W9-ADH-413

In the
NAME *of* SCIENCE

ST. MARTIN'S PRESS
NEW YORK

My thanks and appreciation to Tim Brent, my editor at St. Martin's, and his outstanding colleague Julia Pastore, Mark Fowler, our legal consultant, Robert Berkel, this book's production editor, Noah Lukeman, my agent, and everyone else who in one manner or another has helped in contributing to the book and bringing this project to completion.

 For information, address St. Martin's Press, 175 Fifth Avenue, New York, N.Y. 10010.

www.stmartins.com

Design by Nancy Singer

Library of Congress Cataloging-in-Publication Data

Goliszek, Andrew.

In the name of science : a history of secret programs, medical research, and human experimentation / Andrew Goliszek.

p. cm.

Includes bibliographical references (p 407) and index (p 433).

ISBN 0-312-30356-4

1. Human experimentation in medicine. 2. Medical sciences—Research—Moral and ethical aspects. 3. Science—Experiments. I. Title.

RA1231.R2G57 2003

619'.98—dc21

2003040649

First Edition: November 2003

10 9 8 7 6 5 4 3 2 1

To my wife, Kathy, who stands with me in whatever I do and who makes everything I accomplish worth it.

CONTENTS

Appendices

INTRODUCTION

A biting wind shook the razor wire surrounding the medical barracks as newly arrived children were examined one by one and divided into groups according to the experiments they'd be assigned to. Some of the youngsters were younger than six, yet they were old enough to have an idea of what was to become of them. The two-day trip across Poland was made worse by the asphyxiating conditions in boxcars so full that everyone had to stand or lie in their own waste. Many of the children had already been separated from their parents, who were either dead, serving as laborers for factory work, or placed in holding pens for extermination. The doctors, their cold eyes studying heads, eyes, torsos, and limbs as if looking at animals in a pet shop, were satisfied that this batch of little guinea pigs would provide enough body parts to last several weeks.

Human experiments at places like Auschwitz were efficient, systematic, and incredibly cruel. From practice surgeries without anesthesia to massive infections that led to gangrene and amputations to excruciating high-pressure, freezing, and heat experiments that often ended in death, the Nazis, along with the Japanese, took human experimentation to a new level. In the history of medicine and science, the period right before and during World War II was unlike any other. The few who survived could only describe it as the closest thing to hell on earth.

In the course of history, human beings have, in one way or another, always used their own kind for human experimentation. Although sporadic, vivisection was practiced by the ancient Greeks and Romans to augment their knowledge of science and medicine. In the third century B.C., vivisection was performed on condemned criminals. Persian kings also allowed physicians to experiment on criminals, as did Egyptians who made great advances in medicine, no doubt because of human experimentation. In fact, Jewish law articles about Nazi medicine describe experiments by Cleopatra to determine how long it takes to form a male versus a female fetus. According to the Alliance for Human Research Protection, Cleopatra would force her condemned handmaidens to become impregnated and then subjected them to periodic operations to open their wombs at specific times of gestation.

The Middle Ages were rife with such barbarism that torture and sometimes human experimentation was simply a part of life. Physicians who wanted to practice their craft, or students who needed to learn, used cadavers, animals, and even human beings as subjects. Prisoners, heretics, and witches provided a steady stream of bodies at a time when abject cruelty was as much a part of life as disease and poverty. Writing in the *Review of Medicine in Chile,* R. Cruz-Coke notes that, "there was no cruelty in performing vivisection on criminals, since useful knowledge for the progress of medicine and relief of diseases was obtained."

The Renaissance, though an age of enlightenment, had its share of human suffering. In her book on the history of science and ethics, Joanne Zurlo states that anatomists during this period were sometimes accused of practicing vivisection. Most of us probably never heard of Fallopius, the Italian physician after whom the fallopian tube is named, who received permission from the grand duke of Tuscany to conduct any experiment he wanted on criminals. The centuries that followed left their own legacies of inhumanity, which should have provided those of us in later centuries with a guide for better living. Unfortunately, the axiom that history repeats itself was never truer then in the case of human experimentation.

During the past hundred years, science and medicine have had a phenomenal run, in part because of experiments and research involving human subjects. From risky and sometimes deadly medical procedures to life-threatening chemical and radiation exposures to the administration of mind-altering drugs and dubious vaccines, modern human experimentation has added its own chapter to an already long and nefarious history. Interest in drugs, mind control, and human behavior is centuries old. Potions, tonics, herbal medicines that alter behavior, drugs that induce personality disorders, and hypnosis as a means of controlling human behavior have always fascinated scientists and will continue to be the focus of research for decades to come. The greatest difference now is that we have the technology and wherewithal to perform experiments previously considered infeasible.

Earlier research in the United States was often limited to health and medical experiments, and designed to target diseases that affected mainly large portions of the population. In most cases, research was done with treatments or cures in mind; in others, treatments were denied or studies ignored either because the disease in question was limited to black populations or poor immigrant groups, or so that researchers could follow the progression of an untreated disease from beginning to fatal end. The purpose in both instances was simply to add to the body of knowledge, regardless of the consequences, or to answer questions addressed by basic research.

The years following World War I were a transition period, marked by rapid progress in technology and a growing interest in chemical and bacteriological research. Industrial productivity and advances in medicine increased during the 1920s as researchers broke new ground in virtually every field of science. By the 1930s, science and medicine were on the verge of major discoveries. X rays and radiation experiments became more common, new molecules were being uncovered or produced each day, breakthroughs occurred faster than anyone thought possible, and science, despite the warnings, was at the forefront of a brave new world. It was an exciting and rewarding time to be a biomedical researcher.

Once again, with increased threats from enemies abroad, especially during and after World War II, the focus of the nation's research efforts shifted dramatically. In what we now know to have been an all-out effort to develop new classes of chemical, biological, and psychochemical agents, the government established one of the greatest cooperative military-scientific ventures in history. For more than thirty years, top-secret programs flourished, fueled by Washington's paranoia and driven by a simple goal: to stay one step ahead of our counterparts in Germany, Japan, and the Soviet Union.

More recent incidents and revelations leave no doubt that the use of chemical and biological agents is becoming a dangerous reality. Russia's use of a secret gas to free eight hundred hostages in a Moscow theater has shown the world exactly how effective a chemical attack could be. New evidence has come to light that may implicate both Saddam Hussein and Fidel Castro in producing West Nile virus as a bioweapon. Iraqi and Cuban defecting scientists have admitted working on weaponized versions of West Nile. Also, recently uncovered documents expose the U.S. military's intent to develop a new generation of biological and chemical agents in possible violation of international treaties. Using advanced techniques of biotechnology, scientists in virtually every country now have the ability to make the most horrible weapons that mankind has ever seen.

At no time in history have we witnessed the progress in science and medicine that we've witnessed this past century. Both in the private sector and in government, an explosion in technology and information has occurred that is, without question, unparalleled. We've learned more, discovered more, and invented more in the last fifty years than we have in all the years since the beginning of mankind. But with such progress came a terrible price. In many instances, life has been exploited for the benefit of medical and scientific progress or sacrificed in a zealous effort to preserve the American way of life.

From early human medical research to CIA-sponsored mind control programs and radiation exposure experiments, *In the Name of Science* introduces readers to the terrifying world of chemical and biological warfare, genetic engineering, ethnic weapons, secret virus cancer programs, and AIDS research. Some believe that unless we are vigilant it can all happen again. With advances in medicine and recombinant DNA, and with the completion of the Human Genome Project, we're on the brink of discoveries that some fear will make real the threat of population control, gene warfare, ethnic cleansing, or worse. The best way to ensure that such history is not repeated is to disclose the truth and learn from the past. By detailing the secret programs, medical scandals, and shocking events that have made this past century one of the darkest in scientific history, I hope to have accomplished just that.

In the
NAME *of* SCIENCE

1 THE CHEMICAL REVOLUTION: BRINGING BAD THINGS TO LIFE

Nathan Schnurman, a seventeen-year-old sailor recruited to test U.S. Navy summer clothing in exchange for a three-day pass, never thought he would be gasping for air inside a gas chamber instead. The instructions he received were simple, and he didn't think much of it at the time when he was ordered to put on a mask and some special clothing. During the experiment, the mustard gas and lewisite he was exposed to seeped through his mask, making him first nauseous and then violently ill. He demanded to be released, but was refused because the scientists conducting the experiment told him that it was not yet completed. Shortly after his second demand, he passed out. When he regained consciousness, he found himself lying outside the gas chamber and thinking how lucky he was to be alive.

"I called to the corpsman via an intercom and informed him of my condition and what was happening, and requested I be released from the chamber, now," Schnurman testified before a judiciary committee. "The reply was 'no,' as they had not completed the experiment. I became very nauseous. Again I requested to be released from the chamber. Again permission was denied. Within seconds, I passed out in the chamber. What happened after that, I don't know. I may only assume that when I was removed from the chamber, I was presumed dead."

Another serviceman, Lloyd B. Gamble, had dedicated more than seven years of his life to the U.S. Air Force. When he volunteered for a special program to test new military protective gear, he was offered various incentives, including a liberal leave policy, family visitations, superior living and recreational facilities, and letters of commendation to be made part of his permanent record. During the first three weeks of testing. Gamble was given two or three water-sized glasses of a liquid to drink. He soon developed erratic behavior and even attempted suicide, but what he didn't learn until eighteen years later was that what he'd received as a human subject was LSD. Even after he found out, the Department of Defense (DOD) denied that he'd participated in the experiments, although an official publicity photo shows him as one of the servicemen volunteering for a "special program that was in the highest national security interest."

Both Schnurman and Gamble were victims of a massive organized program that used both the military and civilians to carry out human experiments involving chemicals and chemical agents. All participants had been sworn to secrecy, like eighteen-year old Rudolph Mills, who discovered forty-six years after his own gas chamber experiments that four thousand other servicemen were essentially human guinea pigs for the Chemical Warfare Service (CWS). Though his health began to deteriorate while still in the navy, Mills did not learn that his lifelong physical health problems were likely related to mustard gas exposure until more than forty years later. According to a September 28, 1994 General Accounting Office report, the DOD and other national security agencies used hundreds of thousands of human subjects in tests and experiments involving hazardous and often deadly substances.

This kind of duplicity doesn't begin or end with the military, however. For decades, scientists working for corporations have been hiding research results, relying on flawed or fraudulent studies, or disregarding the health effects of chemical products in order to ensure a steady stream of profits. Because even a small change in data can often have a major effect on the findings of a study, it is sometimes difficult to determine whether or not researchers have acted ethically. Take the case of two scientists who had published a mortality study comparing cancer rates of workers exposed to a hazardous substance with those who were not and then later placed four exposed workers in the unexposed group. This simple switch increased the death rate in the control group while significantly decreasing the death rate in the exposed group. While the researchers contended that the reclassification was done in good faith, the incident triggered a dispute within the FDA as to whether an ethics investigation should or should not have been conducted.

In some cases, there was widespread use of toxic chemicals on humans simply because no one knew how dangerous the chemicals were. After DDT (the potent insecticide that replaced lead arsenate) was developed, the U.S. government dusted millions of soldiers to prevent malaria and typhus. This miracle chemical that killed hundreds of different pest species was made famous in a 1948 *Life Magazine* photograph of a teenaged girl eating a hot dog surrounded by a cloud of DDT. What DuPont scientists did not realize until decades later was the extent to which their altered molecules and synthetic chemicals would accumulate in the environment and continue to show up in the blood of virtually every American twenty-five years after its ban.

By just taking a look at the world around us, we quickly realize the impact chemicals have had on virtually every aspect of our lives. We're literally surrounded by a sea of organic and inorganic compounds. Our bodies are composed of thousands of chemicals, each made from billions of molecules that react with one another and assemble into complex forms to make life possible. We eat chemicals, drink chemicals, breathe chemicals, put chemicals on and in our bodies, and take chemicals whenever we're sick. From the moment we're born to the day we die, we are so dependent on chemicals that we wouldn't know what to do without them.

Over the last hundred years, that dependence has become an addiction. Natural recipes handed down for centuries have been replaced by products promising everything from clean kitchen counters to cancer cures. Along comes the chemical industry, and we now have more than fifty thousand synthetic compounds—many of them unregulated, some of them miracles of humanity, and others more deadly than anything nature could come up with. If we've learned anything from history, it's that natural products can often be deadly. When man gets into the act, they can become even deadlier.

Chemical Warfare Agents

In 1978 London, Georgi Markov, a Bulgarian exile, stood patiently on a street corner and watched the stop-and-go of traffic while awaiting the next bus. The sky was overcast, and the steady stream of commuters made him less likely to think that anything out of the ordinary was about to take place. Perhaps he was thinking of his family back home or about what he had to do that day. But as he looked at the passing cars, he suddenly felt dizzy, lost consciousness, and collapsed. Within a few days he stopped breathing and died. His mysterious death remained a mystery until the autopsy was done, when investigators discovered a tiny pellet beneath his skin containing ricin, a chemical six thousand times more toxic than cyanide. The Bulgarian, they eventually learned, was a former agent murdered by the communist Bulgarian government with an umbrella gun supplied by the KGB and fired unnoticed in a crowd of passersby who never suspected that chemical warfare had been waged so easily.

The use of natural chemicals has been reported for more than two millennia. As far back as 600 B.C., when the Athenians poisoned with helleborus root a river used by its enemy as drinking water, chemicals have been used as a means of waging war. In 200 B.C., Carthage defeated one of its enemies in a battle by leaving behind casks of wine tainted with mandragora, a root that produces a narcotic-like sleep. After enemy soldiers drank the wine, the Carthaginians returned and killed them. In one of the more bizarre examples, Hannibal, in a naval battle against Eumenes II of Pergamum, lobbed venomous snakes onto the decks of enemy ships to defeat the Pergamum sailors. In addition, as we know from historical records, arrows tipped with poison chemicals have been used for nearly as long as there have been bows to shoot them.

Limiting the use of chemicals as weapons was suggested as far back as 1675, when a French-German agreement was signed in Strasbourg prohibiting the use of poison bullets. But within two centuries, large-scale development of chemical weapons had begun. In 1874, to stem the fear of chemical warfare, the Brussels Convention was adapted prohibiting the use of poison weapons. Twenty-five years later, an international peace conference held in The Hague led to a worldwide agreement outlawing the use of projectiles filled with poison gases. These agreements, it was hoped, would put an end to the development of weapons thought too horrible to be used against human beings. It didn't.

Modern chemical warfare actually started in the nineteenth century with incendiary arsenic bombs that sent plumes of poison smoke across enemy battle lines. Soldiers exposed to the smoke died a grisly death. Muscle spasms and severe vomiting were followed by cardiovascular collapse and death within a few hours of inhalation. The twentieth century proved no less civilized. After rumors of a new and deadly weapon invented by the Germans early during the First World War, the German

Army bombed British forces in Neuve-Chapelle with dianisidine chlorsulfonate. A few months later, they attacked Russian forces with xylyl bromide. Both incidents were merely learning experiences and a prelude to what was to be the first large-scale chemical attack on April 22, 1915.

That day, two hours before sunset, the Germans covered themselves head to foot in protective suits and released nearly two hundred tons of chlorine gas from canisters toward the French troops. The greenish mist was taken by a light wind, and within minutes began sinking into the four miles of trench lines where the soldiers experienced something for which they were not prepared. Panic ensued as the men began choking and gasping desperately for breath. When the battle had ended, more than five thousand soldiers had died from asphyxiation. It didn't take long before both sides recognized the impact of chemical warfare and began using chlorine gas on each other while developing even more efficient and practical means of waging war.

Phosgene, a choking gas like chlorine but ten times as toxic, was the next agent to be used. Blister agents were introduced in 1917 and have been used ever since, notably in 1980 to 1988 during the Iran-Iraq War. By the end of 1918, more than one-fourth of all artillery shells fired contained chemical weapons that killed about one hundred thousand people and injured more than a million. In the late 1930s, Germany first developed the G-series nerve agents, such as sarin. In 1936, mustard gas was used by Italy against the Abyssinians. Spanish troops used it in North Africa between the world wars. The Japanese killed large numbers of Chinese from 1937 to 1943 with lewisite, mustard gas, and various biological agents. In the 1950s, England developed even more lethal nerve agents, the V series, which includes the best-known nerve agent, VX.

One of the most secret chemical weapons facilities of all was located near the Russian town of Podosinki. Code named "Tomka," its mission was to develop poison gases to be delivered by artillery, aviation, and special gas projectors. Another Soviet poison facility called "Lab X" was operational as far back as 1937. According to Pavel Sudoplatov, deputy director of foreign intelligence (precursor of the KGB), the lab was used to develop poisons for assassinating enemies both inside the country and abroad. It's not known how much of the research was shared with rogue nations such as Iraq, but evidence gathered during the Gulf War suggests that there was a good deal of cooperation between the former Soviet Union and Saddam Hussein, who had no fewer than five chemical and biological weapons factories when the United Nations (UN) inspected Iraq after the Gulf War.

According to investigations by the *Weekly Mail and Guardian*, in the 1980s a company called Protechnik in South Africa was allegedly the largest chemical, biological, and nuclear laboratory in Africa and carried out secret bizarre experiments to test special bullets and heat-resistant clothing. According to a 1989 Central Intelligence Agency (CIA) report, scientists at the facility worked closely with Israel in the 1980s to develop a chemical warfare capability. (The company has since come under new ownership and is no longer said to be engaged in such research.)

Overall, more than three thousand chemicals have been tested for possible use as toxic weapons. In many cases, the agents were first developed as pesticides composed of organic molecules known as organophosphates and then adapted for use on human beings. According to the Chemical Weapons Convention (CWC), ratified on April 29, 1997, there are now five recognized classes of chemical warfare agents:

Nerve agents: After contact with the skin and lungs, these highly toxic organophosphorous chemicals kill by disrupting metabolism and blocking nerve transmission. The first nerve agent, tabun, was developed in 1936 as a pesticide. VX is so toxic that a single drop the size of a pinhead on bare skin may cause death. Symptoms include seizures, vomiting, convulsions, muscle paralysis (including the heart and diaphragm), loss of consciousness, and coma. Death may occur in one to ten minutes. Examples include sarin (GB), soman (GD), tabun (GA), and VX.

Blister agents and lewisite: Also called vesicants, blister agents are absorbed through the lungs and the skin, burning lung tissue, skin, mucous membranes, the windpipe, and the eyes. There are few deaths from blister agents, but a large number of casualties. They damage the respiratory tract and cause severe vomiting and diarrhea. Examples include nine sulfur mustards (HD), three nitrogen mustards (HN), phosgene oximine (CX), and three lewisites.

Blood agents: Distributed by the blood to various tissues and body parts, these agents destroy blood tissue, thereby disrupting oxygen flow to the heart and causing suffocation. Examples include hydrogen cyanide and cyanogen chloride.

Choking agents: Absorbed through the lungs, choking agents cause fluid buildup in lung tissue, preventing the victim from breathing. Essentially, these chemicals cause drowning by inducing alveoli within the lungs to secrete a steady flow of fluid. Examples include phosgene (CG), diphosgene (DP), chlorine (Cl), and chloropicrin (PS).

Toxins: These chemicals are extracted from living organisms. Ricin, a protein extracted from the castor oil plant, is ounce for ounce more toxic than nerve agents. It acts by blocking the body's synthesis of proteins. Saxitoxin, an organic chemical produced by blue-green algae and accumulated in the mussels that feed on it, acts on the nervous system.

More recently, and despite a series of treaties and agreements, chemicals and toxins have been used widely as both offensive and defense weapons. After World War II, the fear that some countries would actually use these weapons of mass destruction initiated secret research programs and prompted a series of open-air tests involving human subjects. Some of the chemicals and biological agents tested, referred to as "simulants" by the military, were released over populated areas and cities. Accounts of these tests are detailed in the next chapter.

Neither international conventions nor worldwide outrage has mitigated the growing research and development thought critical in maintaining an advantage over rogue nations. Buttressing the argument is evidence that chemical weapons have been used during the past two decades in Afghanistan, Iran, Iraq, Southeast Asia, Mozambique, and Azerbaijan. Today, with state-sponsored terrorism and experts willing to sell their knowledge to the highest bidder, it's hard to know exactly who is sitting on large stockpiles of poisons, plagues, and lethal gases. According to the CIA, more than twenty countries are either developing or already have chemical weapons.

The list is a "who's who" of enemies, rogue states, and nations simply trying to keep up with threats by its neighbors. Besides the thirty thousand tons in the United States and at least forty thousand tons in Russia, stockpiles around the world are growing. Egypt was the first Middle Eastern country to use chemical weapons when it employed phosgene, mustard, and nerve agents against Yemeni Royalist forces in the mid-1960s. Israel began its program in the 1970s in response to the Arab chemical threat. Syria developed its own weapons in response to Israel. Iran's program was started after Iraq's use of chemical agents during the 1980–1988 war. Libya, which received its first chemical weapons from Iran, used them against Chad in 1987. Not to be outdone, Saudi Arabia got into the chemical weapons business and is now suspected of having its own arsenal. In response to regional tensions, China, India, Pakistan, Burma, North and South Korea, Vietnam, and Taiwan have also developed programs they claim are strictly defensive.

Unfortunately, at the height of what had been world paranoia about chemical agents, experts agreed that the only way to know the physiological effects of these agents was to use human subjects. A recent U.S. Senate staff report prepared for the Committee on Veterans' Affairs acknowledged that in the 1940s alone approximately sixty thousand military personnel were used as human subjects to test two chemical agents, mustard gas and lewisite. Most of the subjects were not informed of the nature of the experiments, never received medical follow-up after their participation in the research, and were threatened with imprisonment at Fort Leavenworth if they discussed the research with anyone, including their wives and parents. In fact, not only were discharged personnel forbidden to talk about their experiences, they could not even describe their exposures to family doctors who tried to determine the cause of severe respiratory illnesses.

Rudolph Mills, the eighteen-year-old navy seaman mentioned earlier, was one of many such individuals who testified about his experience before the Senate Committee on Veterans' Affairs. "I had on an experimental mask and the navy was trying to determine if people wearing these masks could communicate with each other," he recounted. "I was enticed to sing over the intercom. . . . No one ever told me that the mask became less and less effective against the gas with each use. . . . We were sworn to secrecy. . . . At the age of 43, I underwent a long series of radiation and later surgery to remove part of my voice box and larynx. . . . It didn't occur to me that my exposure to mustard gas was responsible for my physical problems until June 1991, when I read an article in my hometown newspaper."

The harrowing tales had one theme in common: All told of veterans convinced that they had been lied to about the nature and dangers of the experiments. Testimony by fellow subject John William Allen was also chilling. Exposed to sulfur mustard several times in clothes that had become impregnated with toxic chemicals from previous experiments, he was removed from further exposure after passing out in the gas chamber and receiving many wounds as a result of the chemical. In a written testimony, he states, "The government has lied to us for fifty years over and over again. If I would have been shot on the front lines at least I would have had it on my record and would have received medical treatment."

The 1953 Wilson Memorandum (Appendix II), which adopted rules from the Nuremberg Code (Appendix I), was supposed to protect individuals from such harm and inform them of risks before they were to provide consent. But again, between 1958 and 1975, thousands of volunteers were recruited and experiments carried out as if those rules did not exist. Take the case of Ken Lamb, an airborne soldier who volunteered for an experiment under the new rules because all he wanted to do was collect on the promise of a three-day pass to see his fiancée in New York.

Lamb recalls the day his commanding officer made the offer and his enthusiasm when he arrived at Fort Detrick, Maryland. He remembers sitting in a sterile, hospital-like room and watching as a researcher in medical garb placed a drop of liquid on his forearm. He immediately became nauseous and dizzy, and it took a while for him to recover from the numbness that spread through his arm and into his body. Before returning to his unit, he was ordered not to discuss the experiments with anyone and was never told what the liquid was. Not until he developed inoperable cancer thirty years later did he learn that army scientists had exposed him to VX. Recently, the Office of Veterans Affairs rejected Lamb's claim for disability, citing no evidence of a link between his cancer and the experiments.

A decade later, from 1962 to 1971, U.S. servicemen would be purposely subjected to a chemical that experts knew at the time to be one of the most toxic known to man. Agent Orange was an herbicidal 50:50 mixture of 2,4-D and 2,4,5-T, which contained dioxin, a contaminant that does not occur naturally. Unlike the dioxin used by civilians, the military version was undiluted and sprayed at a rate of three gallons per acre in concentrations up to twenty-five times the manufacturers' suggested rate. According to the Veterans Administration, as many as 4.2 million U.S. soldiers could have made contact with Agent Orange as a result of "Operation Ranch Hand."

Reaching a peak in the mid-1960s, the bulk of Agent Orange was sprayed from fixed-wing aircraft to defoliate the dense jungles where enemy soldiers could hide. Smaller amounts were released from helicopters, trucks, riverboats, and even by hand. Dr. James Clary, a former government scientist with the Chemical Weapons Branch of the Air Force Armament Development Laboratory, said, "When we [military scientists] initiated the herbicide program in the 1960s, we were aware of the potential damage due to dioxin contamination in the herbicide. We were even aware that the military formulation had a higher dioxin concentration than the 'civilian' version due to the lower cost and speed of manufacture." In all, nineteen million gallons of undiluted Agent Orange were dumped on Indochina, with an impact on the environment and human health that is felt to this day.

Soon after Operation Ranch Hand began, reports surfaced of health problems and significant increases in human birth defects. It wasn't until April 15, 1970 that the U.S. surgeon general warned that use of 2,4,5-T "might" be hazardous to our health. But despite concerns by scientists, health officials, politicians, and the military itself about the toxicity of Agent Orange, the spraying program continued unabated until 1971. Recent studies by the Environmental Protection Agency (EPA) and the National Cancer Institute (NCI) have proven that there are no safe exposure levels to dioxin and that exposed humans have a 60 percent greater risk of dying from cancer.

Today a major Agent Orange scandal is brewing in Thailand. According to Thailand's science, technology, and environmental minister, documents released by the U.S. ambassador reveal that the U.S. and Thai militaries secretly tested chemical weapons, including Agent Orange, from 1964 to 1965, then dumped the toxic remains in an area that was subsequently unearthed during construction of an airport runway. Soil samples sent to U.S. and Canadian laboratories found high levels of both 2,4-D and 2,4,5-T present. Because dioxins can spread so easily throughout the food chain, there's now fear that the land, which has since been used for farming, may also end up being a toxic killing field.

The United States was not alone in its use of soldiers for human experiments. A recent search of British documents has found that as many as twenty thousand soldiers might have been used as guinea pigs at England's Porton Down testing station from 1939 to 1989 to test chemical agents. According to Alan Care, a lawyer representing a group of former servicemen, unwitting volunteers were tricked into participating as test subjects and were exposed to nerve gas, mustard gas, and LSD. "Most of the men," said Care, "believed they were going to Porton Down for the purpose of common cold research and were in fact gassed with sarin."

Sarin, recall, is the same gas that killed twelve people and contaminated three thousand in the Japanese subway system, and was used by Saddam Hussein against his Kurdish population.

During World War II, Porton Down, a top secret chemical weapons center in Wiltshire, geared up to counter the top priority menace of chemical warfare. Patrick Mercer, who had gone through the facility as an army officer, said, "There were a series of bunkers to which you were thrust from time to time to be gassed and to go through ghastly exercises underground wearing a gas mask." Another soldier, Ronald Maddison, died after exposure to sarin. The whole time he was being gassed, he thought he was taking part in a program to find a cure for the common cold.

Although British subjects were being gassed, researchers had known since the 1920s that mustard gas was absorbed through the skin and affected every organ in the body. However, they played that down so the military experiments could proceed. Professor David Sinclair, a Porton medical officer, described one experiment as follows: "When the grenade exploded or the armor piercing shot was fired (I always hoped it was properly aimed), shrapnel used to bounce angrily off the furniture, and after it had subsided I would push down the metal plate and the crew would take up their positions and attempt to drive off. I was the lucky one who had a respirator on, and I had to observe the reactions of the unfortunates who had not. The immediate effects included a feeling of grit in the eyes, followed by severe pain, lacrimation, and spasm of the eyelids."

The latest evidence linking exposure to chemical agents and health effects has shown up in Gulf War veterans, many of whom experienced unexplained neurological symptoms after coming home. As many as four hundred thousand U.S. soldiers were ordered to take the investigational nerve agent medication pyridostigmine bromide every eight hours for days, weeks, or months. A study by researchers at the University of Texas in Dallas found that even low levels of exposure to nerve gas and pesticides, when combined with pyridostigmine, may cause irreversible brain damage. Pyridostigmine also happens to be a nerve agent. The results confirm an earlier 1993 study by Dr. James Moss, a scientist at the U.S. Department of Agriculture, who found that the medication given to Gulf War soldiers caused the common insect repellent deet (diethyltoluamide) to become seven times more toxic than when used alone. Coincidentally, deet and other repellents were widely used during the Gulf War as protection against sand flies, scorpions, and other pests.

Even more troubling is the fact that researchers had evidence as early as 1978 that neostigmine, a close molecular relative of pyridostigmine, causes profound physiological, electrophysiological, and microscopic disruption of nerve endings and muscles. Based on the published reports, some of these changes increase in severity over time with continued treatment. It was because of this concern that the Human Subjects Committee reviewing the studies considered the possibility of mentioning the possibility of death in the informed consent form. After some deliberation, it was decided that such a warning was unnecessary because death, they said, was not likely.

That didn't seem to matter a great deal to military officers, who forced personnel under their command to take pyridostigmine, whether they became intensely sick or not. For example, Carol Picou was a nurse who had been stationed in the Gulf for five months when she started taking the drug. By the third day, she developed incontinence, blurry vision, and uncontrollable drooling. The side effects became worse one hour after she took a pill but stopped after she refused to continue taking them. Her commanding officer ordered her to resume taking the pills for fifteen days, even watching to make sure she swallowed them. Currently, Carol Picou has permanent medical problems, including incontinence, muscle weakness, and memory loss.

Similarly, Lieutenant Colonel Neil Tetzlaff had immediate side effects when he started taking pyridostigmine bromide on the plane ride to Saudi Arabia. His nausea and vomiting became so severe that he needed emergency surgery to repair a hole in his stomach. When he became ill, the military doctor told him to continue taking the pills because the doctor had no idea that nausea and vomiting were known side effects. According to Tetzlaff's sworn testimony, the doctor acted as if the pyridostigmine was as safe as a cough drop. Other soldiers and pilots experienced respiratory arrest, loss of consciousness, abnormal liver function, irregular electrocardiograms, joint pain, sensitivity to chemicals, and anemia.

Nurse Picou's case was especially troubling because virtually every pyridostigmine study done up to that time excluded women. Scientists believed, based on other studies, that women would be affected differently; that women on birth control pills had different levels of neurotransmitters that would interact with nerve agents; that women in different stages of their reproductive cycles might respond more intensely; and that reactions might be unique in women who are menstruating or who have breast cancer. None of this kept the military from forcing women to take pyridostigmine, even when they exhibited symptoms indicative of serious health effects.

Civilians participating in the Gulf War were also exposed without informed consent or information about potential side effects. DOD contractors and news media, for example, were given pyridostigmine without being told that the drug was experimental or that it was being administered in a regimen not proven effective or safe. Journalists and other nonmilitary personnel began experiencing serious medical problems similar to the Gulf War veterans, whose illnesses were categorized as Gulf War syndrome.

The chief researcher of the University of Texas study, Dr. Robert Haley, has warned of "an epidemic of Parkinson's disease coming out of Gulf War syndrome." His colleague, Dr. Frederick Petty, adds that "some of these veterans are beginning to show early, subtle symptoms of brain disease. The question is whether, over time, these overstimulated brain cells will wear out and die. If so, these patients could develop degenerative brain diseases such as Parkinson's." According to the researchers, as many as eighty thousand Gulf War veterans may exhibit these symptoms within twenty years.

If the evidence is substantiated, then in the final analysis military scientists and officers had ignored the Nuremberg Code, the Declaration of Helsinki (Appendix IV), and the requirements of informed consent that virtually all government agencies are subject to, and knowingly participated in human experimentation. Dr. Arthur Caplan, director of the Center of Biomedical Ethics at the University of Minnesota during the Gulf War, testifying before a senate committee, said that "these agents were used, as we have heard, in large populations for purposes other than those for which they were originally designed in some cases, and circumstances under which they had never before been tried out in the desert. This seems to me to cinch the case that what took place fell into the category of experimental, innovative, and investigational, and that makes them research."

Interestingly, the DOD has been desperately trying to make permanent the waiver of informed consent, arguing that "to not finalize it provides an arguable defect under the Administration Procedures Act and leaves both the DOD and FDA open to greater liability." But liability is only part of the rationale. If the request is granted, it would also give the

DOD unrestricted use of investigational drugs by military personnel, even though investigational status means that efficacy and safety have not been proven. As of now, the U.S. Food and Drug Administration (FDA) has not acted on the DOD's request.

Recently published findings by the National Academy of Sciences and the Department of Veterans Affairs conclude that the exposure to chemical agents has resulted in long-term effects, disabilities, and even death for participants. For decades the Pentagon had not only denied this but denied that chemical agent research ever took place. And though much of the secret testing has since been exposed, many documents remain classified to this day.

In Harm's Way: Civilian Guinea Pigs and Chemical Weapons

Some eighty miles west of Salt Lake City lies a stretch of flat desert as serene and picturesque as it was when the Mormons first settled there in 1848. Every spring, cactus and wildflowers erupt with blossoms in a rainbow of colors that glimmer beneath the deep blue sky. The browns and grays of the rugged landscape are often stark but always beautiful in this section of the United States, which houses the Dugway Proving Grounds.

Outside the million acres, much of it spread across the Great Salt Lake Desert, an ominous sign reads: WARNING: DO NOT HANDLE UNIDENTIFIED OBJECTS. REPORT THEIR LOCATION TO SECURITY. A single access road leads to the entrance of Dugway's Proving Grounds, the vast expanse of its isolated 210-mile border patrolled mostly by air. It is this kind of isolation from the rest of Utah's population that gives these facilities an air of mystery and kept what happened on March 13, 1968 a secret for more than twenty years.

That morning, streaking across the sky, an F-4E Phantom jet locked on to its predetermined target and released more than a ton of the nerve agent VX along a narrow strip of ground. The next morning, farmers twenty-five miles downwind of the release, in an area called Skull Valley, noticed that their sheep were dying. By the time the count ended, more than six thousand animals lay dead from what was officially reported as unknown causes. And though the army paid one million dollars in restitution to the farmers, it didn't admit for decades to open-air testing of VX, which experts testified had evaporated so slowly that it remained on the ground for days.

Recently declassified documents reveal that 1,635 field tests were conducted with VX, GA, and GB from 1951 to 1969. In some tests, thousands of pounds of nerve agents were dropped but only a small percentage reached the ground. For instance, a March 1964 report revealed that in one spray only 4 percent of the VX dropped reached the test grid; in another test, recovery of VX from a high-wind speed test was less than 40 percent. Other open-air trials included more than 55,000 chemical rockets, artillery shells, and bombs and 134 tests to determine the hazard to personnel downwind. Some of the deadly agent drifted off the range and made its way to unsuspecting residents nearby. All in all, a half-million pounds of nerve agents was spread across the Utah desert.

One of the more disturbing pieces of evidence that could link these tests with disease is the dramatic rise in neurological disorders such as multiple sclerosis in counties either close to or that included Utah's Dugway Proving Grounds. Besides tests of chemical agents such as VX, 328 open-air germ warfare tests were conducted, as were 74 airborne tests that spread radioactive particles. A study by the University of Utah concluded that Tooele County, where many of these tests had occurred, has a multiple sclerosis rate seven times higher than the national average. Utah itself has a rate twice that of the United States.

Ray Peck, one of the residents of Skull Valley at the time, remembers the morning after the VX was dropped. According to the *Deseret News*, he woke up to find a layer of freshly fallen snow covering his land. It was so pretty and clean, he said, that he walked out into it, picked up a handful, and ate it. The tranquility of the moment was suddenly broken when he turned his head and saw some dead birds nearby and a rabbit on its side twitching for several minutes before dying. He didn't know what to think, but suspected that something terrible had happened when the sheep began dying and an army helicopter landed in his yard carrying medical personnel to take blood samples from his family. Peck recalls coming down with violent headaches, numbness, and burning in his arms and legs after that day. His daughters have suffered from similar symptoms, as well as an unusually high number of miscarriages, which he attributes to VX exposure.

The total number of square miles experts believe remains contaminated with unexploded bombs, rockets, and artillery shells, some of which still contain deadly agents, is roughly the size of Rhode Island. But despite the evidence of contamination and health problems, secret new military testing continues today. Scientists at Dugway Proving Ground are currently experimenting with toxic chemicals in the Melvin Bushnell Materiel Test Center, a thirty-million-dollar laboratory for simulating chemical attacks and testing protective clothing. Still being considered is construction of an entire mock city, complete with homes and subway systems, to facilitate practice of chemical warfare scenarios.

Beneath the Surface: Oceans As Toxic Military Waste Dumps

Shortly after World War II in 1945, a British merchant ship set sail toward the Baltic Sea. But instead of merchant marines, its crew consisted of British sailors; and rather than its usual fare, the creaking vessel was loaded with captured German nerve gas, phosgene, and arsenic-containing compounds. The destination, somewhere along the coast of Norway, was kept secret. The mission was to blow up the ship and send its deadly cargo to the bottom of the sea. Twenty such ships were conscripted into Her Majesty's Service, each one laden with poison gas and sunk in the cold waters of the North Atlantic.

The following year, the United States launched Operation Davy Jones's Locker. For the next two years, naval ships sailed into Scandinavian waters on five occasions and dumped some forty thousand tons of chemical agents. Kyle Olson, senior member of the Arms Control and Proliferations Center, explained that those who dumped chemicals were simply following standard procedure. "Sea dumping was thought to be the quickest and best way to do disposal because the materials would be dissipated at sea," he said. But according to Alexander Kaffka of the Russian Academy of Sciences and Chairman of the Conservation for Environment International Foundation, "Rules were often broken, which led to the most dangerous kind of dumping at shallow depths, in straits, and in areas of active fishing." In fact, one of the major dump sites in the Baltic Sea has a mean depth of only 170 feet.

The mass dumpings by U.S. and British forces continued unabated throughout the 1950s, 1960s, and 1970s. After the British unloaded thirty-four shiploads of chemical and conventional weapons in the Norwegian Trench by 1949, they began Operation Sandcastle, a secret program in which cyanide, sarin, phosgene, and mustard gas were loaded onto merchant ships and scuttled eighty miles off the northeast coast of Ireland. At the same time, the United States began Operation Chase (Cut

Holes And Sink 'Em), in which more than 50,000 nerve gas rockets were dropped 150 miles off the coast of New York and then later off the coast of Florida. Other U.S. sites included waters near California and South Carolina.

In one of the last sea dumpings, the U.S.S. *Corporal Eric Gibson* was loaded with VX and sarin nerve gas canisters and towed two hundred miles from the shores of Atlantic City. As the naval ship sailed off, an explosion was set off aboard the *Eric Gibson.* Within minutes, the toxic munitions settled under 7,000 feet of water where they remain today. Most New Jersey residents or Atlantic City visitors looking out at the waves lapping the coastline have no clue what lies beneath.

Between 1945 and 1970, more than one hundred sea dumpings occurred in virtually all of the world's oceans except the Arctic. Many of the sites have been documented, but others remain unknown. Also unknown are the conditions of canisters, which may be deteriorating after more than fifty years and leaking potentially deadly chemicals into sensitive marine ecosystems. The last inspection was made in 1974, and because of the cost involved, the army says it has no plans to do any more. Since the CWC does not provide a legal basis for actions taken before 1985 or for chemical weapons that remain submerged in the ocean, there is no responsibility or obligation to locate and clean any sites containing toxic chemicals.

The secret dumpings, claim many experts, are an ecological and humanitarian time bomb. During the past forty years, mustard bombs have been found along German and Polish beaches, and Danish fisherman have hauled in rusted containers with chemical agents. There is even evidence that during transport, some of the munitions were packaged in wooden crates and thrown overboard en route, where they remained floating away from the intended dumping area. If the canisters were not packaged securely and have begun to shift with the currents and tides, they could also become damaged, which would create a toxic flow with consequences beyond the immediate fish stocks.

How long this issue remains unresolved depends on the nations involved in the original dumping and the nations that allowed it to go on along their shores. The question comes down to accountability and the enormous costs involved. Admitting to the secret sea dumps would mean remediation and a massive clean-up effort that would necessarily involve both civilian experts and the military. For the time being, with so many other domestic and global issues, politicians are content to leave that issue to future generations.

Does Russia Have a Secret Chemical Weapon?

Standing before an enthusiastic crowd on the outskirts of Moscow, the Russian extremist Vladimir Zhirinovsky proclaimed that his country has a secret weapon capable of destroying the West. At the time, few people knew what this state secret could be. However, based on recent evidence, it is suspected that Zhirinovsky was referring to a class of chemical agents developed since the late 1970s under the collective name Novichok, which means "newcomer."

Following World War II, both the United States and the Soviet Union used German and Japanese research to improve their own chemical weapons programs. However, in an effort to disrupt the Soviet program, the United States began a disinformation campaign to convince the Kremlin that it had achieved far greater chemical weapons successes than it really had, especially in developing a super nerve gas called GJ. Rather than disrupting it, the deception produced the opposite effect.

The Soviet Union's intense efforts to keep up with the West led to the development of several new agents that were not technically banned by the CWC because they were binary (benign when kept separate but lethal when combined). While denouncing U.S. research on binary chemicals, the Soviets were pouring vast resources into developing their own. The agents went by the code names Substance 33, A-230, A-232, A-234, Novichok 5, and Novichok 7. Most of them were at least as toxic as the nerve agent VX and some purportedly ten times as toxic. For instance, A-232 is so lethal that a microscopic amount can kill a person. Vladimir Uglev, the Russian scientist who personally developed A-232, revealed its existence in an interview with the magazine *Novoye Vremya* in 1994 and admitted that it was specifically developed to circumvent the CWC. If the agents are truly as toxic as suspected, forty thousand tons—an amount not very difficult to produce—would be enough to kill all human life on earth.

According to Michael Waller, a senior fellow with the American Foreign Policy Council, Novichok agents can be more toxic than any chemical weapon known, cause diseases like biological agents, and alter human genes, thus causing birth defects and infant disorders for generations. This was corroborated by Vil Mirzayanov, a 26-year veteran of the Soviet chemical weapons program, who spoke out publicly about Novichok and the creation of an entirely new class of deadly binary chemical agents. Mirzayanov was arrested in 1992 for "revealing state secrets."

On May 25, 1994, Mirzayanov, in a *Wall Street Journal* exposé, blew the lid off the Soviet lie. Safely in the United States, he freely told the *Journal* how the CWC would actually help, not hinder, Russia's production of chemical weapons because of loopholes that the Clinton administration ignored. He went on to describe the intensity of research into binary weapons such as Novichok, how they are easily disguised as common agricultural chemicals, and how difficult it would be for inspectors to identify the compounds because the formulas were kept in such secrecy.

The most blatant loophole the Russians had taken advantage of is the CWC omission that if a weapon is not "specifically" listed it cannot legally be banned or controlled. Because the West has no idea what these compounds are, Russia is free to continue the secret program and produce as much of these chemicals as they want. According to Mirzayanov, fifteen thousand tons of Substance 33 alone have been manufactured in the city of Novocheboksarsk.

To this day, Russia has not officially acknowledged Novichok. When asked about the secret weapon, General Stanislav Petrov, commander of Russia's radiation, chemical, and biological defenses, has said, "No, it does not exist." Too many experts who have actually seen it say otherwise, and they worry that the formula in the wrong hands could make nuclear weapons seem irrelevant.

But on October 27, 2002, the Russians gave the world a frightening glimpse into how effective and fast acting a chemical attack would be after using a secret gas to end a hostage crisis in which Chechen rebels held more than eight hundred people in a Moscow theater. The unidentified chemical, thought by some experts to have been an experimental opiate, was so potent that Chechen suicide fighters passed out in their chairs before they had a chance to move their fingers and detonate the explosives strapped to their waists. Rescuers entering the theater witnessed some of the victims paralyzed with their mouths agape, others already dead, and still others convulsing or gasping for air and slumping over from asphyxiation. Dr. Andrei Seltsovsky, Moscow's top physician and chairman of the city's health committee, reported that all 117 deaths except for two resulted from gas poisoning.

When asked specifically about the gas, the Russian government would only say that it was a "special substance." According to sources, it was developed by the FSB Security Service, successor to the KGB, and so secret that Russian authorities initially refused to provide an antidote for fear of having the substance identified or being accused of violating the CWC. Even Dr. Viktor Fominykh, the Russian president's chief medical officer, admitted that the substance's precise composition was kept secret from him. In fact, so closely guarded was the nature of the gas that the Russians were reluctant to transport the injured and dying to other hospitals for fear that outside examiners would alert experts to what the world was beginning to suspect: that Russia's ongoing chemical weapons program was alive and well and could very well be developing new classes of agents not covered by the CWC.

No one is sure if Russia will ever reveal the exact nature of the gas it used. But the loss of more than one hundred lives in a matter of minutes was a stark reminder of just how lethal chemical agents can be. But another grim reality that may have been overlooked is the possibility that the Russian military had used the hostage crisis as a golden opportunity to test the effective use of a supposedly nonlethal weapon. Since the motto of the former Soviet Union, which applies even in today's modern Russia, is "The state is more important than the individual," it's possible that a cold predawn morning on October 27 was the perfect moment to perform a human experiment while rescuing more than eight hundred hostages. Fortunately, most survived the onslaught. More importantly, Russian scientists answered whatever questions they had about a secret chemical that will surely be used again if it benefits the state.

Recently Declassified Chemical Weapons Programs

In October 2002, the Pentagon reluctantly admitted to having held open-air chemical tests over naval ships in the Pacific and over land in Alaska, Hawaii, Maryland, and Florida for more than a decade. From 1962 to 1973, the tests were done in order to develop defenses against weapons such as sarin and VX, two of the most deadly nerve agents known to man. According to declassified DOD documents, 150 separate projects were conducted under the code name Project 112, which was directed from the Deseret Test Center in Fort Douglas, Utah.

One set of thirty-five trials was conducted near Fort Greely, Alaska between June 7 and December 17, 1965 as part of the Elk Hunt tests, designed to measure the amount of VX nerve agent picked up on the clothing of personnel moving through contaminated areas and minefields, the amount deposited on personnel contacting contaminated vehicles, and the amount of VX vapor rising from VX-contaminated areas. Human subjects wore rubber outfits and M9A1 masks, and afterward were decontaminated with wet steam and high-pressure cold water hosing. Other tests at the Gerstle River test site in Alaska, code-named Devil Hole I and Devil Hole II, involved the release of sarin and VX from rockets and artillery shells filled with the lethal agents.

Another 1965 test, code-named Big Tom, involved the spraying of bacteria over Oahu, Hawaii during May and June to simulate a biological attack against an island complex. *Bacillus globigii* was disseminated from a high-performance aircraft, an Aero spray tank mounted on a U.S. Navy A-4 aircraft, and a Y45-4 spray tank mounted on a U.S. Air Force F-105 aircraft. The next year, from April to June 1966, M138 bomblets filled with BZ were exploded in the Waiakea Forest Reserve southwest of Hilo, Hawaii. BZ is a code name for an ester of benzilic acid, which affects the human mind, rendering contaminated subjects unable to perform an assignment or less able to resist for a short period of time.

Open-air tests over navy ships were also part of Project 112. Code-named SHAD (Shipboard Hazard and Defense), the DOD conducted a series of tests to determine the vulnerability of naval ships to chemical and biological agents. The SHAD program was planned and conducted by the U.S. Army's Deseret Test Center and used live toxins and chemical poisons against U.S. servicemen.

In one such test, named Flower Drum, the USS *George Eastman* (YAG-39) was sprayed with sarin nerve agent, sulfur dioxide, and methylacetoacetate from a gas turbine mounted on the bow of the test ship and by direct injection into the air supply system. The following year, in a test code-named Fearless Johnny, VX nerve agent and diethylphthaline mixed with 0.1 percent fluorescent dye DF-504 were used to measure interior and exterior contamination and to demonstrate the effectiveness of the shipboard water washdown and decontamination systems. The USS *George Eastman* (YAG-39) was once again the test subject vessel for all the trials, conducted in August and September 1965 off the coast of Honolulu, Hawaii. A second ship, the USS *Granville S. Hall* (YAG-40), was assigned to Fearless Johnny as an escort and laboratory vessel.

Many of the men involved complained of negative health effects at the time and say they're now suffering from severe medical problems as a result of their exposures. After forty years, the DOD is trying to locate and assist qualified veterans. But here's the catch. Many of the veterans have already died, and although the Veterans' Administration (VA) says it will accept information provided on the location, dates, units, ships, and substances involved in the exercises, it also says that it cannot verify the accuracy of that information. In other words, veterans should be prepared to have concrete proof of their involvement or else be given short shrift when applying for disability claims. Veterans who believe they have legitimate claims should contact the VA Health Benefits Service at (877) 222-8287.

The Chemical Industry: Causing More Harm Than Good?

Nearly 140 years ago, Lewis Carroll wrote *Alice's Adventures in Wonderland.* As one of his characters, he created the Mad Hatter, an English term that also refers to someone who behaves in an irrational, bizarre, or delusional way. What Lewis Carroll did not know when he penned his famous book was that nineteenth-century hat makers would sometimes behave in odd ways because of simple chemical poisoning. Before beaver pelts were sent to England from the United States, they would first be treated with arsenic, lead, or mercury—three elements commonly found in the environment and in many contemporary products. The English hatters would lick the skins to make them soft and pliable as they worked them over, and would consume the toxins, which eventually caused bizarre speech and personality changes.

We don't know how many of today's physical and mental health problems are the result of chemical exposure, but of the seventy thousand or so chemical products currently on the market, quite a few have been linked to serious health effects. More than three hundred have been designated by the National Institutes of Health as carcinogenic and, according to experts, more people than ever are developing cancers of the lung, bladder, skin, brain, pancreas, and soft tissues as a result of exposure to toxic chemicals. In comparison with men born in the late 1800s, those born in the 1940s have twice the rate of non-smoking-related cancers. Women fare no better. Those born in the 1940s have a 50 percent higher cancer rate, including breast cancer, than women born just eighty years earlier. Overall, from 1950 to 1998, there has been nearly a 60 percent increase in both males and females of all cancer cases.

Even the most common chemicals, such as nitrates (the main ingredient in fertilizer and food preservatives), are associated with rising cancer rates. A 1996 National Cancer Institute (NCI) press release warned that dangerous nitrate levels in drinking water, particularly in rural areas, have led to a significant increase in the incidence of non-Hodgkin's lymphoma. The greatest increases have been in groups that consume the highest levels of nitrates and in farmers who use fertilizers on a regular basis. Biochemical studies in humans have also shown that when nitrates combine with water, they form *N*-nitroso compounds, many of which are known carcinogens.

Because there are two ways to report cancer trends—incidence rates, which refer to the number of age-adjusted cases per 100,000, and death rates per 100,000—there seems to be a disconnect between the positive statements we hear about winning the war against cancer declared by Richard Nixon in 1973 and the fact that cancer incidence rates are higher than ever. Although mortality rates may be lower because of advances in detection and treatment, and despite claims that cancer is on the decline, in most types of cancer we are actually losing the battle and, in some cases, losing it significantly.

According to the NCI's SEER Cancer Statistics, incidence rates for many common cancers have increased dramatically from 1950 to 1998, some by almost 500 percent! The greatest increases have been in breast (63 percent), testicular (125 percent), kidney (130 percent), thyroid (155 percent), liver (180 percent), non-Hodgkin's lymphoma (185 percent), prostate (194 percent), lung (248 percent), and skin melanoma (477 percent). Taken together, even with cancers that have seen a decrease, the overall increase in cancers during the past fifty years has been 60 percent. While the mortality rates may be lower due to better treatments, such a significant rise in cancer incidence is seen by many experts as an indication that something is terribly wrong.

The overall increase in cancer rates isn't simply the result of an aging population or better detection. In the last twenty years, childhood cancer rates have risen by 20 percent, making cancer the second leading cause of death after accidents. Incidence of the most common childhood cancer, leukemia, has increased by about 17 percent in that period, whereas the incidence of brain cancer has risen more than 25 percent. Combined, these two cancers, which have been linked to chemical exposure, account for half of all childhood cancer cases. One study found a strong relationship between brain cancer and chemicals like chlorpyrifos (trade name: Dursban) used against fleas and ticks. Other studies show leukemias and other cancers four times greater when children live near oil refineries, automobile factories, and chemical facilities. In England, it was observed that when children changed addresses, those who succumbed to cancer were more likely to have lived near hazardous facilities before or shortly after birth. Sadly, since chemicals are not tested for effects on fetuses, infants, or children, we simply have no idea whether exposure to even smaller doses than allowed are harmful.

Parents themselves may be poisoning their children without realizing it. A recent pesticide study found that residues last much longer than previously thought and may become concentrated at very high levels. For example, after homes were treated with chemicals, residues were present for as long as two weeks on dressers, carpeting, and toys that children put in their mouths. In one home, the concentration was six to twenty-one times higher than the recommended "safe" dose. Because sunlight is important in breaking down pesticide molecules, poisons disbursed indoors settle into materials and do damage for much longer periods.

For adults, the chemical revolution that started in the 1940s has had a different effect. Rather than altering normal brain or nerve development and function, many of today's chemicals attack the reproductive and endocrine systems and disrupt the sensitive balance of hormones that guide virtually every other system in the body. An adult born in the United States who was sampling for chemicals in his or her blood would probably find at least fifty industrial toxins known as endocrine or hormone disruptors. Many of these disruptors have been linked to cancer of the testes, prostate, breast, ovary, and uterus; and because of their accumulation in body fat and other tissues, they can be passed from generation to generation. More frightening is the fact that artificial hormone disruptors, because of "biomagnification" in the food web, are often present in concentrations millions of times higher than are natural hormones.

This is not a problem specific to the United States. The increase in global distribution and use of hormone-disrupting chemicals over the past thirty years has caused a disturbing increase in cancer rates throughout the industrialized world. In the United States, testicular cancer has increased by more than 40 percent, in England and Wales by 55 percent, and in Denmark by an astounding 300 percent. Since testicular cancer disproportionately strikes young men, the increase cannot be attributed solely to a rise in the aging population. At the same time cancer rates are increasing, sperm counts since 1960 have been steadily decreasing. In Europe alone, sperm counts have declined at a rate of three million per milliliter per year, with those born most recently having the lowest sperm counts.

Taking average sperm counts around the world, scientists have shown a drop of more than 50 percent from about 160 million per milliliter of semen to about 66 million per milliliter—roughly one-third as much sperm as is produced by a hamster. Also coincidental with the increasing use of hormone-disrupting chemicals is the ratio of male to female births, with female births outnumbering male births in many industrial nations. The chemical effect hypothesis is supported by data from Seveso, Italy, where a major dioxin spill resulted in the birth of a total of twelve daughters and no sons to nine couples with the highest dioxin exposures.

Hormone-disrupting chemicals also wreak havoc on women's reproductive organs. Since the chemical revolution that was supposed to make life easier for women who chose to stay at home, breast cancer in the United States has increased by 1 percent per year; in Denmark it has risen by 50 percent since 1945. The United Kingdom has seen significant increases as well. Could this be a matter of inheritance? According to epidemiologists, the increases are too great to be attributed to genetics or aging. As an example, researchers looking at Japanese women who emigrated from Japan, where the prevalence of breast cancer is only one-fifth that of the United States, observed that within a single generation the prevalence of breast cancer among Japanese women had become as high as it was in their new homeland. Today an American woman's lifetime risk of getting breast cancer is one in eight compared to one in sixteen in 1940.

The same chemicals that trigger cancer are also causing girls to enter puberty earlier than in previous years. Two centuries ago, North American women reached puberty at age seventeen. Today it's closer to twelve. Those five extra years of estrogen exposure create an entirely new set of health issues, since early puberty has been linked to an increase in reproductive problems and cancers later in life. Dr. Patricia Whitten of Emory University in Atlanta, Georgia looked at two hundred years of medical records for clues as to what is triggering such early puberty in industrial countries. Her conclusion is that hormone-mimicking chemicals are the culprits and that even though some, such as DDT, have been banned since the early 1970s, they remain everywhere in the environment and are still being used heavily in many areas outside the United States. One recent study found that women with blood DDT levels as low as twenty billionths of a gram per milliliter have a fourfold greater risk of breast cancer than women with levels of two billionths of a gram.

One would think, therefore, that the chemicals consumed and used in so many products by so many people would be regulated to at least some degree. In reality, the federal government does not screen chemicals for safety before they go to market, and if they do it's only after questions have been raised or incidents reported. Of the thousands of pesticides used, for example, only about 150 are formally registered with the EPA, which no doubt explains why hazardous and even deadly products can remain on the market for decades before their dangers are discovered.

Even when chemicals are known to be deadly, they continue to find their way into products used every day by millions. As parents watch their children climb the wooden beams of a playground set, they probably have no idea that the pressure-treated lumber used to manufacture the equipment was injected with so much chromate copper arsenate (CCA) that a twelve-foot section contains enough of the poison to kill 250 people. The parent who spends a weekend building a deck for his or her family might not realize that the U.S. wood products industry uses half of all arsenic produced in the world today. According to Renee Sharp, principal author of a report by the Environment Working Group, "In less than two weeks, an average five-year-old playing on an arsenic-treated playset would exceed the lifetime cancer risk considered acceptable under federal pesticide law."

Such contaminants as arsenic in wood products, lead in water, cadmium and manganese in soil, and pesticides in virtually every part of the home and environment could very well account for the behavioral changes in children over the last generation. That statement is based on new research by scientists at Dartmouth College, which shows that toxic pollutants cause people to behave with increased aggressiveness, commit violent crimes, develop learning disabilities, and lose control over impulsive behavior. Recent studies estimate that between 80 percent and 95 percent of females in the United States use pesticides. Tests of indoor air in Jacksonville, Florida showed pesticides in 100 percent of the homes studied.

Combined with poverty, social stress, drug abuse, and genetics, chemicals may be the principal ingredients for future social disaster.

Roger Masters, one of the Dartmouth researchers, has proposed a controversial idea that he calls the neurotoxicity hypothesis of violent crime. What Masters and his colleagues found is that low-level poisoning by toxic pollutants such as lead is associated with homicide, aggravated and sexual assault, and robbery. When lead uptake occurs at age seven, for instance, it is linked to juvenile delinquency and increased aggression. In one of the largest studies done, which examined behaviors of one thousand black children in Philadelphia, lead was associated with both the number and the seriousness of juvenile offenses.

Another pollutant, manganese, was also found to affect brain development. While lead in the brain damages nerve cells associated with inhibition and detoxification, manganese lowers brain levels of serotonin, a neurotransmitter associated with impulse control and mood regulation. When serotonin levels decrease as a result of manganese poisoning, there's an increase in mood swings, aggressive behavior, and impulsiveness. The most widely prescribed antidepressant, Prozac, works by increasing the amount of time serotonin remains in nerve synapses. Chemicals such as manganese could interfere with or even alter the biomechanisms specifically targeted by Prozac.

The argument for limiting these kinds of toxins is based on several factors, all of which have a profound effect on brain development and function, especially in children. For example, infants and children absorb up to 50 percent of the lead they ingest, compared with only 8 percent for adults, so that even amounts considered too small to be dangerous for adults can be deadly for children. The highest levels of lead and mercury uptake are reported in groups most likely to commit violent crimes, such as inner-city minority youths. Lead uptake is increased among individuals having diets low in calcium, zinc, and essential vitamins. Since calcium deficiency greatly increases the absorption of manganese, undernourished children are much more severely affected. In poor minority communities, which tend to have a disproportionate number of landfills, chemical manufacturing facilities, and toxic pollutants, black teenaged males consume on average 65 percent less calcium than whites. Alcohol increases the uptake of toxic metals such as lead and manganese. Since the poor consume more alcohol and less calcium, the combination of calcium deficiency, which increases manganese absorption, and alcohol abuse, which increases lead and manganese uptake, is equivalent to living in a toxic waste dump.

To test their neurotoxicity hypothesis, the Dartmouth scientists looked at FBI and EPA data for various socioeconomic groups. After controlling for factors that would abnormally skew their results, the group found that counties having the greatest lead, manganese, and alcohol consumption also had violent crime rates three times the national average. In what could turn out to be one of the most significant findings about pollutants and behavior, the Dartmouth study sheds new light on why the chemical industry, whose very survival depends on a continuous and ever-increasing supply of chemicals and toxins, is so worried.

The government's answer to the growing problem of dangerous compounds, specifically pesticides, was a 1996 law called the Food Quality Protection Act (FQPA), which places sharp limits and sometimes bans on the production of a number of these chemicals. In response, the chemical industry took a giant step in protecting itself and getting around EPA safety standards and FQPA regulations by doing what would seem unthinkable: replacing animal with human experimentation. What had been fed for decades to lab rats, mice, and guinea pigs was now, in some cases, actually being fed to human beings.

Since the United States has a more restrictive set of guidelines regarding use of human subjects, most of the recent studies have been conducted with paid volunteers in the United Kingdom. For example, Amvac Chemical Corporation of California funded researchers at the Medevel Laboratories in Manchester, England to test dichlorvos, a neurotoxic pesticide used in flea collars and pest strips under the name "No-Pest" and "Doom." As part of the experiment, adult men were given a mixture of dichlorvos dissolved in corn oil in order to measure acute health effects.

In a similar study, the French chemical company Rhone-Poulenc gave thirty-eight men and nine women orange juice laced with the highly toxic insecticide aldicarb. Used mainly on crops such as potatoes and cotton, aldicarb in humans causes nausea, diarrhea, and neurological symptoms. More recently, Inveresk Clinical Research Ltd., an international research laboratory in Scotland, gave human subjects oral doses of azinphos-methyl, a neurotoxic pesticide banned by the EPA because of its health effects in children. In many of these studies, the subjects had experienced adverse physical reactions and the experiments had to be stopped.

More recently, Loma Linda University, funded by military contractor Lockheed Martin, conducted the first large-scale human experiment to test the harmful effects of the drinking water contaminant perchlorate, which is one of the toxic components of rocket fuel. Researchers paid one hundred individuals one thousand dollars each to eat perchlorate every day for six months. What was known about perchlorate at the time was that it damages the thyroid, causes cancer, and prevents normal development in fetuses and children. Still, according to the Environmental Working Group, the human subjects were fed up to eighty-three times the safe dose of perchlorate set by the State of California.

But why do human experimentation when lab rats are cheap, expendable, and provide a reliable living system for testing chemicals? The answer, for better or worse, is economic incentives. By eliminating the safety factors that are applied when animals are used for testing, companies can legally increase the concentrations of chemicals to be used on crops and added to water and air. To understand how this is accomplished, we have to look at NOAELs and the EPA's method of determining what is safe and at what dose.

The EPA has a long-standing methodology for setting human exposure levels based on animal studies. A two-step safety protocol is implemented. In step one, animals are given incrementally smaller doses of a chemical until a dose with no effect is identified. Once this NOAEL (no observable adverse effect level, or the amount of chemical that could be administered without triggering a biological response) is established, the EPA adds a tenfold "interspecies" safety factor in case humans are more sensitive than the lab animals being tested. In step two, an additional tenfold safety factor is then added in case variations within human species (intraspecies effects) exist that make some individuals, especially children, more sensitive to the chemical than others.

Since recent EPA studies have found that some people are ten thousand times more sensitive to certain air pollutants than the average person, regulatory standards are often set at levels a thousand or more times lower than those considered toxic. In essence, the chemical being tested is significantly diluted from the time it enters the animal test stage to the day it's brought to market. The final concentration is the reference dose (Rf D), defined as the dose of chemical that the most sensitive human can consume safely every day for a lifetime of seventy-five years.

Increasingly, however, chemical companies have been eliminating the animal phase and going directly to human testing in order to reduce or eliminate the interspecies uncertainty factor. Their claim is that the NOAEL is too high or not needed and that it keeps some pesticides off the market altogether. By going directly to human testing, the standard tenfold animal safety factor is bypassed in favor of humans being used as the initial lab rats. The new emerging strategy substantially reduces the cost and time of testing while allowing as much as ten times more chemicals on food or in the water.

The real victims of this kind of manipulation by chemical companies are not necessarily adults. Since the toxicity of a compound is often much greater for fetuses, infants, and children, and since tissues and organs such as the brain are more vulnerable to toxins at an earlier age, the implications for society are staggering. Pound for pound, children get higher doses of chemicals than adults because they drink seven times as much water per pound, eat four times more food per pound between the ages of one and five years, breathe twice as much air, and have a greater surface area to volume ratio, which makes their bodies absorb chemicals more quickly through their skin. According to the National Research Council (NRC), "exposure to neurotoxic compounds at levels believed to be safe for adults could result in permanent loss of brain function if it occurred during the prenatal or early childhood period of brain development. This information is particularly relevant to dietary exposure to pesticides, since policies that established safe levels of exposure to neurotoxic pesticides for adults could not be assumed to adequately protect a child less than four years of age."

Like adults, children are also the victims of endocrine-disrupting chemicals, but in a different way. They are exposed in the womb during the critical period of development and in early life when their bodies are more vulnerable to absorption of contaminants. Every pregnant woman in the world has endocrine disruptors inside her that attack during narrow windows of fetal development and cause irreversible changes in her child's brain structure and function, which then leads to behavioral, intellectual, and social abnormalities. Some of these chemicals become toxic only after they go through the liver, and some that are not toxic to the mother may be very toxic to the embryo or fetus.

According to the Washington, D.C.–based Environmental Working Group, pesticide companies, farm groups, and food processors claim that there will be an increased reliance on direct human studies in order to avoid the tenfold interspecies uncertainty factor. Currently, six organophosphate insecticides, found to be toxic to brain and nervous tissue, and two carbamate insecticides have been submitted to the EPA for registration and regulation. Both rely on test results from human studies. A review is underway to determine the merits and ramifications of experiments relying solely on adult human tests.

One of the best kept secrets that the chemical industry wants desperately to keep from the general public is the danger of chemical mixtures. We know that certain chemicals can be highly toxic by themselves, but no one fully knows the real dangers of multiple exposures because of the cost and impossibility of testing the tens of thousands of chemicals now on the market and the thousand or so new chemicals added each year. To test just one hundred chemicals in combinations of three for a single effect would require more than 150,000 tests. Multiply that by the number of effects or diseases for each organ system and one can see why no one is even suggesting it. The fact is that while testing a chemical individually may not yield a statistically significant effect, mixing it with other chemicals can increase its potency a thousandfold, dramatically intensify its negative effects, and prevent it from ever getting to market.

Although there are only about seven hundred different active ingredients (an active ingredient is the chemical in a product that is principally responsible for the effect) in pesticides, in reality these are mixed with each other and with other chemicals to produce the tens of thousands of toxic formulations currently available. Once a "tolerance" level is set (the amount of toxic residue on a crop that a consumer can eat but that can still kill a target pest), the reference dose is set, which is the safe amount that can be eaten directly. The problem is that tolerance levels and reference doses are meaningless when pesticides are not used properly or when they are sprayed illegally in high concentrations. For example, the FDA data show that 25 percent of all peas contain illegal amounts of pesticides. The same has been shown for pears, blackberries, onions, and the apple juice mothers often give to their children because they assume it's healthier than soda. Furthermore, tolerance levels do not take into account exposure to a range of chemicals, which has the same effect at low levels as exposure to a single chemical at a much higher level.

Another well-kept secret is that chemicals are almost never studied for toxic effects at low doses. The rationale here is twofold. First, there's a common assumption that the higher the dose, the greater the effect. Second, for statistical purposes, higher doses produce better statistical results. The flaw in this strategy is that we are neglecting an entire body of studies showing that low doses of certain chemicals on some organ systems can actually be worse than or have the opposite effect as high doses. For example, the developing brain, nervous system, and endocrine system are especially sensitive to low doses of certain chemicals and hormone disruptors.

At this point you may be wondering how a system with so many rules and regulations to prevent fraud and protect citizens can allow this to happen. But it's exactly because bureaucracies have gotten so large and have had to deal with so many issues and individuals that the whole process is tailor-made for corruption. As an example, let's look at an investigator's account of what happened at IBT, a toxicology laboratory responsible for nearly half of all the consumer products, pesticides, and drugs submitted to the EPA and FDA.

Adrian Gross, a pathologist for the FDA, was doing what he'd done every day. On his desk in front of him were growing stacks of papers, reports, charts, and results from completed studies waiting to be checked and shuffled over to the next reviewer in the system. Normally, the procedure was routine enough that a quick perusal would have been sufficient. But on that particular day, when Gross examined a rat study of the arthritis drug Naprosyn, his gut told him something was wrong. At first glance, the data and results just didn't look right, so he decided to dig a little deeper. His initial instincts proved correct because after further review his team of investigators discovered that scientists had faked data by switching around sick and healthy rats or by inventing data for nonexistent rats.

"IBT is the worst anyone's ever seen," said Dowell Davis, one of the investigators. "They were hell-bent on providing their clients with favorable reports. They didn't care about good science. It was all about money. They really had what was almost an assembly line for acceptable studies." Further investigations into other companies' research found that data were sometimes omitted or simply made up in order to improve statistical significance; and some animal deaths were deliberately ignored in final reports to conceal potential dangers and side effects.

A principal testing lab for DuPont and Monsanto, Craven Laboratories of Austin, Texas, committed similar acts. Fifteen of its employees were charged with fraud in 1990 after investigators uncovered phony studies on twenty pesticides. When the EPA's inspector general later looked into problems with oversight, it found that the agency had audited just 1 percent of the more than two hundred thousand studies done by eight hundred pesticide labs in the United States. Of the studies investigated, many were audited only after the pesticides were on the market.

Although this kind of behavior is not pervasive, scientific fraud has been increasing, not only because of the tremendous amount of money involved in research, development, and potential future revenues, but also because scientists themselves fear for job security. Grant renewals are often contingent on positive results. A drug company that awards grants to university scientists, whose jobs often depend solely on grant money, may seek out individuals willing to do what they can to ensure those results. One researcher told me that when he was offered a position at a major midwestern university he was given two years to obtain major outside funding. If he failed to do this, his third year could be spent looking for another job or a new line of work. That kind of stress and pressure to bring money to a research institution, especially on individuals with families to support, can do more to institutionalize scientific fraud than anything else.

Some years ago, I personally testified to NIH investigators about what I had witnessed as a researcher at a major medical university. That story, which I include in chapter 7, is revealing in that it illustrates the lengths to which some scientists will go to chase the ever-shrinking piece of the grant pie.

Sometimes it's about money; sometimes it's about national security or vital national interest. But it's always about people and how they are affected by governments and organizations that often care little about the consequences of their actions. Chemicals are only the tip of the iceberg. The biological toxins and agents discussed in the next chapter are what experts really fear most.

2 NATURE'S WEAPONS: MAN AND BIOLOGICAL WARFARE

The Tatar army gathered outside the walled city of Kaffa along the southern border of Russia in preparation for its final, desperate assault. On the scorched earth in every direction, besieged troops lay dying, some from fatal wounds, many from bubonic plague carried along the Venetian and Genoese silk trade routes. The infected, covered head to foot in festering boils and vomiting blood so violently they could hardly breathe, were abandoned and left to drag themselves away from the main army where they would collapse and die. It was on the eve of that final battle in 1347 that the corpses were collected and thrown onto large wooden catapults for what was to be one of the first documented cases of systematic biological warfare in history.

What ensued must have been nothing short of gruesome. Catapulted cadavers flew into the walled city one after another, crashing like foul stones against buildings and along cobbled streets. The air thickening with contamination, the water poisoned with the infecting agent, a growing mountain of decaying flesh filled the air with a stench so putrid that even birds must have fallen from the sky. By all accounts, the ploy was so effective that terrified city dwellers fortunate enough to have survived the onslaught fled Kaffa in fear for their lives. Many escaped in rat-infested sailing ships to various Italian ports that subsequently served as centers for the spread of disease. Carrying the plague to distant cities such as Constantinople and Venice, infected travelers caused the spread of "Black Death" throughout western Europe, thus eliminating a third of the population.

Perhaps the battle of Kaffa was an omen of things to come, because biological agents have been used as a means of eliminating populations ever since. The crusaders similarly left plague-ridden bodies in the camps of infidels. At the battle of Carolstein, bodies of plague-stricken soldiers plus two thousand cartloads of excrement were hurled into the ranks of enemy troops. In the fifteenth century, Pizarro is thought to have infected South American natives with smallpox-contaminated clothing. During the French and Indian War, Sir Jeffrey Amherst, commander of the British forces in America, ordered that smallpox be spread among Indian tribes, writing in a 1762 letter, "You would do well to try to inoculate the Indians by means of blankets as well as to try every other method that can serve to extirpate this execrable race." Within months of his order, an epidemic ensued and killed a large portion of the Indian population. Back in Europe, Napoleon tried to force the surrender of enemy troops by infecting them with swamp fever. A hundred years later, Dr. Luke Blackburn, the future governor of Kentucky, used biowarfare against Union troops by infecting clothing sold to soldiers with smallpox and yellow fever. Also General Johnston, retreating from Vicksburg during the height of the Civil War, infected ponds and lakes with the bodies of decaying sheep and pigs.

Since the dawn of civilization, man has witnessed nature wage its own biological war against itself. Epidemics from bubonic plague to influenza to smallpox to AIDS are a grim testament to the awesome power that life's tiniest microbes have over the planet's most dominant organism. It should not surprise anyone that it would only be a matter of time before someone decided to use that power and change forever the way we look at nature's bioweapons.

Unlike nuclear weapons systems, which are expensive to build and maintain and require an extensive personnel network, biologics are much less complex. Biological weapons make use of microorganisms, such as bacteria or viruses, and are either sprayed, dropped by bomb, or carried by vectors such as insects to infect the targeted enemy and cause disease. Arsenals include, among others, agents of anthrax, botulism, brucellosis, cholera, plague, smallpox, typhoid, and yellow fever. They are more difficult to remove from the environment than chemicals, cover a much larger area in comparison, and can produce stunningly gruesome effects. We can only imagine what new classes of microorganisms will be produced through advances in genetic engineering.

However, bioweapons needn't be delivered by bomb or during war to be effective or deadly. They can be introduced surreptitiously in small towns and local communities, and can potentially spread infectious diseases to epidemic proportions. Take the case of the Bhagwan Shree Rajneesh cult that did exactly that in late September 1984 in a small northwestern town in Oregon.

It was a typical fall afternoon that day in Oregon when one of the townsfolk decided to quit work and walked the two blocks to the local diner where he spent most of his lunch breaks. He greeted friends, joked with the staff, and studied the chalkboard menu beneath the mounted head of a deer. Not particularly hungry that day, he paused over the salad bar, picking through the trays of lettuce, tomatoes, and condiments before moving on. When his plate was half full, he ladled on salad dressing, then stopped at the coffee urn to fill his cup. A broad smile across his face, he handed the cashier who'd been there as long as he could remember a ten dollar bill and never suspected that he was about to be the town's first target of biological warfare. In the restaurants across the street and around the corner, similar scenarios were being played out.

That evening, several people developed fever and nausea. Within a week, doctors received thirty complaints of illness, which by then was identified as salmonella (food poisoning). In two weeks that number rose to two hundred, and it grew to nearly one thousand by the time the epidemic was contained. How did this happen? Who was behind it and why? As it turned out, the Rajneeshis, who lived on the outskirts of town, had an underground tunnel that led to a secret laboratory where they cultivated salmonella they'd purchased from a Maryland bioresearch supply house. Judging by the success of the epidemic, the cult members were very adept at what they were doing. After growing the bacteria in a medium, they suspended it in liquid and transported it in tubes to local restaurants where they poured it on salads and into salad dressing and coffee creamers. Their rationale was to make people so sick that they would not be able to make it to polling places to vote against cult interests in a local election. Similar incidents, such as the failed attempts to release botulism and anthrax at eight Tokyo locations in the 1990s, have been reported around the world since then.

Biological agents differ from chemicals in that they not only disrupt or destroy physiological systems but can also reproduce, thereby staying in the environment for years, even decades, and can actually become more toxic over time. In that respect, biological weapons pack an enormous lethal punch. An individual bacterium, for example, can produce billions of offspring in less than twenty-four hours. In the case of anthrax, a bacterium can literally survive for centuries without nutrients, in its spore form, until it finds a host to infect. If Japan had used anthrax at Pearl Harbor, the entire area would still be contaminated today. Gruinard Island, a picturesque island off the coast of Scotland, was off limits to humans from 1941 to 1987 because of anthrax contamination by British scientists. It would still be contaminated had the soil not been soaked with hundreds of thousands of liters of formaldehyde. The incredible stability and survivability of anthrax spores, as well as their propensity to increase rapidly in soils soaked in blood, make it the bioweapon most desired by nations doing biological warfare research.

So what would an attack scenario using inhalable anthrax be like? It would probably come unexpectedly, by way of a strategically placed canister programmed to open once the terrorist was gone, or perhaps by a bacterial bomb containing insects, infected feathers, or balls of cotton batten. In most cases, you would not even be aware that a cloud of deadly pathogen had engulfed you. The released weapons-grade spores, one to five micrometers in size, are small enough to be unnoticed as you inhale them deep into moist lung tissue where they reconstitute and begin to reproduce almost immediately. The warm, conducive environment stirs the bacteria into a frenzy of activity that causes them to release toxins, which attack every cell in your body. Within twenty-four hours, you become nauseous and short of breath. Your chest aches as though something inside of you is trying to find its way out. Less than a day later, you're burning up with fever, fluid builds inside your lungs, your bladder spasms uncontrollably, and soon you're vomiting a vile mixture of yellow fluid and white foam. By now antibiotics and vaccines are useless because the bacterial toxin coursing through your body is a massive tidal wave of poison that is virtually unstoppable. As your brain cells are literally eaten away, you become delirious and go into shock. That's when your heart, spleen, and other organs hemorrhage. As blood oozes from openings in your body, you gasp for air and die one of the most horrible deaths imaginable. The entire unspeakable ordeal, triggered by as few as five thousand microscopic spores, will take no more than three days.

A single bomb with a biological agent such as anthrax does far more damage than a single bomb or canister filled with chemicals. Ten grams of anthrax, for instance, could affect as many people as a ton of the nerve agent sarin. Another way to look at it is that the aerosolized release of anthrax spores upwind of the Washington, DC area could kill more people than a hydrogen bomb. Former Defense Secretary William Cohen said that a quantity the size of a bag of sugar could wipe out half the city's population. According to scientists and experts in the field, biological agents cover an area ten thousand times greater than any chemical agent developed, and it's this tremendous ability to proliferate that makes biological agents much more potent overall.

A CIA report made public in October 2002 states that "Iraq admitted producing thousands of liters of the biological agents anthrax, botulinum toxin, and aflatoxin." Intelligence officials, who told U.N. weapons inspectors that Iraq is hiding as much as 7,000 liters (1,800 gallons) of anthrax that was supposed to be destroyed in 1995, are worried that Saddam Hussein could easily have hidden the agents unless totally unfettered access to virtually all military and civilian facilities is allowed.

If anthrax is at or near the top of everyone's short list, smallpox must run a close second. By the time it was declared officially eradicated in 1980, smallpox had killed, maimed, or disfigured more than 10 percent of all humankind who had ever lived (more than any other infectious disease, including the Black Death of the Middle Ages). Explosively contagious, smallpox can infect a host who inhales just a few particles, then spread like wildfire through air, a sneeze, or even a conversation between people. Ten days later, a fever develops, followed by vomiting and red spots all over the body. The spots become pustules, which enlarge and split at the dermal level beneath the outer skin, where nerve endings are located. When splitting occurs, it is excruciating. Death is caused by shock, cardiac arrest, or collapse of the immune system. In the worst case—called extreme or black pox—the virus ravages the membranous linings of the throat and digestive system from the mouth to the rectum and can destroy the body's entire layer of skin. This form is close to 100 percent fatal.

Unfortunately, since routine smallpox vaccinations ended decades ago, few individuals retain immunity to the disease. Other than the U.S. Centers for Disease Control and Prevention (CDC), the only other known stockpile of smallpox virus exists in Vector, the Russian State Research Center for Virology and Biotechnology in Koltsovo near Novosibirsk, formerly the Soviet Union's top secret biowarfare facility, which experts believe still conducts bioweapons research and development. Even small viral stocks finding their way into the hands of terrorist organizations would become, in short order, a threat to the entire world. Recent intelligence suggests that Iraq, North Korea, and France also have stockpiles of smallpox.

Besides the proliferation aspect, one of the most attractive characteristics of biological weapons to poorer countries and terrorists is that, in comparison with other weapon systems, they are easy to transport, longer lasting, and relatively cheap to manufacture. In fact, biological weapons have been referred to as the "poor man's atomic bomb" because, dollar for dollar, conventional weapons cost several thousand times more to inflict the same number of casualties as would a single biological agent. Kathleen Bailey, former assistant director of the U.S. Arms Control and Disarmament Agency, said that a major biological arsenal could be built in a 225-square-foot room on a budget of ten thousand dollars.

Almost immediately at the end of the World War I, the Japanese Army began the study of biological warfare, the rationale being that bioweapons would be the most economical means of waging war for a country so poor in natural resources. The program grew over the next two decades and became established as a major military operation. Intense research into biological agents for the United States began in earnest when reports emerged that both the Japanese and Germans were actively developing biological weapons. Military leaders naturally felt compelled to respond. President Roosevelt ultimately agreed, and in 1942 Secretary of War Henry Stimson established the biological warfare program at Camp Detrick, Maryland under the auspices of the War Research Service (WRS) and headed by George W. Merck, president of the Merck Pharmaceutical Company. Camp Detrick, home to the second largest scientific project after the Manhattan Project (which created the atomic bomb), was later renamed Fort Detrick, which today is the center of continued biological weapon, viral, and cancer research. One of its main sections is the U.S. Army Medical Research Institute of Infectious Diseases (USAMRIID).

Biological agent experiments, including the CIA's mind control projects, were the most sensitive of all the government research programs because they involved the exposure of human beings to an array of drugs, toxins, and microorganisms known to have potentially serious effects. Nonetheless, for nearly three decades, the U.S. government, together with civilian research scientists, conducted a massive research effort involving thousands of human subjects. Were it not for the inadvertent discovery of top secret documents that exposed these projects, many of the experiments might have continued.

Classification and Characteristics of Biological Agents

There's certainly nothing new about biowarfare. Ever since man began observing what nature can do to human beings, it seems that diseases and natural products have been used to inflict death and destruction. What is different about the twentieth century is the intensity of bioweapons research and development and the advances in technology that have allowed the production of biological agents in quantities that could literally wipe out the world's entire population several times over.

Tactically, biological agents have three characteristics that make them ideal weapons. (1) They are highly infectious and toxic. One of the goals of bioweapons makers is to produce the greatest effect using the smallest number of organisms. For example, enough anthrax spores to fit on the head of a pin may be lethal. (2) They can be extremely contagious. An effective biological agent is not only lethal to the individual but to other individuals who come in contact with the infected victim. For instance, smallpox is one of the most contagious diseases in the world and can spread through a population with relentless speed. (3) They can be genetically altered to make them more virulent and less responsive to antibiotics. One of the real fears experts have is that nations such as Russia may have developed "supergerms," which are resistant to anything we have.

Biological agents are typically classified as *pathogens*, which are disease-producing microorganisms such as bacteria and viruses; *toxins*, which are poisons that are produced naturally by living organisms such as plants, microbes, and animals; and other agents of biological origin, such as bioregulators and small proteins. Of the many types of biological agents, only a few have weapons potential. Here are the leading candidate diseases caused by these agents:

Anthrax: An acute bacterial zoonotic (communicable from animals to humans) disease that usually affects the skin (cutaneous anthrax) and digestive system but in the most severe cases is inhaled (pulmonary anthrax). In order for anthrax to be weaponized, the spores must be less than five micrometers in diameter and are usually coated with another chemical so as to be more easily aerosolized and dispersed. With pulmonary anthrax, the number of spores inhaled is critical. In the worst-case scenario, an infected individual may quickly develop symptoms resembling an upper respiratory infection, go into shock, and die within forty eight hours. A recently published article in the journal *Vaccine* claims that Russia has already developed a vaccine-proof anthrax.

Botulism: A severe toxic condition produced by the botulinum bacteria, the most common source is contaminated food. Symptoms of botulism include double vision, difficulty swallowing, vomiting, constipation or diarrhea, and paralysis that progresses downward through the body until the entire body is affected. Neurological symptoms can appear within twelve hours. Unless intravenous and intramuscular antitoxins are administered promptly, death ensues as a result of respiratory failure.

Cholera: This is an acute and severe diarrheal disease transmitted by water and food. The cholera toxin attacks the mucosal epithelium, causing vomiting and rapid dehydration. One of the most rapidly fatal illnesses known, incubation can be a few hours to a few days, and the symptoms must be managed promptly to prevent cardiovascular collapse. A healthy person may become hypotensive within an hour of symptom onset and die within two to three hours if not treated. More typically, the disease progresses from liquid stool to shock within four to twelve hours, with death following in as little as eighteen hours.

Ebola: One of the least talked about candidates for a bioweapon, Ebola virus has the potential to kill hundreds of thousands of people if released into a congested area, such as the New York subway system. Ebola virus is highly contagious, has no known effective treatment, and causes severe hemorrhaging and fluid leakage by attacking blood vessels and triggering release of cytokines and anticoagulants. Within a few days, an infected individual literally becomes a sack of blood, fluid, and bones before dying of shock and organ failure.

Plague: The disease that killed one-third of the European population in the fourteenth century, plague (even more so than smallpox) is the most lethal and invasive disease known to man. The pathogen remains viable for weeks in water, soil, and grains, and can survive for months to years at near-freezing temperatures. The most common vectors are rodents, squirrels, and prairie dogs, but humans typically are infected by fleas that live on rats. Unless treated, plague can progress to three different forms: (1) bubonic plague, which causes fever, headaches, nausea, vomiting, diarrhea, and liver damage; (2) septicemic plague, which attacks the bloodstream and causes fever, chills, hypotension, and blood clotting in the limbs, with necrosis and gangrene; and (3) pneumonic plague, which infects the lungs and may lead to respiratory failure and shock within eighteen hours.

Rickettsia (Query or Q fever): An acute disease characterized by fever, sudden headache, sweats, chills, and malaise, rickettsia is usually spread by airborne dissemination of excreta by domestic livestock or by direct contact with infected animal products such as wool, straw, or milk. More debilitating than it is lethal, rickettsia typically lasts for two to fourteen days.

Smallpox: Eradicated in 1979 by the World Health Organization (WHO), the only known sample of this ideal killer is at the CDC in Atlanta, Georgia and its counterpart in Moscow. However, it's suspected that other countries, such as Iraq and North Korea, may also have stockpiles. Even in its original form, the smallpox virus is easily cultured, extremely contagious, and hardy enough to survive an aerosolized suspension that can spread over a fifty-square-mile area.

Staphylococcal enterotoxin B (SEB) poisoning: One of several exotoxins (toxins found outside the bacterial cell) produced by *Staphylococcus aureus*, SEB, when ingested, produces the same symptoms as food poisoning. Inhaled SEB causes fever, chills, muscle pain, and difficulty breathing within three to twelve hours. In the worst-case scenario, fevers of 106 degrees Fahrenheit may last several days and lead to shock and hypoxia.

Tularemia: This is a zoonotic bacterial disease that produces a variety of clinical symptoms, such as ulcers at the site of infection, enlarged lymph nodes, abdominal pain, diarrhea, vomiting, and pneumonia-like illness. The most common transmission is through animal vectors, especially insects and arthropods. It is thought that a highly resistant strain has been developed that destroys the respiratory system.

Unit 731: Japan's Infamous Biowarfare Factory

The harvesting of subjects was becoming almost routine, even when the locals knew that enemy soldiers were in the area. But that's exactly why the Japanese military had moved the center for bacteriological research in 1936 from the Army's Medical College in Tokyo to Ping Fan in northern Manchuria, where the supply of Chinese and Russian peasants would be virtually unlimited. The head of the most infamous human experiment facility, its name changed to Unit 731 in 1941, was General Shiro Ishii, a graduate of Kyoto University and a rabid proponent of biological weapons research. As one of twenty-six known killing laboratories in China, Unit 731 housed a variety of military and medical personnel who made no secret of their loathing for the Chinese people. As one veteran of Unit 731 openly admitted, "They were logs to me. Logs were not considered human. They were either spies or conspirators. They were already dead, so now they die a second time. We just executed a death sentence."

Twenty kilometers southwest of Harbin, Unit 731 was constructed for the sole purpose of human experimentation and the study of biological agents. Located on thirty-two square kilometers of uniformly barren plains, the 150-building facility stood like an isolated penal colony, a combination prison, medical laboratory, worker dormitory, army barracks, and museum of horrors. A former worker at Unit 731 testified that he saw six-foot-high glass jars containing humans cut in pieces and pickled in formaldehyde. Smaller jars were packed with internal organs, hands, feet, and heads, all labeled as Chinese, Korean, Mongolian, English, French, and American. A common practice at the facility, according to other workers, was to lock up diseased prisoners together with healthy ones to measure how readily the disease would spread and how quickly the men would die. The most gruesome secrets, by far, were the vivisections, carried out as callously as if the subjects were nothing more than expendable lab animals.

The stories told by Unit 731 doctors and medical assistants at the war criminal trials were grotesque enough to be almost unbelievable. Imagine, if you will, listening in shock to horrific accounts of cruelty so unspeakable that they make your stomach turn. Then try to envision what it must have been like to catch sight of an approaching squadron of Japanese soldiers and know at once that life for you is virtually over.

It had been several hours since the Japanese Imperial Army dropped canisters of plague-infected fleas onto a village near Ningpo in the Chechiang province on October 29, 1940. Everyone in the farming village thought it may have been a reconnaissance flight or perhaps a training mission, because it didn't seem as if the munitions had done any damage. However, within days they knew better because they suddenly came down with fever and got sicker by the hour. By the time plague had taken hold, soldiers arrived, covered in airtight suits and facemasks to protect them from the bacteria while they rounded up a group of frightened men and locked them into a specially constructed transport vehicle.

The trip to Unit 731 took several hours. On the way past smaller villages and open fields, the prisoners remained silent. They thought of their families left behind, wondering if they'd been taken as well or if they would simply die from disease within the next few days. The men had all heard rumors, but they'd seen others die from cholera, plague, typhoid, and even anthrax. Yet they accepted their fate and continued to stare almost trancelike at the metal walls inside the vehicle as it lurched forward and ground to a sudden stop. When the doors opened, two soldiers directed the men out and ushered them toward a brick building. Once inside, the infected men were separated and confined to small, cagelike cells adjacent to the experiment rooms.

The following morning, a shrill rattle echoed through the empty corridor when one of the cell doors was opened. Squinting through the dimness of his surroundings, one of the newly arrived prisoners barely made out the shape of a guard who stepped into the cage and grabbed him by the arm. The young man was forced out and escorted to a room where a doctor and his team of medical assistants were waiting next to a bed. Stripped of his clothing, each of the man's limbs was then tied to the bed as he looked up and searched the grim faces above him for any clue of what was about to happen. Assuming that he was to be given anesthesia, he didn't struggle until he saw the doctor pick up a scalpel and place the blade in the middle of his chest.

In a futile effort to dislodge the bindings that were holding him down, the prisoner grimaced and cried as the sharp edge, which sent a chilling shudder through every inch of his body, was pressed against his sternum. The initial incision triggered a blood-curdling scream. The medical staff, some of which had witnessed the procedure dozens of times, observed without emotion. They shifted this way and that for position, stretching their necks to get a better view of the live dissection, which they were told was necessary to demonstrate how the disease had ravaged the subject's internal organs but which had to be done without anesthesia so that the blood vessels and organs to be examined were not affected. As the scalpel imbedded itself down against the bone, the doctor sliced through the soft flesh effortlessly until he reached the lower abdomen and exposed the stomach, liver, pancreas, and intestines. The wailing and screams of agony intensified and then dissipated as shock set in. All that the man could offer was a guttural moan before he took his final breath and died. A former medical assistant, who insisted on anonymity, described this kind of vivisection almost nonchalantly, saying, "This was all in a day's work for the surgeons."

According to some estimates, some twelve thousand individuals died at the hands of the Japanese, either as a direct result of medical experiments or because they were no longer needed and therefore were summarily executed. Another two hundred and fifty thousand were exterminated in "field tests" when entire areas would be contaminated with plague-infected fleas or sprayed with infectious germs. Inside the labs, subjects typically survived for a few days to a few months depending on the type of experiments done. Those selected for practice surgeries by doctors traveling to China in order to hone their skills on live humans were marked for death that day. In these cases, anesthesia was given, and the victims would be used for amputations, appendectomies, tracheotomies, and other surgical procedures before being killed by injection. The less fortunate were those who participated in medical experiments and tests before being infected with plague bacteria or other pathogen.

For example, to determine how best to treat frostbite, a prisoner would be taken outside in subfreezing temperature and left with his forearm exposed, which was drenched in water until it froze solid. After a few days, the forearm would be amputated and the prisoner taken out to perform the experiment with his upper arm. This was repeated with the other arm and both legs until only the prisoner's head and torso were left, at which time he would be infected with a pathogen such as plague and subsequently vivisected without anesthesia.

Testimonies by former Unit 731 workers include accounts of burn experiments with flamethrowers, bullet and shrapnel tests, pressure chamber experiments to determine the amount of pressure the body can withstand before the eyes pop out of their sockets, centrifuge tests in which prisoners would be spun to death, gassing experiments, intravenous injections with seawater to test its effectiveness as a substitute for sterile saline, and bizarre surgeries that included removal and reattachment of organs and amputated limbs. None of these seemed especially cruel to the perpetrators, some of whom are not remorseful to this day. As an aging Japanese farmer and former Unit 731 worker said recently, "There's a possibility this could happen again, because in war you have to win."

As the war neared its end, Unit 731 personnel proposed a suicide attack on the United States. Code-named Cherry Blossoms at Night, the plan was to use kamikaze pilots to infest California with plague-infested fleas, much as they had done throughout northeastern China. The plan, whose target date was September 22, 1945, was never carried out because of the Japanese surrender in August. During those final days, Unit 731, along with the other human experiment facilities was blown up in an effort to destroy evidence.

Although the United States had known about Japan's biological research and human experiments years before the end of the war, prisoners of war captured in the South Pacific, as well as Japanese naval sources, revealed the true extent of the program. Yet despite the evidence of war atrocities, the importance of the data to American military scientists was paramount. Colonel Sanders, who interrogated many of the Unit 731 leaders and scientists, recommended to General Douglas MacArthur that no one involved in biological warfare research be prosecuted as a war criminal. General MacArthur, realizing the significance of Japan's research efforts, not only agreed to immunity in exchange for data but also ordered Colonel Sanders to keep silent about the human experiments.

As early as January 1946, the U.S. Army's unofficial newspaper, *Stars and Stripes,* claimed that Americans were part of Ishii's human experiments. Because of MacArthur's grant of immunity, however, nothing was done, even after it was learned that as many as 1,174 servicemen were used as guinea pigs in medical research, some of whose livers were purportedly eaten by the experimenters after they had killed them. The Allied war crimes tribunal on March 11, 1948 found twenty-three defendants guilty of war crimes. Five were sentenced to death. As supreme commander of the Allied forces, General MacArthur reduced the sentences in 1950. Eight years later, all of the convicted, included those sentenced to death, were free men.

The fact that General Ishii's experiments were the only known source of data collected on the effects of biological agents on living human beings was enough for the U.S. government to overlook one of history's most heinous crimes against human rights in exchange for scientific information. It was later discovered that the CIA had hundreds of thousands of pages of records and documents, including the operational records of Unit 731. Other than the few medical workers who were prosecuted in 1948, no members of the Japanese biological warfare group were ever indicted or prosecuted. For his part, General Shiro Ishii was offered not only protection and immunity but also received a rather generous retirement package until his death in 1959 at the age of sixty-nine.

The legacy of Unit 731, as well as the shameful decision by the U.S. government to keep secret the details of human experiments in exchange for biowarfare data, haunts us today. If nations could hide such atrocities and give blanket immunity to such butchers, what else have they hidden from their past or will they forgive in the future? Exposing such sins will at least make us realize that even civilized nations, given the right circumstances, can become uncivilized before our very eyes.

Fort Detrick: 60 Years of Biowarfare Research and Human Experimentation

There was an especially dire sense of urgency at the height of World War II. Intelligence knew the extent of Germany's biological warfare program and feared that the rockets and bombs raining down on European cities might easily be converted to bioweapons. That urgency gave birth in 1942 to Fort Detrick, a ninety-two-acre tract chosen for its remote location and proximity to Washington, D.C. and Edgewood Arsenal, home to the U.S. Chemical Warfare Service. Touted at the time as the world's largest and most sophisticated biological warfare facility, it employed nearly five hundred scientists, many of them microbiologists and plant pathologists. Secretary of War Henry Stimson conveyed U.S. determination to counter whatever Germany had when he wrote:

> The value of biological warfare will be a debatable question until it has been clearly proven or disproven by experiences. The wide assumption is that any method which appears to offer advantages to a nation at war will be vigorously employed by that nation. There is but one logical course to pursue, namely, to study the

> possibilities of such warfare from every angle, make every preparation for reducing its effectiveness, and thereby reduce the likelihood of its use.

The purpose of Fort Detrick was twofold: to develop defensive measures against biological weapons attack, and to research and develop weapons that the United States could use to respond "in kind" if attacked by an enemy using biological agents. One of the key figures advising President Franklin D. Roosevelt was George W. Merck, president of the pharmaceutical corporation that still bears his name. As Lieutenant Colonel Richard Clendenin, author of *Science and Technology at Fort Detrick, 1943–1968,* wrote, "This was an enormous task. . . . It was literally without precedent and had to be prosecuted with all possible haste. . . . The effort was cloaked in the deepest war time secrecy, matched only by the Manhattan Project for developing the Atomic Bomb."

In May 1942, the Federal Security Agency (FSA) was assigned the task of leading the biowarfare effort in order to obscure its existence and its real intent. George Merck was named director of the War Research Service (WRS) in August 1942; its mission was to oversee construction of the laboratories and the actual establishment of the biological warfare program. An important element of Fort Detrick research was to investigate how diseases were transmitted, whether by inhalation, digestion, or through skin, and to establish methods for creating effective and virulent microbes while developing protective measures against them. Critical to that end were experiments involving human subjects.

So secret was the biowarfare research that the American public didn't learn about it until January 1946 in a report released by the War Department. One of the goals, according to that report, was to investigate offensive possibilities in order to learn what measures could be taken for national defense. Perhaps the most famous and ubiquitous product developed was 2,4,5-T, the dioxin-containing herbicide used in Agent Orange and so indestructible that it remains in the environment decades after its ban.

By the time President Nixon announced the end of the U.S. offensive biological warfare program in 1969, tens of thousands of soldiers, seaman, and civilians, with and without knowledge, participated in experiments involving biological agents. Although few died, many became ill, and a significant number have developed diseases in much higher numbers than could otherwise be explained. And because these individuals were sworn to secrecy and had their medical records secured for the sake of national interests, it's impossible to know the true account of their health and mortality.

October 19, 1971 marked the dawning of a new era for Fort Detrick. On that day, President Richard Nixon, in a ceremony at Post Headquarters Building 812, declared his war on cancer and with it the creation of the Frederick Cancer Research and Development Center as part of the National Cancer Institute. The seventy buildings that had once been home to the most dangerous microbes on earth were now the Frederick Cancer Research and Development Center. Today it is known as the National Cancer Institute at Frederick, and the work conducted there involves research not only on the fundamental biology and genetics of cancer but also in virology, immunotherapy, and retrovirology (the study of special viruses such as HIV). It also houses the U.S. Army Medical Research Institute, which conducts research on vaccines, drugs, and countermeasures for biological warfare.

What else had been going on that would have justified an increase in the biowarfare budget a year after Nixon pledged to end offensive biological warfare research? The offensive biowarfare program may have "officially" ended, but defensive biowarfare projects have continued unabated; and many doubt that research on defensive capabilities is possible without concomitant research on offensive weapons. There has also been declassification of a top secret "Special Virus Cancer Program" (SVCP) designed to discover and even create cancer viruses. More on this in chapter 6. But just as the mysteries of Fort Detrick have been dying down, a recent development has people concerned anew about its activities and fearing a new kind of biological warfare.

In June 2000, the U.S. Senate approved a 1.3 billion dollar aid package for Colombia under the condition that it would step up its use of chemical pesticides to eradicate drug crops. During the previous eight years, despite dumping almost three million liters of Monsanto's herbicide Roundup on 350,000 acres of coca and 110,000 acres of opium poppies, coca production tripled while nontarget plants were destroyed and water supplies contaminated. There was a desperate call for more action in the war on drugs. But what the U.S. Department of Defense tried to keep under wraps was the addition of a new and deadly weapon in its biological arsenal.

Developed at Fort Detrick's Agricultural Research Service (ARS) branch, *Fusarium oxysporum*, a fungus known as EN4, can make treated soil unfit for coca production for up to forty years by releasing a mycotoxin that kills plant roots. After years of molecular genetic manipulations, plant pathologists, working with the DEA, had finally come up with a strain so virulent that its use is prohibited in the United States. By DNA sequence encoding, ARS scientists had literally transformed *Fusarium oxysporum* into a fungus with enough pathogenicity to wipe out coca and opium production.

Although the potential of Fort Detrick's latest research effort is enormous, *Fusarium* has been shown to cause disease and even death in humans, especially in those with depressed immune systems. As it does in plants, once *Fusarium* infects humans, it releases mycotoxins that dissolve cell membranes, enter the cell, and reproduce. From there, they invade more cells in a progressive fashion, causing weakness, fever, skin lesions, painful ulcerations, necrosis, and, in some cases, cancer. One medical study found that 76 percent of *Fusarium*-infected patients with lowered immunity from other illnesses died. In areas with high rates of HIV, malnutrition, and other health problems, the use of *Fusarium oxysporum* would be equivalent to legalized biowarfare.

In an effort to keep the program alive, Madeleine Albright, former U.S. secretary of state, had urged the UN Drug Control Program (UNDCP) in 1999 "to find more support for fungal eradication, and to solicit funds from other governments in order to avoid the perception that this is solely a U.S. Government initiative." The rest of the world has not responded very favorably, citing mounting evidence that America's newest biological agent may have long-term effects on human health and biodiversity. Nonetheless, the United States is pressing on at full speed, despite more and more studies showing the dangers to humans, animals, and nondrug crops. It has continued research and development, and has been conducting field tests in Central Asia with the goal of targeting millions of acres in Asia and Latin America.

On September 4, 2001, the *New York Times* uncovered three projects that until now had been kept secret from the public: (1) Project Clear Vision, funded by the CIA to reconstruct a Soviet-designed biowarfare bomblet and test its dispersion characteristics; (2) Project Jefferson, funded by the Defense Intelligence Agency (DIA) to produce a genetically modified strain of anthrax that would be antibiotic resistant and to test its effects against anthrax vaccines used by the government; and (3) Project

Bacus, in which the Defense Threat Reduction Agency (DTRA) attempted to purchase all of the necessary components and to covertly construct a small biowarfare production site. These recent reports add fuel to the fire of speculation that our intelligence agencies have been a virtual beehive of activity in the field of biowarfare and have no plans to slow down any time soon.

Operation Whitecoat

From 1953 to 1975, the United States experimented with a variety of human and animal diseases and toxins, as well as crop diseases it hoped could be used to destroy enemy food supplies and other vegetation. Many experiments that tested biological agents on human subjects, referred to collectively as Operation Whitecoat, were carried out at Fort Detrick in the 1950s. As part of its vast entomological program, military scientists bred mosquitoes infected with yellow fever, malaria, dengue, cholera, anthrax, and dysentery. Nothing was out of bounds if it meant developing weapons that could incapacitate quickly and effectively.

Planning for the exposure of volunteers to microorganisms was begun in the summer of 1953 by U.S. Air Force medical officers assigned to Fort Detrick's Chemical Corps Biological Laboratory. The main obstacle was obtaining enough volunteers for research that would involve tests with Q fever or rickettsia, yellow fever, hepatitis A, plague, Venezuelan equine encephalitis, Rift Valley fever, and certain intestinal disease–producing agents. Part of the problem was solved by using civilian prison volunteers from the Maryland House of Corrections in Jessup, but a large number (twenty-three hundred) of volunteers were subsequently recruited from the ranks of Seventh-Day Adventist military personnel who were conscientious objectors. Before assignment to the project, each volunteer was required to sign a statement indicating that he or she would be used to develop preventive measures against infectious disease–producing organisms and be injected with new, experimental vaccines.

The success of the project was virtually guaranteed by church officials who forged an agreement with the U.S. Army to establish church membership as a potential source of volunteers. Without the assistance of the Seventh-Day Adventists, according to Lieutenant Colonel W. D. Tigertt, commanding officer of Fort Detrick's medical unit, the necessary information of the highest importance about the nation's health could not be obtained. In a 1954 meeting, an official statement was issued by the church approving the project as planned. On November 3, 1955, the church newspaper openly endorsed the program and colorfully described the contribution of each volunteer in terms of service to his country and his valuable performance beyond the call of duty. Because of the church hierarchy, where power is still concentrated at the top and flows down to a congregation that looks almost blindly to its leaders for guidance, the joint venture became a constant and reliable source of human subjects who would have little or no Sabbath conflicts.

Operation Whitecoat was unique in that it recruited almost exclusively Seventh-Day Adventists sent to the U.S. Army Medical Training Center at Fort Sam Houston, Texas. "Whitecoat" referred to the coats worn by medical personnel. The future army medics were selected for two reasons: (1) As Seventh-Day Adventists, the men prided themselves on humanitarian service and saw this as an opportunity to volunteer for something that would benefit mankind. (2) They were pacifists who were classified 1-A-O (noncombatants) and feared being shipped to Vietnam as field medics. As one soldier admitted, "We were told that if we did not volunteer, we would receive combat duty overseas." Another also claimed that he'd volunteered for the experiments so that he would not be shipped out, saying, "The truth is, we were getting killed pretty good over in Vietnam. There's not too many of us that wouldn't have gone to Vietnam if we hadn't volunteered." Thus, the military had a highly homogeneous control group and was able to take advantage of individuals who could easily be coerced into volunteering for human trials.

The initial experiments were conducted as both field tests and laboratory studies. After being flown from Fort Detrick to the Dugway Proving Grounds in Utah, a group of volunteers would be transported to an isolated test location where they were ordered onto wooden platforms at various levels. When the atmospheric conditions were right, the experiments began. Medical officers conducting the tests put on their gas masks and radioed to overhead aircraft to commence dispersing the infectious agent onto the test site. Within minutes, the planes took on the role of biological crop dusters, spraying clouds of Q-fever virus on the volunteers. After becoming infected, the soldiers were flown back to Fort Detrick for monitoring and observation.

Upon their return, soldiers were left to develop fever for three days before antibiotic therapy was initiated. Some reported getting seriously ill from the tests. Rickettsia or Q fever causes sudden onset of headaches, high fever, weakness, and severe sweats, chills, and malaise. One volunteer lost consciousness and woke up in an ice bath. Those who did not go to Dugway would be taken to a sixty-by-sixty-foot lab building enclosed by a three-story gastight sphere. Only one of a few such spheres equipped for human subjects, this was where the vaccination and exposure experiments took place. Damaged by fire in 1975, it remains standing today, a constant reminder of Fort Detrick's past.

Before the volunteers arrived, hermetically sealed canisters of biological agents were opened to monitor the speed and dispersion pattern of the aerosol. Animals were vaccinated and tested first to ensure that they would survive the biological assault. The human test subjects were next. After a period of vaccination, the Whitecoat volunteers were locked in the sphere and told to step into a special portal where they breathed in the released biological agent. For the next few weeks, they would be watched closely for signs of any reactions from either the experimental vaccine or the biological agent.

The Operation Whitecoat experiments occurred at the same time as other open-air tests, and it's not known if volunteers were cross-contaminated with biological agents. For example, Venezuelan equine encephalitis (VEE), a brain virus that had only been identified in the rat population in Florida, was suddenly discovered in animals around Dugway. Experts believe that VEE was only one of the agents used in biological open-air releases and transferred to the nearby animal population.

Once the Q-fever experiments ended in 1958, the next set of tests involved such exotic diseases as tularemia and yellow fever. Tularemia infections would cause enlarged lymph nodes that sometimes ruptured and leaked, abdominal pain, diarrhea, vomiting, and pneumonialike illnesses. Yellow fever in its mildest form caused fever, nausea, and vomiting, but allowed to go untreated produced hemorrhage of mucous membranes, vomiting of blood, degeneration of the liver, and subsequent jaundice, thus giving the disease its name. Some volunteers later claimed that they'd been given injections of plague and rabbit fever.

When the military draft ended in 1973, recruitment terminated and the project officially ended in 1975. The Seventh-Day Adventist (SDA) church has downplayed its role years ago in the recruitment of young men as human subjects but proudly admits to its contribution to public health and national security. In fact, the *Seventh-Day Adventist Encyclopedia* states, "Another example of noncombatant heroism while in service of their country is Operation Whitecoat, a project involving medical experimentation, staffed entirely by SDA volunteers." What the defenders of Whitecoat have difficulty recognizing is that the project, besides discovering defenses and developing vaccines against airborne diseases, was also an essential part of biowarfare research.

Open-Air Vulnerability Testing

Soon after the end of World War II, a growing fear spread that the Soviet Union's biological warfare program was developing so rapidly that U.S. national security would be threatened. In response, the U.S. government granted immunity from war crimes prosecution to the Japanese scientists who had been involved in bacteriological warfare research and who had performed deadly human experiments on captured prisoners of war.

Although the biological warfare budget was cut back following the war, the possibility that a rogue nation would use biological agents was enough to keep weapons research alive and well. On October 5, 1948, the Committee on Biological Warfare, headed by Ira Baldwin, issued a report concluding that the United States was particularly susceptible to covert attack and that current biological warfare research did not meet the requirements necessary to prepare defensive measures against special biological warfare operations. In response to this lack of preparedness, the committee suggested infecting ventilation systems, subway systems, and water supplies with microorganisms to test the extent to which subversive dissemination of pathogenic biological agents was possible.

By 1950, Fort Detrick was fully operational and included research units to test aerosols on humans and animals and to study paramilitary and covert biological warfare activities. Projects included not only biological agents, such as anthrax and botulism microbes, but also viruses, fungi, parasites, and arthropods, such as insects and spiders, that could be used as vectors to transmit disease. Delivery systems tested included aerosol sprays, bomblets, feather bombs (germ-containing feathers dropped by plane), rodents, fleas, flies, and other carriers.

It didn't take long for the DOD to act on the Baldwin committee's 1948 findings and suggestions. Testing on human vulnerability began in earnest in 1950 and included several major cities in the United States. The objectives were to determine how vulnerable U.S. cities were to biological agent attack and whether or not there would be any residual effect. Ostensibly, none of the biological organisms used were pathogenic, although there were reports of illnesses following all of the tests. None of the reports mentions the health effects on humans. From 1951 to 1970, thousands of open-air tests were conducted at Dugway Proving Ground, a military facility eighty miles from Salt Lake City, using bacteria and viruses that cause disease in humans and animals. In one 1968 study, sixty-four hundred sheep died following intentional release of a deadly nerve gas from a plane. The DOD denied any responsibility for the accident, but autopsies on the sheep disclosed the nerve agent VX.

According to *Special Report No. 142. "Biological Warfare Trials at San Francisco, California,"* six experimental biological warfare attacks with *Bacillus globigii* and *Serratia marcescens* were launched from September 20 through September 27, 1950. The ships from which the bacterial aerosols, along with zinc cadmium sulfide fluorescent particles, would be released positioned themselves at various distances offshore. In the heart of the San Francisco Bay area, forty-three sampling stations were set up with special filters and collectors to measure the incoming microorganisms and, according to the special report, "test the offensive possibilities of attacking a seaport city with biological warfare aerosol, measure the magnitude of the defensive problem, and gain additional data on the behavior of a biological warfare aerosol as it is borne downwind."

Each of the releases lasted thirty minutes, and the six trials took into account such factors as wind velocity, direction and speed of the ship, atmospheric conditions, and the bacterial counts at the collection sites. What was not considered were the toxic effects of the bacteria or the adverse health effects of the people exposed.

Page twenty-four of the report details what happened during the September 25 experimental release. On that day, one of the ships released *Bacillus globigii* as it cruised along the San Francisco shoreline. The size and shape of the aerosol approached the ideal theoretical distribution of a two-mile-long cloud as it was borne downwind. Respiratory exposures were relatively high, even on the eastern side of the bay at stations within the cloud pattern. The maximal distance of effective travel extended inland to Station 43, approximately twenty-three miles from the aerosol source. It was later revealed that some of the bacteria were detected as far away as fifty miles inland.

On shore, hundreds of thousands of unsuspecting residents breathed in the bacilli as they engulfed the city. Based on the report, nearly everyone in San Francisco inhaled five thousand or more fluorescent particles each day for the duration of the tests. The conclusion drawn was that a biological warfare attack by ships or other sources located some distance offshore was entirely feasible. The eight hundred thousand residents of San Francisco, some of whom had inhaled millions of bacteria as they shopped or walked to work every day, were never told that they had been part of a massive biological warfare experiment or given the opportunity to monitor their health.

The army's rationale for using live *Serratia marcescens* was that it was a biological simulant and tracer not normally capable of causing infection. However, since 1913, when the first cases of infection in humans were described, there have been isolated reports that classify the bacteria as pathogenic. So for at least forty-seven years before the San Francisco open-air tests, scientists had known about the potential health effects but chose to release *Serratia* anyway.

According to Leonard Cole, author of *Clouds of Secrecy*, a patient at Stanford University Hospital came down with an infection caused by *Serratia marcescens* within four days of the sprayings. During the months that followed, ten more individuals became infected, one of whom died. A contentious Senate hearing before the Subcommittee on Health and Scientific Research in 1977 exposed the fact that *Serratia* had been used even after evidence of its pathogenicity was documented.

The bitter exchange that occurred at that hearing, with Senators Kennedy and Schweiker grilling Brigadier General William Augerson, Assistant Surgeon General for Research and Development, and Lieutenant Colonel George Carruth, Staff Officer, Chemical and Nuclear Biological Chemical Defense Division, left little doubt that the army was caught red-handed using biological agents that it knew would cause health problems. Senator Schweiker, looking General Augerson in the eye, stated, "I believe, notwithstanding the safety officer and notwithstanding the *AMA Journal* report about SM, you ran these tests up through 1968, some sixteen years after the Fort Detrick safety officer had determined he felt there was a serious problem and seventeen years after the AMA article said they caused death. That is what I have the most trouble with."

Shortly after the San Francisco tests, in the summer and fall of 1952, Dugway Proving Ground became the site for experiments to determine how *Brucella suis* and *Brucella melitensis* spread throughout the human population. In June and September of that year, military scientists, without public knowledge, tested dispersal methods and the effects of infection. Today some experts claim that as a result of those tests, most, if not all, of us are infected with these microorganisms.

In 1953, both St. Louis, Missouri and Minneapolis, Minnesota were targeted with zinc cadmium sulfide, a chemical used to test dispersal patterns and the efficiency of detection devices. In the Minneapolis experiment, according to *Joint Quarterly Report No 3: The Spraying of Minneapolis*, the goal was to test the strategic use of biological agents against target cities. Experiments included determining street level dosage patterns, testing the penetration of aerosol clouds into homes and schools at various distances from the site of release, and observing the lingering effect of the test clouds within buildings. As with the San Francisco releases, the report makes no mention of the effects on human health.

The spraying of Minneapolis occurred over a three-month period and involved sixty-one separate releases of zinc cadmium sulfide fluorescent particles. Tests were conducted between 8 P.M. and midnight or between 1:30 P.M. and 5 P.M., when individuals were either at work or commuting home or children were at school or playing outdoors. Experimenters would operate continuous blower-type aerosol generators from trucks or rooftops, after which samples were taken outside windows, from roofs, on the ground, and from inside buildings to measure the extent of penetration.

In St. Louis, residential, commercial, and downtown areas were sprayed with zinc cadmium sulfide from April through June. *Joint Quarterly Report No. 4* states that of the thirty-five releases, which comprised afternoon, predawn, and nighttime operations, two were made on a citywide scale. The focus of the sprayings was the same as it was in Minneapolis except that the tests were conducted in poorer sections of town where residents would not be as likely to raise questions and concerns or complain to authorities. Findings showed penetrations not only into residences but also into banks, office buildings, and medical facilities where concentrations inside were as much as fourteen times greater than they were outside. Similar tests were conducted in Iowa, Nebraska, South Dakota, and Virginia.

The military had no qualms about using its own for open-air vulnerability testing as well. Robert Bates, a naval crewman aboard the USS *Navarro* (APA 215), was ecstatic when his ship was ordered to Hawaii in

1966. "It was basically an R&R cruise," he told CBS News. Until they actually got to Hawaii, that is. Once there in the blue waters of the Pacific, 4-C jet aircraft began swooping down and spraying clouds of *Bacillus globigii* in front of the eleven ships to simulate biowarfare attacks. Code-named Autumn Gold, the ships were attacked nine times that month. According to a May 15, 2000 CBS news interview, Bates claimed, "There were people with chemical suits on the ship with some kind of apparatus apparently monitoring what was going on. They wouldn't talk to you. You'd try to carry on a conversation, try to find out what was going on, but they just flat out ignored you. It always bothered me."

His fellow crew member, George Arnold, added, "I remember an airplane flying over and I could see it sprayed something and then a little later I felt this mist on my face. They were doing that just to see how much they could get stuff absorbed into our body, probably in the amount it would take to kill us if they were to use something like anthrax." CBS News obtained documents referring to the sailors as "test subjects," who were ordered to give throat swabs. Those selected to wear gas masks were the "control group."

Although the subjects were members of the military, they had every right to be fully informed about the tests and given the option of whether or not they wanted to take part. Because they weren't, the government had violated its own policy, which states, "The voluntary consent of the human subject is absolutely essential." Most of the details of the month-long operation remain classified, but a December 1, 2000 directive from the Veteran Affairs undersecretary for health shows that more than one hundred secret biological warfare tests were done with biologics and chemicals that, other than *Bacillus globigii*, included *E. coli*, sarin, and VX nerve gas, trace amounts of asbestos and radioactivity, as well as "other chemicals." The expanded list of code names includes Copperhead, Project Shipboard Hazard and Defense (SHAD), Eager Belle, Flower Drum, Fearless Johnny, Half Note, Purple Sage, Red Beva, Scarlet Sage, and Shady Grove. As late as 1996, the Pentagon denied having information about any of the tests. Two years later, fifteen bound volumes relating to just one of the tests—Autumn Gold—suddenly appeared. The Department of Veteran Affairs is still deciding what and how much will be released.

Perhaps the most brazen display of open-air testing occurred below ground in the New York City subway system. It must have been stiflingly hot at the Twenty-third Street Station of the Seventh Avenue line on June

6, 1966, when the first New York vulnerability test began—so stifling, in fact, that commuters were probably more concerned with getting to where they were going than with the strange goings on around them. Army scientists and technicians, trying not to attract attention during peak travel hours, waited unobtrusively on the station platform, hiding the specially made light bulbs filled with 175 grams of the bacteria *Bacillus subtilis* variant *niger* and thirty grams of charcoal particles. The Fort Detrick report states that each light bulb contained eighty-seven trillion bacilli and that the charcoal was used as a darkening agent to make the bacterial deposits less noticeable in the subway tunnels. It also notes that test personnel were given letters identifying them as members of an industrial research organization as cover in case they were questioned. *Bacillus subtilis* was chosen because of its similarities to anthrax spores.

The experimental procedure was simple. When a train pulled in, the scientists prepared themselves for attack. They walked across the platform, positioned themselves next to the train, and waited. When the train pulled out, they tossed the bulbs onto the tracks, allowing the aerosol cloud to be pulled down the tunnel after it. In some cases, the bulbs were thrown onto the tracks as a train was arriving, thus engulfing it and the passengers completely in the cloud of bacteria. On other days, the scientists shattered infected bulbs on walkways or into ventilation grates that opened into the subway system. Passers-by who found themselves engulfed in the clouds simply brushed off their clothing and moved on, never suspecting that they had just been the targets of an army biowarfare attack.

The tests were conducted over a five-day period from June 6 through June 10 to study the vulnerability of a subway system to covert attack. Midtown New York City lines were selected because of their heavy traffic and the number of lines available for testing. Following each test, scientists measured dissemination patterns of the bacteria, penetration into train cars, length of exposure to which passengers were subjected, and concentration of bacteria in various stations. The army report concluded, among other things, that (1) more than a million commuters breathed in countless trillions of bacteria, (2) on the uptown platforms, people were inhaling almost one million bacteria per minute, and (3) not a single tester was questioned about his actions, thereby reinforcing the contention that large populations are susceptible to attack by terrorists. Like the other experiments, no mention was ever made of possible health effects of exposure to the bacteria.

How safe was the *Bacillus subtilis* used in the experiments? Microbiologists, as well as several textbooks, state that *Bacillus subtilis* can cause infections, invade the blood stream in certain diseases, and even serve as a carrier for pathogenic viruses that may lie dormant for a while, then suddenly cause disease with no apparent cause. The scientists conducting the human experiments should have known that the bacteria was not harmless and probably did, since research papers as early as 1960 describe *Bacillus subtilis* inhalation experiments and raise concerns about safety and health effects. Still, as recently as 1986, the army, in acknowledging that it was using *Bacillus subtilis* for open-air testing, claimed that the bacteria were nonpathogenic to man. With that attitude, there's no reason to think that somewhere in the United States or elsewhere the military is not continuing that practice today.

Biopreparat: Russia's Top Secret Bacteria and Virus Program

Surrounded by the Ural Mountains in Russia, the town formerly known as Sverdlosk had been frozen for more than six months. When the spring thaw came in late April 1979, the Siberian air was still crisp, the north wind biting into the weathered faces of men and women who never suspected that an accidentally released plume of anthrax had just floated ten miles south from Military Compound 19 and settled like an invisible mist over the entire city.

The first sign that something had happened occurred several days later when people began complaining of high fever and severe burning in their chests and stomachs. One by one the ambulances came. Within days, more than a hundred infected patients lay dying in hospital wards, gasping for breath, convulsing and writhing in pain as doctors and nurses looked on helplessly. The final death toll was estimated at between two hundred and one thousand, though an official count has never been reported and could have been much higher. For years, the Soviet Ministry of Health blamed the outbreak on contaminated meat, until the summer of 1992 when Russian President Boris Yeltsin finally acknowledged the accident.

Though experts had suspected all along that an anthrax outbreak occurred, it wasn't until March 1998 that tissue samples taken by Russian pathologists finally confirmed not one but four different strains. Reports by defecting scientists are chilling. They tell of a vast and isolated complex of laboratories at the heart of the Soviet biowarfare program that was staggering in its ability to produce biological agents. Sometimes referred to as "Black Biology," the official name that still sends shudders through anyone familiar with it is "Biopreparat."

Founded in 1974, Biopreparat was a collection of forty clandestine biowarfare facilities spread throughout the former Soviet Union. At the height of the Cold War, more than thirty thousand scientists and staff were employed in the research, development, and production of anthrax, Ebola (Marburg variety), plague, smallpox, tularemia, and a variety of viruses. The virus research was conducted at the Vector laboratory complex in Koltsovo. In 1990, almost five thousand scientists still worked at the Vector labs. Today some fifteen hundred people are still there and involved in top secret research, which means that a large number of scientists are unemployed, working at other labs or other jobs, or have left Russia to sell their expertise to whoever is willing to pay.

Shortly after its start, Biopreparat achieved astonishing success in producing weapons-grade biological agents. Stockpiles of smallpox were stored in bunkers near SS-18 intercontinental ballistic missile silos that were programmed to target at least one hundred of the largest cities in the United States. The freeze-dried powder was manufactured so that it could be loaded into special warheads, each one capable of carrying one hundred pounds of smallpox. Experts suspect that the Soviets may also have had as many as four hundred plague warheads as well. Each missile could be fitted with ten such warheads containing as many as five different biological agents.

According to testimony before the U.S. Senate, the missiles had special cooling systems to keep viruses alive during the heat of reentry and parachutes so that the warheads could drop down over a city, burst apart at a certain altitude, and fire bomblets full of smallpox in all directions. Once the finely powdered smallpox, plague, or Marburg organisms dispersed, they would become virtually invisible and spread quickly for miles throughout the population. A recent inspection of Russian biowarfare facilities has left experts convinced that this kind of research and development continues today.

Dr. Kenneth Alibek, former director of Biopreparat and author of *Biohazard*, says, "In the Soviet's view, the best biological agents were those for which there was no prevention and no cure. For those agents for which vaccines or treatment existed—such as plague, which can be treated with antibiotics—antibiotic-resistant or immunosuppressive variants were to be developed." He goes on to describe the Soviet emphasis on creating genetically combined strains of two or more viruses and transforming harmless microorganisms into pathogenic ones through molecular biology and genetic engineering. By the 1980s, not only did the Soviet program catch up with the United States, it surpassed it to become the most sophisticated biological weapons program in the world.

In 1990, with a budget of one billion dollars, Biopreparat scientists could produce two tons of weaponized anthrax a day and were said to have successfully spliced plague and a neurotoxin to create a new super-weapon. The bank of ten thousand viruses, including more than one hundred strains of smallpox, may be the largest in the world. Dr. Byron Weeks, an officer with the U.S. Air Force Medical Corps, has testified that both Russia and Iraq have created genetically modified plague, anthrax, and tularemia pathogens, and that Russian strains of weaponized anthrax are resistant to penicillin, tetracycline, and probably most other antibiotics. Substantiating that claim, Dr. Alibek told a congressional house subcommittee on national security, veterans affairs, and international relations that the Soviet Union actually began developing antibiotic-resistant biological agents in the 1970s and made significant breakthroughs in the 1980s. "At first, it was three antibiotics," he said, "then five, and then finally we developed a strain that was resistant to ten antibiotics, including Cipro and quinolines."

Testifying before the U.S. Congress, Alibek claimed that the Soviet Union did everything in its power to transfer smallpox virus from the Ivanovsky Institute in Moscow to Vector. There, scientists were to explore the smallpox genome as fully as possible to facilitate genetic engineering operations, identify closely related viruses, and perform genetic engineering work for the purpose of inserting genes of other viruses into the smallpox virus to create new organisms. By 1990, the Russians had the ability to produce tons of smallpox virus in as little as a few days. After defecting in 1992, Alibek wrote, "I don't remember giving a moment's thought to the fact that we had sketched out a plan to kill millions of people." That statement was no exaggeration. According to a *New York Times* report, officials visiting some of the bioweapons plants in Stepnogorsk, Kazakhstan found ten twenty-ton fermentation vats four stories high, capable of holding twenty thousand liters and producing sixty thousand pounds of anthrax spores in a little more than six months.

Alibek also believes that Russian scientists have been able to genetically engineer smallpox virus with VEE, as well as smallpox with Ebola. These monstrous combinations, he says, are especially dangerous because there are no treatments and because the kill rates approach 100 percent. He also suspects that Soviet know-how made its way to Iraq, which had tested anthrax and botulism in rockets, aerial bombs, spray tanks, and SCUD missiles. Based on inspection reports following the Gulf War, Iraq had produced nineteen thousand liters of concentrated botulism and eighty-five hundred liters of anthrax.

How much of this material has been allowed out and how many scientists are selling their secrets is not known. In 1994, however, General Anatoli Kuntsevich, the man in charge of dismantling Biopreparat, was charged with shipping eight hundred kilograms of toxic chemicals to a Middle Eastern country thought to be Syria. Also having close ties with former Soviet scientists are the likes of Iraq, Libya, North Korea, and Cuba, which is thought to have developed its own Novichok agents with the assistance of its former ally. Of even greater concern are independent terrorist organizations like Al Qaida, which have been trying for years—probably with some success—to acquire weapons of mass destruction from individual scientists or from the Russian mafia.

But Alibek believes that research long kept secret may have been revealed in, of all places, readily available scientific journals. "When you read Russian scientific journals, that's what scares me to death," he says. "If you take Russian scientific journals from 1992 to 1998, start reading what kind of articles they published throughout this period, you will be able to find everything. How to create genetically engineered anthrax, antibiotic-resistant anthrax, how to develop protection of the virus using simple techniques, how to manufacture the virus using simple techniques, and so on and so forth. It is available, unfortunately, now . . . I would say that the number of publications is huge and if somebody is interested in finding some new ways to develop biological weapons, this information is available. You can go to any library in the U.S., and I believe any library around the world, and get this information."

Is there still a Biopreparat? A Vector? Almost everyone, including Soviet defectors now working in the United States, believe so. It may not be the vast system of complexes and facilities it once was, but it doesn't need to be. The technology to create microorganisms that can literally wipe out the human population has been common knowledge for some time. What remains a mystery and a deep concern is who has it and how and when they will use it. We can only hope that the idea of unleashing something that could cause the extinction of the very nations that would even think of using it is enough to prevent such actions.

Modern Biowarfare: Did Bioweapons Cause a Pandemic of Gulf War Illness?

When I received a letter from Dr. Garth Nicolson who, at the time, was chairman of the department of tumor biology at the University of Texas M. D. Anderson Cancer Center and a Nobel Prize nominee, I didn't know what to expect. His letter began with the startling claim that "about one-half of the Gulf War illness patients have an invasive mycoplasma infection." From what I understood about *Mycoplasma*, one of the microbes used in the development of germ weapons, I knew that it burrows itself deep into cells, exits, then travels to another area of the body such as synovial joints. Since one of the first symptoms of Gulf War illness is severe joint pain, I realized why there'd been such an effort to discredit the individuals trying to investigate the source and link the two.

"What is interesting about these mycoplasmas," Nicolson continued, "is that they contain retroviral DNA sequences such as the HIV-1 envelope gene, suggesting that they may have been modified to make them more pathogenic and more difficult to detect. It's also interesting that we have been working with a support group of Texas Department of Corrections employees that were apparently exposed to the same unusual mycoplasma, possibly during a Defense Department–supported vaccine testing program in selected state prisons here in Texas. One of the biotech companies involved had U.S. Army contracts to study mycoplasmas and has been named in lawsuits as selling or supplying chemical and biological weapons to Iraq."

My initial reaction, based on mechanisms of HIV and *Mycoplasma*, was that Dr. Nicolson's claim makes sense. For more than a decade, HIV has been studied for its effectiveness in a new role: as a retroviral package and vector to deliver genes to cells. Dr. P. O. Brown, in the 1990 issue of *Current Topics in Microbiology and Immunology*, states that retroviruses like HIV have been widely used as vectors for genetic engineering and are likely to be the first vectors used for introducing foreign genes into cellular chromosomes. Moreover, the journals *Cell* and *Virology* have published detailed articles describing gene therapy techniques in which parts of the HIV-1 organism are packaged for delivery of DNA into human cells. According to Dr. Nicolson, a mycoplasma with the HIV envelope gene could never have originated in nature but only through genetic engineering.

Could parts of HIV have been used to facilitate entrance of mycoplasmas through cell membranes? Armed with a new technique called *gene tracking*, Dr. Nicolson, together with his wife, a cell biologist and president of the Rhodon Foundation for Biochemical Research, proved that the particular strain of *Mycoplasma* found in Gulf War vets had incorporated into it as much as 40 percent of the HIV protein coat, making it extremely pathogenic. With gene tracking, blood is separated into red and white cells and fractionated into nucleoproteins that bind to DNA. The purified nucleotides are then probed to determine the presence of specific *Mycoplasma* gene sequences. The more I examined the materials Dr. Nicolson sent me, the more convinced I was of the biological agent scenario as opposed to the theory that chemical exposure may have been the culprit.

The probability of a biological source is strengthened by three facts. First, exposure to chemicals cannot account for so many sick soldiers (at last count, as many as one hundred thousand soldiers have Gulf War illness). Not all soldiers were stationed near areas where chemicals had been released, and many had left the Iraqi war zone before the war started. How could individuals who were not exposed be affected? Second, chemical exposure does not explain how so many spouses of returning veterans have developed the illness. Only biological agents are contagious. Chemical exposure only affects the individual exposed and cannot be passed on. Finally, the United States had no capability to detect biological agents in the Persian Gulf, so how could anyone discount the possibility so quickly unless they did not want the issue raised in the first place?

In his letter, Dr. Nicolson went on to say, "We have been able to assist thousands of soldiers recover from a life-threatening disease that is caused by invasive mycoplasma infections. We have learned that over 6,000 U.S. soldiers have died of infectious diseases and chemical exposure in Operation Desert Storm. I suspect that this is being hidden from the American public for political, economic, and legal reasons."

Speaking out like that has meant nothing but trouble. Since his discovery and public announcements, Dr. Nicolson has been pressured by the CIA and DOD to limit or abandon *Mycoplasma* research and to curtail public statements, his mail and packages have been intercepted and some have disappeared (he'd sent me information via Federal Express to prevent this), and on two occasions there have been attempts to destroy frozen Gulf War disease blood samples. He has since left the University of Texas and founded the Institute for Molecular Medicine in Irvine, CA, where he continues Gulf War illness research and treatment of veterans and their families.

When I talked to Dr. Nicolson by phone shortly after receiving his letter, I became even more convinced. He'd told me how *Mycoplasma* had been tested on prisoners and death row inmates in Huntsville, Texas in the 1980s prior to the Gulf War. He went on to describe how guards had contracted it from inmates and then passed it on to their families. As he spoke to me, I sensed a great deal of concern and urgency in his voice, especially when he talked about the thousands of young veterans who, he said, had died of "unusual" diseases and cancer.

But were it not for his stepdaughter, a Blackhawk helicopter crew chief in the 101st Airborne Division, Drs. Garth and Nancy Nicolson would never have sacrificed their health and careers to uncover what may be a modern cover-up of enormous proportions. "Practically everyone in her unit came down with Gulf War illness," said Nicolson. The Nicolsons themselves were infected after handling Gulf War veterans' blood samples. That's when Dr. Nicolson immediately suspected an infectious agent and decided to get to the bottom of it. It wasn't easy. It seemed that no one was willing to cooperate. Military logs crucial to Gulf War veterans' health were missing and couldn't be found. Despite that, the Nicolsons persevered and were able to manage their illness with antibiotics targeted against *Mycoplasma*, proving that the source of everyone's illness was biological, not chemical.

A few months after I spoke to him, reports began coming in from the Middle East, especially from countries in or near the war zone, that as many as 25 percent of the civilian population was infected and suffering with Gulf War illness. Studies of wind patterns showed that periodic dust storms kicked up germs and spores throughout the region and may have deposited them on distant soils. Many returning U.S. veterans started out with flulike symptoms that worsened to debilitating joint pain, chronic fatigue, nausea, gastrointestinal problems, memory loss, vision problems, and severe headaches. A few years after Desert Storm ended, the number of rare illnesses and cancers has risen dramatically.

In the Middle East, unconfirmed death tolls, based on figures supplied by the International Red Cross, included as many as two hundred fifty thousand Iraqi soldiers and perhaps as many civilians. According to Dr. Nicolson, more babies are born deformed or with birth defects than are born normal. And health officials reported that child mortality increased by fifty-five thousand immediately after the war. On August 31, 2001, a survey in the *British Medical Journal*, taken by the School of Medicine in London, showed that nine thousand British service personnel believe they have Gulf War syndrome. In the United States, that number is much higher; in fact, a congressional hearing cast serious doubts about the military's explanations, concluding that veterans were more than likely the subjects of accidental or purposeful foul play.

While investigations continue, so does the spread of Gulf War illness. As early as 1994, the Senate Banking, Housing and Urban Affairs Committee reported to Congress that approximately 77 percent of the spouses and 65 percent of the children of Gulf War illness patients were showing signs and symptoms of the illness. Shortly after, a team of researchers at the University of Texas Southeast Medical Center at Dallas reported a link between three primary syndromes of brain and nerve damage and Gulf War veterans. And as recently as December 10, 2001, Anthony J. Principi, Secretary of Veterans Affairs, said that preliminary evidence shows that veterans who served in Operation Desert Shield and Desert Storm are twice as likely as veterans who were not deployed to Southwest Asia to develop amyotrophic lateral sclerosis (ALS), also known as Lou Gehrig's disease. He said, "They believe that there was an association between service in the Gulf and ALS—and preliminary evidence indicates that they were correct." Since ALS typically strikes adults fifty to seventy years old, the fact that veterans as young as nineteen have been diagnosed is significant.

In testimony before the U.S. House of Representatives on January 24, 2002, Dr. Nicolson said that few infections can produce the complex chronic symptoms found in Gulf War patients, but infections caused by *Mycoplasma* and *Brucella* can. The fact that 40–50 percent of Gulf War illness patients have such infections compared to only 6–9 percent of nondeployed individuals is troubling. Also troubling is Dr. Nicolson's own study showing that almost all ALS patients (about 83 percent, including 100 percent of Gulf War veterans with ALS) are infected with *Mycoplasma*. Perhaps we'll never know definitively what it was that triggered Gulf War syndrome. Many experts now agree that it was probably a combination of chemical exposure and infection with biological agents. However, until an admission is made and a legitimate effort is launched to combat it, Gulf War illness and its effects may be with us for some time.

West Nile Virus: An Omen of Things to Come?

If terrorists have learned anything, it's that biological agents needn't be prime weapons of mass destruction like anthrax or smallpox. A simple and slower-acting agent can be just as effective. Some in the U.S. government believe that outbreaks of a disease such as West Nile virus, a mosquito-borne illness unknown in the United States until 1999, could very well be the test run in a next phase of terrorism. Senate Judiciary Committee Chairman Patrick Leahy of Vermont, in a September 2002 radio interview with WKDR of Burlington, Vermont, questioned whether the astonishing increase in West Nile virus was coincidental or whether our defenses were being tested against a biological attack. Senator Leahy's concern was based on evidence that both Saddam Hussein of Iraq had overseen and Fidel Castro of Cuba still oversees bacteriological programs that include the weaponization of West Nile virus.

A recently declassified 1995 letter from former Surgeon General David Satcher to Michigan Senator Donald Riegle detailed the shipment of various pathogens to Iraq, including a shipment of West Nile virus on May 25, 1985. Needless to say, the revelations caused a firestorm. But when the accusations over who was most responsible for arming Saddam with bioweapons had subsided, the revelations became even more frightening, and the pieces to a horrible puzzle were beginning to fall into place. We began to think back to the late 1990s, when top Iraqi defectors claimed that Saddam Hussein had bragged about wanting to use West Nile virus against the United States and that Iraqi scientists had developed the ability to mutate and weaponize all sorts of viruses. Just two years after Saddam's threat, West Nile virus first erupted in Queens, New York and in Florida and then, according to the Maryland State Health Department, migrated inexplicably along the Interstate 95 corridor rather than spreading randomly as it should have in nature.

The question experts grappled with was why West Nile virus? The answer may have come in a 1999 *New Yorker* interview with Richard Preston, author of *The Hot Zone*, who quoted Mikhael Ramadan, one of Saddam's former bodyguards. "In 1997, on almost the last occasion we met, Saddam summoned me to his study. Seldom had I seen him so elated. Unlocking the top right-hand drawer of his desk, he produced a bulky, leather-bound dossier and read extracts from it. The dossier holds details of his ultimate weapon, developed in secret laboratories outside Iraq. Free of U.N. inspection, the laboratories would develop the SV1417 strain of the West Nile virus, capable of destroying 97 percent of all life in an urban environment. He said SV1417 was to be operationally tested on a Third World population center. The target had been selected, Saddam said, but that is not for your innocent ears."

Preston went on to report a conversation he'd had with an FBI agent who said anonymously that "if I was planning a bioterror event, I'd do things with subtle finesse, to make it look like a natural outbreak. That would delay the response and lock up the decision-making process." Adding to those concerns, Dr. Kenneth Alibek, former director of Biopreparat, the Soviet Union's biowarfare program, recently stated at a congressional hearing that the West Nile virus outbreak was indeed suspicious. The Pentagon had known for some time that Soviet scientists were working with West Nile virus as a possible biological agent. They also believed that one of the Soviet Union's principle allies, Cuba, had an advanced biological weapons program and was working actively to weaponize West Nile virus.

A September 2002 UPI report stated that Undersecretary of State for Arms Control and International Security John Bolton has been repeatedly frustrated at the unwillingness of U.S. intelligence to admit to or disclose information about Cuba's bioweapons program, which probably includes anthrax, smallpox, and strains of encephalitis such as West Nile virus. The report claims that defecting Cuban scientists told U.S. authorities that the strain of West Nile virus infecting the United States can be traced to birds infected at Cuban bioweapons labs because some of Fidel Castro's experiments involve the use of animals as carriers of weaponized germ agents.

"Castro's biological front," says Carlos Wotzkow, a Cuban ornithologist who defected in 1999, "was extended to the Institute of Zoology in 1991 to develop ways of spreading infectious diseases, including encephalitis and leptospirosis, through implantations in migratory birds." Another defector, Dr. Roberto Hernandez, added that scientists worked on viruses resistant to insecticides and that "military officers running the labs ordered us to trap birds with migratory routes to the United States with the idea of releasing contaminated flocks which would be bitten by mosquitoes which, in turn, infect humans."

Both claims are corroborated by Colonel Alvaro Prendes, former vice chief of the Cuban air force, who told officials that a compound the size of two football fields in eastern Havana houses giant tanks for toxic gases and has its own water supply as well as backup generators. Ken Alibek revealed in his 1999 book, *Biohazard,* that Castro obtained bioweapon technology from top-ranking Biopreparat scientists who made repeated trips to Cuba during the late 1980s and early 1990s. "We knew that Cuba was interested in biowarfare research," Alibek says. "We knew that there were several centers, one of them very close to Havana, involved in military biotechnology."

Alibek also made another startling accusation, this time about the Cuban-Iraqi connection. According to Alibek, Saddam Hussein learned how to conceal his acquisition of bioweapons technology from Cuba. "The model was one we had used to develop and manufacture bacterial biological weapons," he explains. "Like Cuba, the Iraqis maintained the vessels that were intended to grow single-cell proteins for cattle feed. What made the deals particularly suspicious were additional requests for exhaust-filtration equipment capable of achieving 99.99 percent purity—a level we only used in our bioweapons labs."

In light of shocking revelations that there's likely a massive biological weapons facility ninety miles from the shores of the United States, what has the U.S. response been? Incredibly, instead of pursuing the possible terror link, the Clinton administration's National Security Council team claimed there was no evidence to support any of the allegations, despite even testimony by Castro's own bioweapons scientists. Later, on June 5, 2002, John Ford, head of the current State Department Bureau of Intelligence and Research, testified at a congressional hearing that "Cuba does indeed have an offensive weapons research program."

Apparently, none of the evidence or testimonies has been enough to take seriously even Special Negotiator for Chemical and Biological Weapons Donald Mahley's warning that Cuba has been selling dual-use biotechnology to Islamic countries with close ties to terror groups. The biotech systems include biological agents, pathogens, and technology that could be used to weaponize bacteria. Castro's recent trips abroad included meetings with leaders in Iran, Libya, Syria, and the United Arab Emirates, whose banks have been implicated in dealings with Al-Qaida while they maintained training camps in Afghanistan.

But the mother of all neglected bioweapons red flags was apparently overlooked by the CIA and FBI. The UPI report ends with the revelation that U.S. investigators ignored two Cuban intelligence agents indicted in Florida on August 4, 2001 while they worked for the U.S. Postal Service. During their interrogations, they told the FBI that they'd been ordered by Fidel Castro to obtain jobs in the U.S. Postal Service to study post office security for possible bioterror attacks. The first anthrax death in the United States was exactly two months later, in October 2001. While the FBI's anthrax investigation focused almost exclusively on domestic terrorists, the real culprits may have been foreign agents living among us who had discovered how easy it is to deliver terror through the mail.

Germ Warfare in the Twenty-first Century

During the past decade, and despite the 1972 Convention on the Prohibition of the Development, Production, and Stockpiling of Biological and Toxic Weapons, the United States has been heavily involved in the research and development of biological agents. Funding for the next generation of these agents has been steadily ongoing, with as many as 120 universities engaged in some form of biological agent research. From 1980 to 1986 alone, the budget for biowarfare research increased from 160 million dollars to more than one billion dollars. A major reason for the heightened interest was the proliferation of biological weapons around the world. In 1989, at least ten countries were actively developing toxic agents, according to former CIA director William Webster. By 1995, that number had grown to seventeen and included such nations as Iran, Iraq, Libya, North Korea, Cuba, China, and Russia, which is purported to have the largest and most sophisticated bioweapons arsenal in the world.

One of the more surprising revelations of the past year has been that the CIA has taken on a significant and aggressive role in biowarfare programs. Since September 11, 2001, the agency has been given even more power to obtain information regarding biological agents that it needs to carry out activities in the name of national security. A government statement before the U.S. Senate in April 2002 illustrates how terrorists have actually strengthened the hands of those who wanted intelligence agencies to once again get involved in biowarfare programs:

> An area of significant multi-agency homeland security collaboration is in genetic sequencing of microbes with possible terrorist implications. The effort is being coordinated through OSTP's interagency microbe project working group. All agencies (NSF, NIH, CDC, DOE, DARPA, USAMRID, CIA, and Agriculture) doing genetic sequencing are participating and agreeing on what should be sequenced, to what level and quality, and who will do the sequencing. This is a real success story as multiple agencies are pooling their resources to attack a part of the bioterrorism threat.

Thanks to an often clandestine partnership between government and the private sector, advances in molecular biology have allowed scientists to develop synthetic viruses and bacteria and produce deadly mutant strains of existing microorganisms. The same technology used to revolutionize medical therapies is readily transferable to biowarfare research and development. Current arsenals include everything from naturally occurring venoms and toxins to genetically altered combinations of viruses and pathogenic bacteria, spliced together to create entirely unique and vaccine-proof organisms. The insidious side effect of this kind of development is obvious in that biological agents are less costly and more easily transportable than other weapons.

The earliest known biological agents were simple infectious pathogens, microscopic organisms that attack the body's defenses causing physiological breakdown, disease, and, in many cases, death. Since the 1970s, however, the stakes have been raised; the search for more effective offensive capabilities has taken military scientists into an area many hoped they would never be allowed to enter: genetic engineering.

With the help of private biotechnology firms, recombinant DNA research has intensified. Technologies have been developed to produce strains of bacteria resistant to known antibiotics or that produce deadly toxins. By 1983, the DOD had funded twenty-seven recombinant DNA projects, most with outside contractors. By 1985, that number had grown to sixty. And by 1986, with more than three hundred companies actively engaged in the biotechnology and pharmaceutical industries, the DOD had its pick of companies that had exhausted their initial investment capital and were eager to cooperate with the military for a share of the lucrative grant pie, even if it meant participating in various biological and chemical weapons programs.

As an example of how this technology is being used to create lethal versions of existing organisms, consider this. The genes for the anthrax toxin reside on a circular DNA molecule called a plasmid. A scientist can easily cut that section of the DNA out and splice it into another species of bacteria. Once the transfer is complete, the new organism begins to produce the toxin and essentially becomes a killer bacterium. This technique is done routinely in recombinant DNA work and has actually been used commercially to create a bacterial pesticide against gypsy moths. The anthrax gene is transferred from *Bacillus anthracis* to *Bacillus thuringiensis* as a way to control insects in gardens. The fear that experts have is that the genetically altered bacteria may spontaneously transfer the plasmid to other bacteria, some of which could be infectious to humans or other animals.

Even the promising medical breakthrough of gene therapy could be misused for evil purposes. Rather than repairing or replacing defective genes, the technique may be used to introduce viruses that initially lie dormant and become lethal over time, or infect the host with pathogenic genes that trigger disease. "Designer diseases," on the research burner as well, would afflict individuals genetically predisposed because of certain traits or lack of resistance or would kill by using modified animal pathogens that seek human hosts.

Also very real is the prospect of bioregulators as weapons. Present in our bodies in minute quantities, bioregulators determine hormone release and bodily functions such as temperature, sleep, consciousness, behavior, and emotions. Being natural molecules, they could avoid detection, have immediate and devastating effects on physiology, and, through genetic engineering, be far more potent and trigger more severe responses than more traditional agents.

At no time in history has mankind been so vulnerable to biological weapons. With the fall of the Soviet Union and the desperation of scientists seeking work, there is a market for individuals with the expertise to develop and manufacture weapons of mass destruction. And with terrorist organizations becoming wealthier and more sophisticated, the threat of attack sometime in the near future is very real indeed. As long as that threat exists, there will be ongoing research in the name of national security. Whether we know it or not, some of that research will continue to include human subjects.

From 1970 to the present, new generations of technologies, products, and scientific advances have continued to emerge. Genetic engineering or recombinant DNA technology has literally revolutionized the manner in which we conduct science and has allowed researchers to produce molecules that previously could not be produced. By turning living organisms into chemical factories, scientists are able to mass produce not only useful products but also deadly toxins, biological agents, and viruses more deadly than any we've seen. For example, in a 1970 hearing before a house appropriations committee, scientists testifying before Congress claimed that "within a period of five to ten years, it would be possible to produce a synthetic biological agent that does not naturally exist and for which no natural immunity could be acquired." That possibility has already been realized. With the development of techniques that can alter the genes for coding proteins, we can change the structure of viral membranes, induce virulence, and make vaccines against viruses ineffective simply by developing new strains.

Recently disclosed documents reveal that in possible violation of the Biological and Toxin Weapons Convention (BTWC), U.S. Navy and Air Force biotechnology laboratories have explicitly proposed development of offensive biological weapons, including genetically engineered microbes, cluster bombs designed to disperse biological agents, new strains of antibiotic-resistant anthrax spores, and nonlethal bioweapons for use against crowds and in antinarcotics operations. According to one Naval Research Laboratory report, "It is the purpose of the proposed research to capitalize on the degrading potential of naturally occurring microorganisms, and to engineer additional, focused degradative capabilities into genetically modified microorganisms, to produce systems that will degrade the warfighting capabilities of potential adversaries."

In an upcoming paper to be published in the *Bulletin of the Atomic Scientists*, Dr. Malcolm Dando of the University of Bradford and Mark Wheelis of the University of California claim that in July 2001 the United States inexplicably blocked an attempt by signatories of the 1972 Biological Weapons Convention to implement tough inspections by member countries in order to maintain secrecy over its offensive bioweapons program. Dando also makes the claim that U.S. research includes hallucinogenic weapons such as BZ and "calmative" nonlethal agents similar to the gas used by Russian Special Forces to knock out the Chechen rebels and hostages in Moscow. The Pentagon, according to Dando, is getting around the ban on knockout gases because of a loophole that permits chemicals for "law enforcement purposes."

Throughout history, and even in recent times, man has used biological agents as a desperate means of last resort. Today's transportation systems make the spread of microbes from continent to continent fast and reliable, with contamination measured in hours rather than days or weeks. Even the safety of oceans is no longer much of a barrier against the threat of bioweapons. The twenty-first century, with scientific advances and discoveries growing faster than our ability to foresee or comprehend the ramifications, may very well usher in an era in which those bioweapons are the first or only choice.

3 THE EUGENICS MOVEMENT: PAST, PRESENT, AND FUTURE

One by one the delegates streamed in. Some of the notables included Alexander Graham Bell, future president Herbert Hoover, and even Charles Darwin's son, Leonard Darwin. The international affair, hosted by the American Museum of Natural History in New York, was a who's who of scientists, politicians, philanthropists, and leaders of nations from around the world. It was the fall of 1921, and the event was the Second International Congress of Eugenics.

The fact that so many nations were represented shows the extent to which the movement had grown from small pockets of nineteenth century radical activism to a popular ideology that literally gripped the world. Devoted to improving the human species through genetics, sterilization, and controlled breeding, these men and women were on a mission: to eliminate hereditary defects and, thus, to attack the moral problems of decadence, crime, and social ills like alcoholism and venereal disease. Their ultimate goal was to improve the world by creating the best societies possible through better breeding.

An offshoot of social Darwinism, eugenics did not begin in the United States, as some have claimed, but it *did* take shape through such organizations as the American Eugenics Society and the American Eugenics Party. Coined by Charles Darwin's cousin, Francis Galton, and used to describe the science of heredity and good breeding, the term eugenics is derived from the Greek roots *eu* or good and *genics* or origin. It is often grouped into positive eugenics, which aims to improve genetic stock through selective reproduction, and negative eugenics, which uses forced sterilization and euthanasia to keep inferior genes out of the population. To give it a sense of legitimacy, proponents initially presented eugenics as a mathematical science that would predict human behavior through genetic manipulation. By 1927, the principles were so ingrained in American society that one of America's greatest supreme court justices, Oliver Wendell Holmes, in the majority opinion of a landmark eugenics case penned the now famous statement, "It is better for all the world, if instead of waiting to execute degenerate offspring for crime, or to let them starve for their imbecility, society can prevent those who are manifestly unfit from continuing their kind." Those words were the cornerstone of a decision based on a seven-month-old baby girl named Vivian Buck who just didn't "look" normal.

In almost every way, Carrie Buck of Charlottesville, Virginia was a typical seventeen-year-old girl. What wasn't so typical in the 1920s was a teenager giving birth to a child out of wedlock. When Carrie's illegitimate daughter Vivian was born, the usual uproar over promiscuity was heightened by the fact that Carrie's mother, Emma Buck, was at the time a resident at the Virginia Colony for the Epileptic and Feebleminded. The whispers about Carrie's indiscretion progressed to accusations and then to loathing for a girl who'd committed the ultimate act of shame on the upstanding people of Charlottesville.

For almost a year, Carrie and her daughter were left alone, ostracized for the most part by a community that considered her a shiftless, ignorant, and worthless type of antisocial white southerner. But a routine visit by a Red Cross relief worker who meant no harm would change all that and make Carrie Buck one of the most famous names in the eugenics movement. During that visit, the relief worker noticed something about Vivian that didn't seem normal. Since the worker was aware of Carrie's family background, she assumed the worst and reported her observation to superiors. It didn't take long before Carrie was paid a follow-up visit by officials from the Eugenics Record Office (ERO), where files were kept on disabled individuals and those deemed genetically unfit.

By the following week, scientists at Cold Spring Harbor Laboratory in New York were busy at work. Since they already had IQ test scores for both Carrie and Emma Buck, whom they had previously classified as "morons," they concluded that baby Vivian would most likely share her mother and grandmother's defective traits of "feeblemindedness" and "sexual promiscuity." There was no need of further examinations and tests. Three generations of the Buck family having low intellect was enough to mount a legal effort for Carrie Buck's sterilization.

During the trial, a parade of witnesses testified to Carrie's defects. Dr. Albert Priddy, the colony superintendent, swore that Emma Buck had "a record of immorality, prostitution, untruthfulness, and syphilis." An ERO sociologist and a Red Cross nurse who examined her in the midst of the trial testified that she was of below average intelligence and not normal. Following days of tests and sometimes rancorous testimonies, the judge ordered Carrie sterilized to prevent her from giving birth to other defective children. The decision, met with wide approval, was appealed to the U.S. Supreme Court. On May 2, 1927, Associate Justice Oliver Wendell Holmes concluded that the existence of a deficient mother, daughter, and granddaughter justified the need for sterilization and offered one of the most infamous opinions in modern history (Appendix VIII). Before the decision, almost three thousand people had been involuntarily sterilized in America. After the Supreme Court gavel came down, *Buck v. Bell* had become the law of the land, supplying the precedent for the sterilization of more than eight thousand Virginia citizens and more than twenty thousand people in the United States by the mid-1930s.

Sadly, Carrie Buck's ordeal and forced sterilization were based on lies, incorrect diagnoses, and a plot to guarantee that Virginia's newly passed Eugenical Sterilization Act would be upheld. The law, adopted as a cost-saving strategy to relieve the tax burden in a state where mental facilities were growing rapidly, stated that "heredity plays an important part in the transmission of insanity, imbecility, epilepsy, and crime."

As it turned out, Carrie Buck, perhaps from fear or embarrassment, had kept a dark secret locked in her heart. Her daughter Vivian was not the result of promiscuity but of rape by a relative of her foster parents. Carrie, according to school records, was not the feebleminded "slut" she'd been portrayed as but a fairly good student who had been on the honor roll. A subsequent review of the case uncovered a conspiracy between Carrie's defense lawyer and the Colony of Virginia to ensure the constitutionality of Virginia's law.

The desperate attempt to keep eugenics alive at any cost may not have attracted much attention in rural America, but it surely put a smile across the face of every Nazi official emboldened by the Supreme Court decision. In 1933, the Nazi government adopted the Prevention of Hereditarily Ill Offspring Law, based on *Buck v. Bell*, which provided legal cover for the forced sterilization of more than 375,000 people and the banning of marriage and sexual contact between Jews and Germans. Later, it set the stage for expansion of the law to include euthanasia and human experimentation. A little more than a decade after that, lawyers at the Nuremberg trials were referring to America's laws and policies as defense and justification for what Nazi scientists had done during World War II.

The Tuskegee Syphilis Study: America's Most Infamous Eugenics Scandal

Who needed the Supreme Court anyway? Their imprimatur was simply icing on the cake, as far as the eugenicists were concerned, and a vindication for what many believed was the right thing to do for society. But few in America suspected what was about to transpire because even in this feverish climate, where sterilization of the unfit was seen as a noble cause, no one expected it to go beyond that.

In 1925, the philanthropic group of men known as the Advisory Council of the Milbank Memorial Fund gathered over cigars and cognac to discuss such things as birth control, care of the elderly, and how they would distribute the millions of dollars in their coffers. One of the men, Dr. William Welch, director of the School of Hygiene at Johns Hopkins University, stood before the group and asked a remarkable question for a man in his position: "Aren't we just keeping the unfit alive at the expense of the fit instead of letting nature do the weeding?" That simple question triggered a round of discussions about eugenics and set the course for funding of the Tuskegee syphilis experiment four years later.

Once the paperwork was authorized and signed, hundreds of poor black men in various stages of syphilis were recruited for medical examination—this at a time when even physicians believed in the inherent physical and mental inferiority of blacks. In fact, one of the agenda items at the Third International Congress of Eugenics that year was the Negro problem and methods for sterilization to eliminate bad stock. Racial medicine, as it was called, perpetuated the myth that blacks had lower immunity, greater susceptibility to disease, and self-destructive behaviors that led to a decline in health. So here was an opportunity to look at a sexually transmitted disease in a population thought to have a natural tendency for sexual promiscuity and only themselves to blame.

By May, eager black men began to arrive at the Tuskegee Institute and met with a team of physicians and a nurse named Eunice Rivers. The usual papers were filled out, questions asked, and histories documented. Everyone tried to make the men as comfortable as possible, given the clinical atmosphere surrounding the institute. But Nurse Rivers recalls the fear in the men's eyes soon after the initial examinations were done, when they were told that a twenty-gauge needle would have to be inserted into their spines to remove some fluid. Fear turned into terror for some who had to endure the crude lumbar puncture two or three times before the doctors were able to hit the spinal cord just right. Before they left, the only treatment the men received was a little mercury, administered in amounts thought too small to be therapeutic.

In almost every case, Nurse Rivers remembers, the men had painful reactions; a few were laid up for a week with head and neck pains, and one patient had pain until the day he died. None of the men were ever told they had been diagnosed with syphilis, only that they were being tested and treated for "bad blood." In his book of the same name, James Jones writes of one man's account in his own words. According to Jones, one of the men says, "It knocked me out. I tell you I thought I wasn't going to make it. I fainted, you know. Just paralyzed for a day or two. Just couldn't do nothing."

Within a month, the experiment had ended. Or so the men thought, until Dr. Raymond Vonderlehr, the physician who'd performed many of the spinal taps, became director of the Division of Venereal Diseases. From the very beginning, he wanted the study to continue. "Should the cases be followed over a period of five to ten years, many interesting facts could be learned regarding the course and complications of untreated syphilis," he argued. Since a number of the men he'd examined had severe symptoms and exhibited pathologies that could provide him with a wealth of information, Dr. Vonderlehr, knowing that untreated syphilis caused tumors, blindness, deafness, paralysis, and death, nonetheless viewed this as a golden opportunity. The fact that it was Negroes with untreated syphilis made the research project all the more acceptable.

So the decision was made. In the fall of 1932, the federal government, under the auspices of the U.S. Public Health Service and led by a surgeon general who happened to be a member of the American Eugenics Society, took charge of the study and agreed that it would last until the final autopsy was done. The black men who'd been previously diagnosed with syphilis were sought out and began receiving official recruitment letters from the U.S. Public Health Service, written in a way that would entice them to come back for further treatment.

Macon County Health Department
Alabama State Board of Health and U.S. Public Health Service
Cooperating with the Tuskegee Institute

Dear Sir,

Some time ago you were given a thorough examination, and since that time we hope you have gotten a great deal of treatment for bad blood. You will now be given your last chance to get a second examination. This examination is a very special one, and after it is finished you will be given a special treatment if it is believed you are in a condition to stand it.

You will remember that you had to wait for some time when you had your last good examination, and we wish to let you know that because we expect to be so busy it may be necessary for you to remain in the hospital over one night. If this is necessary you will be furnished your meals and a bed, as well as the examination and treatment without cost.

REMEMBER, THIS IS YOUR LAST CHANCE FOR A SPECIAL FREE TREATMENT. BE SURE TO MEET THE NURSE.

Macon County Health Department.

The deception was brilliant; 399 men with syphilis responded. With the addition of 200 controls, Dr. Vondelehr had a living laboratory, a group of unsuspecting human beings whose disease would progress through terrible stages until death. The title selected for the study was "Tuskegee Study of Untreated Syphilis in the Negro Male."

The beauty of the forty-year study, thought researchers, was its simplicity. Once a year, a roundup began in Macon County. Like livestock, the unsuspecting men, induced by small cash payments, were gathered for their annual "treatment" and examination. At the Tuskegee Institute, they were given aspirin and iron tonic they assumed was medicine for their bad blood. The incentive for enthusiastic young doctors to make the trip to Alabama was a chance to learn diagnostics in a real-life clinical setting and to participate in a once-in-a-lifetime experience that would advance their careers. When they returned to their regular duties, they discussed the various cases, wrote scientific papers, and thought nothing more about the men they'd left behind to die.

From the beginning of the study in 1932 until most of the subjects had died, annual exams and blood tests were done. There were also four main surveys carried out in 1932, 1938, 1948, and 1952. The men had no idea of their guinea pig status, cheerfully participating in the annual events while their syphilis steadily progressed in severity. Because the same public health nurse saw the same men each time, they grew to trust her. Nurse Rivers tells of men being picked up by their nurses at gatherings that had become more like social events, and waving to neighbors as they drove to the clinic for their physicals, pills, and "spring tonic."

As the years went by, the men's health worsened. It soon became clear that syphilis not only shortened its victim's life span but did so mainly while the men were still young. But to fully understand the cruelty of the government study, one has to follow the progression of the disease in individuals whose syphilis has advanced from early infection, in which a small sore develops and then spontaneously disappears, to a more symptomatic stage six to eight weeks later, to the latency period in which there are no outward symptoms but in which the spirochete organism burrows itself into a host of body tissues and organs, to the final or tertiary stage.

Despite what they believed were helpful annual treatments, the men continued to experience a gradual increase in eye disorders, headaches, and other vague discomforts that became more intense. Some of the men complained of deep pain in their bones if the spirochetes happened to lodge in the bone marrow, or pain throughout their bodies if the microbes had spread to vital organs. Over time, the pain became unbearable because bones were literally eaten away by the growing spirochetes. In the case of cardiovascular syphilis, the aorta and aortic valves were dissolved, causing the heart to malfunction and the victims to die either of heart failure or suffocation.

In the worst cases, and those that probably intrigued the doctors most of all, syphilis infected the central nervous system (CNS), including the brain. The accounts of victims with neurosyphilis are often harrowing, but for doctors who had anticipated all year what they would find in those particular patients, the annual visit must have been something special. It had to have been especially surprising for patients to see how much interest these government doctors had in their conditions. All they knew was that something was terribly wrong and that the so-called treatment they were getting was not making them better.

After several years of complaining about increasing numbness in their limbs—no doubt caused by degeneration of nerves and sensory receptors—the men began to think that not even the U.S. government could help them. They were absolutely correct, because once the doctors allowed syphilis to invade the CNS, all that could be done was to observe the pathology and learn from it. During the last few years of the study, the neurosyphilis patients no doubt were among the most interesting test cases. However, anyone at all familiar with the progression of syphilis once it invades the CNS would find it hard to believe that human beings were denied treatment in order to observe what happens to a man in the tertiary stage.

The tertiary stage typically begins with inflammation of the meninges (the membranes surrounding the brain and spinal chord), which then moves to the blood vessels, cranial nerves, and spinal chord tissue. Dizziness, double vision, and irregular dilation of pupils follow, triggering bouts of nausea and vomiting. As nerve damage continues, facial tone is affected, causing muscles to twitch and become distorted. Many victims suffer with episodes of excruciating abdominal, rectal, and laryngeal pain, or experience "lightning strikes" in their extremities that feel as if a stream of fire or electricity is surging through them. Because of degeneration of the dorsal nerve, there is loss of sensation and reflexes in the legs and arms, which makes a person feel as if he's paralyzed. Soon there is loss of balance and coordination. At the same time, sphincter muscles stop working, making it impossible to control bladder and rectal functions.

Within a few years of initial symptoms, the untreated patients must wear diapers as the disease finally moves to the brain. If they don't suffer a stroke or total paralysis, they develop seizures, personality disorders, and mental deterioration beginning with memory lapses, depression, schizophrenia, and toxic psychosis and ending with convulsions, total dementia, and death. Family members unfortunate enough to witness the hideous progression report horrible personality changes that reduce their loved one from a maniac having no control over his actions and bodily functions to a helpless vegetable with little or no brain function.

During the forty years of the study, in which 520 of the 600 original participants were followed, the experimental subjects were allowed to grow more and more sick until most had died. Had it not been for a newspaper story in 1972, the study would have continued even longer. The fact that doctors knowingly encouraged this horrible disease to spread, even after penicillin was shown to be an effective treatment in 1943, is one of the most shameful examples of racial medicine in the United States. And though a formal apology was issued in 1997, critics will forever keep alive America's role in the eugenics movement and link projects like the Tuskegee syphilis study with Nazi justification for their own human medical experiments.

Eugenics and Race Purification

The atmosphere of racism and racial hygiene seemed to gain momentum and gather like storm clouds over the twentieth century. The idea that the unfit, based on standards set by the fit, were like a cancer that needed to be excised from the human race was almost infectious. The infection reached an epidemic in 1930s Germany, where the eugenics torch was passed from moderates seeking strict birth control to zealots who wanting nothing less than elimination of some human beings from the population. When asked to look back at the last hundred years and choose the defining moment that changed the world and with it modern history, historians will probably choose the defeat of Nazi Germany as the single greatest contribution made by Western civilization. But while it's true that the 1930s marked the beginning of Hitler's expansion of power, to uncover the roots of evil one has to go back much further.

The primary sources of Hitler's biomedical vision were Reverend Thomas Malthus (1766–1834), Charles Darwin (1809–1882), and the English naturalist Alfred Russel Wallace (1823–1913). In his famous eighteen-chapter essay, "The Principle of Population," Malthus wrote, "Because all animated life tends to increase beyond the nourishment prepared for it, there can never be real progress or happiness for mankind." He argued that while populations increase geometrically, resources do so arithmetically, and concluded that man is doomed to misery and despair unless he controls population growth.

European nations took those principles to heart and agreed that to prevent imminent disaster they had to strictly control their growing populations. According to Dr. Theodore Hall of the Leading International Research Group, members of the ruling European classes gathered to devise ways to increase the mortality rates of the poor. Hall writes of their methods, "Instead of recommending cleanliness to the poor, we should encourage contrary habits. In our towns we should make the streets narrower, crowd more people into houses, and court the return of the plague. In the country, we should build our villages near stagnant pools, and particularly encourage settlements in all marshy and unwholesome situations." This policy toward the poorest in society was clearly evident in nineteenth century Britain, where policies in Ireland led to famines and death in the 1840s.

In America, the eugenics movement began in 1903 when Charles Davenport, director of the Cold Spring Harbor Laboratory, persuaded the Carnegie Institution of Washington to establish the Station for Experimental Evolution. A year later, the movement was given a boost when social engineering programs funded sterilizations of the unfit and discouraged marriages that would produce defective offspring. Field agents would go house to house, as well as to prisons, hospitals, and institutions for the deaf, blind, and mentally ill, to collect data and study health records. Armed with these kinds of records, many physicians performed sterilizations even before such procedures were legally approved, justifying their actions on the belief that certain racial stock was inferior and ought to be rooted out.

By 1907, Indiana enacted the first official eugenic sterilization law. Connecticut followed a year later. When the Eugenics Record Office (ERO) was established in 1910, it became a clearinghouse for family pedigrees, health records, and reports from caseworkers trained to gather specific information, analyze mate preferences, and suggest better eugenic choices among races. A system of measurements was adopted in which eugenicists used physical proportions and IQ tests to determine the superiority of northern and western European races. For the next two decades, ERO caseworkers descended on prisons, hospitals, and public schools in a mass effort to categorize the U.S. population and provide evidence to support their theories.

In 1914, Virginia sought to sterilize what they called the socially inadequate, feebleminded, insane, criminalistic, epileptic, inebriate, diseased, blind, deaf, deformed, and dependent, including orphans, ne'er-do-wells, tramps, the homeless, and paupers. It went even further in 1924 with the Virginia Integrity Act, designed to focus on what lawmakers called "defective persons whose reproduction represented a menace to society," and prohibiting the marriage between a white person and anyone with even a trace of blood other than Caucasian. As if other states couldn't jump on the eugenics bandwagon fast enough, by 1937 thirty-two states required sterilization of citizens viewed as undesirable or degenerate.

But when we look at the timing of the eugenics movement, we have to consider the coincidental link between the rise of racial policies and one of the greatest eras in U.S. immigration. Unlike nineteenth century immigration patterns, in which much of the influx was from northern Europe, many of the new immigrants passing through Ellis Island each year were coming from southern and eastern Europe. By the early twentieth century, Americans were increasingly marrying people of different races and ethnic groups. Since most of the eugenicists were from northern and western Europe, they feared that the new, less desirable immigrants were weakening American stock.

Alarm bells sounded. Warnings were issued that the inferior masses flooding America's shores were outbreeding superior races and lowering the nation's overall IQ. In the midst of the hysteria, eugenics experts were predicting that the social crime of race mixing would ultimately destroy white civilization. Thanks to their efforts, more than half the states passed laws forbidding mixed-race marriage, many of them imposing heavy fines and even prison sentences of up to ten years. All this occurred before Nazi Germany even considered implementing these kinds of drastic measures.

A key figure in the American eugenics movement from the outset was Margaret Sanger, who founded the first birth control clinic in 1916 and later established Planned Parenthood in 1921 (originally called the American Birth Control League). In her controversial 1932 thesis, *Plan for Peace*, which bears a striking resemblance to future Nazi population programs, she wrote of a "stern and rigid policy of sterilization and segregation to that grade of population whose progeny is already tainted." Her ideals for these segregated individuals included apportionment of land

and homesteads for their entire lives, which, in her view, included nearly half the population.

Many historians go so far as attributing some of Hitler's ideology and pronouncements to Margaret Sanger, challenging readers unfamiliar with Sanger to consider her quotes from *The Pivot of Civilization*, written in 1922:

> There is sufficient evidence to lead us to believe that the so-called borderline cases are a greater menace than the out-and-out defective delinquents who can be supervised, controlled, and prevented from procreating their kind. (p. 91)
>
> We prefer the policy of immediate sterilization, of making sure that parenthood is absolutely prohibited to the feebleminded. (p. 102)
>
> We are paying for and even submitting to the dictates of an ever increasing, unceasingly spawning class of human beings who never should have been born at all. (p. 187)
>
> Nearly half—47.3 percent—of the population has the mentality of twelve-year-old children or less—in other words they are morons. (p. 263)

Margaret Sanger, believing that less than 15 percent of the U.S. population was of superior intelligence, proposed immediate sterilization for the rest. She looked down on organized charity, writing that it is "the surest sign that our civilization has bred, is breeding, and is perpetuating constantly increasing numbers of defectives, delinquents, and dependents." At the same time Dr. Ernst Rüdin, head of the Nazi eugenics program, was working with the notorious SS chief Heinrich Himmler to develop plans for Germany's 1933 sterilization laws, Sanger had invited Dr. Rüdin to write an article for her magazine, *The Birth Control Review*, entitled "Eugenics Sterilization: An Urgent Need."

There is little doubt that Dr. Rüdin and Himmler looked to the West during those early years for any sign of encouragement. After all, both the United States and Britain's eugenics movements were in full swing by 1932, and they certainly must have recognized Germany as a future partner because they appointed Dr. Rüdin president of the International Eugenics Federation. When Rüdin was chosen to head up Germany's

Racial Hygiene Society a few months later, he immediately set out to make American eugenics principles the model for his own. So delighted were the American eugenicists at this honor that they actually printed a copy of the Nazi sterilization law in the September 1933 *Eugenical News*.

In Germany, it seemed as if America's actions were paralleled by General F. von Bernhardi, a German socialist who wrote, "If it were not for war, we could probably find that inferior and degenerate races would overcome healthy and youthful ones by their wealth and their numbers. The generative importance of war lies in this, that it causes selection and thus becomes a biological necessity." Bernhardi's influence was profound because his philosophy of "racial cleansing" became part and parcel of Hitler's thinking from the very start. Later Heinrich Himmler equated anti-Semitism with delousing, saying, "Getting rid of lice [the Jews] is not a question of ideology. It is a matter of cleanliness."

Adolf Hitler hailed these statements and ideas as principles to live by. "The state must see to it that only the healthy beget children," he wrote. "It must declare unfit for propagation all who are in any way visibly sick or who have inherited a disease and can therefore pass it on." He looked on Jews, as well as the defective in his own population, much like Sanger looked on blacks in *her* world. She wrote, "The masses of Negroes, particularly in the south, still breed carelessly and disastrously, with the result that the increase among Negroes, even more than among Whites, is from that portion of the population least intelligent and fit." From the day he took the oath of office as chancellor of the Third Reich, Hitler assumed that many in the West shared his philosophy and would agree wholeheartedly with his plan to make sterilization one of the key elements in his plan for Germany.

A student of Darwinism, and certainly of American eugenics proponents, Hitler believed that preserving the best racial stock by all means necessary would eliminate the unfit and ensure a superior race. His focus on race often verged on paranoia. In fact, despite all his ambitions, political or otherwise, what really made Hitler was a pathological zeal for race purification. "Nature is cruel," he said. "Therefore we are also entitled to be cruel . . . should I not have the right to eliminate millions of an inferior race that multiplies like vermin?" Yet early on, even American doctors praised him, the editors of no less than the prestigious *New England Journal of Medicine* writing in 1934 that "Germany is perhaps the most progressive nation in restricting fecundity among the unfit."

In his book *Mein Kampf*, Hitler made no attempt to hide convictions that often border on hysterical rantings against inferior humans. "Any crossing of two beings not at exactly the same level produces a medium between the level of the two parents," he writes. "This means: the offspring will probably stand higher than the racially lower parent, but not as high as the higher one. Consequently, it will later succumb in the struggle against the higher level . . . the stronger must dominate and not bend with the weaker, thus sacrificing his own greatness. . . . If this law did not prevail, any conceivable higher development of organic living beings would be unthinkable. . . . No more than Nature desires the mating of weaker with stronger individuals, even less does she desire the blending of a higher with a lower race, since, if she did, her whole work of higher breeding might be ruined with one blow."

The conclusion Hitler came to was that mankind was at a genetic crossroads, and since the Germanic people had remained racially pure while much of the world was breeding with inferior races, *it* would rise to be master of the continent. All that Hitler admired about Western civilization and culture was based, in his opinion, on the creativity of a few people and one race. Despite his hatred for Christianity, especially Catholicism, he saw himself as a cultural prophet who divided mankind into good and evil, superior and inferior. One of his greatest fears was that continued inbreeding would destroy the world as he wanted it to be. "If they perish," he wrote, referring to the superior few, "the beauty of this earth will sink into the grave with them." The urgency of Germany's sterilization program is illustrated in a letter to Reichsführer Heinrich Himmler on August 24, 1942:

> Dear Reichsführer,
>
> At the orders of Gauleiter Dr. Jury, his staff has hitherto busied themselves especially with the problem of population, racial policy, and antisocial elements. Since the prevention of reproduction by the congenitally unfit and racially inferior belongs to the duties of our National Socialist racial and demographic policy, the present director of the District Office for Racial Policy, Gauhauptstellenleiter Dr. Fehringer, has examined the question of sterilization and found that the methods so far available, castration and sterilization, are not sufficient in themselves to meet

> expectations. Consequently, the obvious question occurred to him whether impotence and sterility could not be produced in men and women by the administration of medicine or injections.
>
> The director of my race policy office points out that the necessary research and human experiments could be undertaken by an appropriately selected medical staff, basing their work on the Madaus animal experiments in cooperation with the Pharmacological Institute of the Faculty of Medicine of Vienna, on the persons of the inmates of the Gypsy camp of Lackenbach in Lower Danube.
>
> SS–Oberführer Gund

Although eugenics did not begin with Hitler, he embraced it as a father would embrace his own child. The reports he'd received from the United States of lectures and exhibits popularizing eugenics and demonstrating the menace of uncontrolled breeding and race mixing only strengthened his resolve. So using the American eugenics movement as a model, he elevated Germany's fledgling eugenics program to a prominence that made the rest of the world take notice. The effort was simple at first, prompting the German geneticist and leading proponent of the Nazi purification program, Fritz Lenz, to complain that Germany's eugenics research could not match that done at institutions such as Cold Spring Harbor Laboratory, where work on race purification was allowed to germinate and flourish. Armed with examples of selective breeding from around the world, especially the United States, the Nazis commingled science and politics to initiate the "final solution" to the Jewish question, the Holocaust.

Work at the Cold Spring Harbor Laboratory during the darkest chapter in its history paved the way for a movement that transformed eugenics from a method of selective breeding meant to improve the Nordic race to a rationale for extermination. There was no compromise. Even President Calvin Coolidge, who signed the 1924 Restriction of Immigration Act into law to maintain strict quotas on immigrants believed to be of inferior stock, stated that "America should be kept American. Biological laws show that Nordics deteriorate when mixed with other races." There is no better example of this kind of racist attitude than the platform written by the American Eugenics Party, the precursor of the movement that would carry on the eugenic ideology in the United States (Appendix X).

It's hard to imagine that Germany chose the United States as a model and an inspiration when instituting its Nazi eugenics policy. However, consider President Theodore Roosevelt's own words: "I wish very much that the wrong people could be prevented entirely from breeding; and when the evil nature of these people is sufficiently flagrant, this should be done. Criminals should be sterilized and feebleminded persons forbidden to leave offspring behind them. The emphasis should be laid on getting desirable people to breed."

Communities around the United States took that advice to heart. For instance, in Winston-Salem, North Carolina, the Bowman Gray School of Medicine became a center for eugenics, gathering records of children with inheritable disorders and forcibly sterilizing anyone having an IQ below 70. In a 1990 interview for the book *George Bush: The Unauthorized Biography*, a former official with the Eugenics Society said:

> I.Q. tests were run on all the children in the Winston-Salem School System. Only the ones who scored really low were targeted for sterilization, the real bottom of the barrel, like below 70. Did we do sterilizations on young children? Yes. This was a relatively minor operation. It was usually not done until the child was eight or ten years old. For the boys, you just make an incision and tie the tube. We more often performed the operations on girls than with boys. Of course, you have to cut open the abdomen, but again, it is relatively minor.

In two of the last "legally" sanctioned sterilization programs in the United States, the Indian Health Service (IHS) initiated a federally funded sterilization campaign and the Department of Health, Education and Welfare accelerated programs that paid 90 percent of the cost to sterilize poor women. The plan was to eliminate members of the population that would place an undue burden on society. In the case of Native Americans, doctors were sterilizing so many reservation women that, according to one account, "one woman was being sterilized for every seven babies born." Throughout the 1970s, countless Native American women continued to have their tubes tied or their ovaries removed by the IHS. No one knows the exact number, but one estimate has the number as high as seventy thousand Indian women sterilized between 1973 and 1976.

However, unlike the United States, which stopped at sterilization and committed the atrocity at Tuskegee, the Nazis used eugenics as a forerunner for mass murder and human medical experimentation. Theirs was a mission of elevating the Aryan people to a level of superiority and world dominance. Everyone else was regarded as nothing more than an inferior animal. It was in this climate that the movement aimed at improving members of the human race became the impetus for instituting a systematic program of Nazi medical research.

Eugenics, Nazis, and Medical Research

The seeds of America's early eugenics movement took hold in Germany as that nation's best hope for the future. In the beginning, it was simply a matter of sterilization. The disabled, mentally ill, and those with genetic diseases were the first targeted. Soon, euthanasia or mercy killings were added to eliminate the unproductive and those who would place an undue burden on the rest of society. The final step in the transformation of eugenics was human experimentation, in which individuals or races thought inferior were used in medical research for the benefit of the superior races. The horror of these experiments was so shocking in its abject cruelty that it eventually drove a stake into the heart of the American eugenics movement.

On a January morning in 1942, the biting wind outside the Polish concentration camp medical facility blew a swath of gray ash across the snow and left in its wake a thin layer of charred flesh. A perpetual stench seemed to cling to the gritty remains, which filled cracks in the wood and mortar and would become as much a part of the soil as the earth itself. On one side of the compound, tall stacks billowed an endless stream of thick smoke. On the other side, the medical facilities stood eerily silent, though when the winds died down the soldiers on guard would flinch at what they knew were the pitiful sounds of human experiments.

We can only imagine what it was like for one of the young men there, no older than twenty, to be led into a sterile room and stripped naked. As he was forced to bend forward, an insulated probe was inserted deep into his rectum. In front of him, a team of doctors looked on as he was lifted by attendants and wholly submerged in a vat of water and ice. Overcome with fear, he tried to resist but then closed his eyes and felt the cold blocks pressing against his skin. Within minutes, the intense shivering gave way to sharp pain, then near-hysteria, and finally numbness. Movement was no longer possible. The sounds he'd heard only moments earlier were now muffled. His vision became clouded and then disappeared as nerves were frozen and the signals between his brain and body disrupted. Soon, he began to lose consciousness as his limbs and internal organs froze and his body temperature approached 25 degrees Celsius.

Although the young man was fortunate enough to be resuscitated, the extreme pain of thawing after being frozen was unbearable. From his forehead to his toes, he felt as though ice picks were stabbing every square inch of his body. And as his core temperature rose, and he began to feel as if his muscles and skin were being ripped from his body, he prayed that tomorrow would be his turn in the poison room, where subjects were killed by various experimental toxins and then autopsied. At least that would mean a quick end. But despite his screams of agony, he thanked God that at least he was not in the adjacent room where even worse experiments were being conducted.

In one of those rooms, similarly frozen victims were placed under sun lamps so hot as to burn the skin, or had water that had been heated to near blistering temperature forcefully injected into their stomachs and intestines. The screams of *those* victims were silenced by the sudden onset of shock followed by cardiac arrest. In another room, a human subject whose intentionally administered wounds had been infected with gangrene culture was delirious with pain as his blood vessels were tied off and shards of glass, mustard gas, and sawdust placed into open wounds to aggravate them further. This test was done to see how long it would take for lethal gangrene to set in.

The medical blocks, as they were known, were areas of Auschwitz where certain prisoners were kept and specific medical research was done. For example, block 10, an especially brutal place of horrors, held many female subjects who served as guinea pigs for gynecological and reproductive research. It was here that Dr. Carl Clauberg's mass sterilization experiments were conducted, in which caustic agents, such as formalin, were injected into the cervix to see how much they would destroy or obstruct the oviducts. Hundreds of women occupied block 10 at any one time, after which they were sent to Birkenau for gassing.

The other blocks were equally frightening. Block 41 was notorious for vivisections, where subjects' limbs were cut open to expose muscles in order to apply various medications, or where prisoners were used for experimental surgeries. In block 28, men had toxic chemicals, such as lead acetate, rubbed into their bodies to cause severe abscess, infection, and painful burns, or were forced to ingest toxic powders for study of stomach and liver damage. Virtually every medical block was manned by SS doctors who viewed their subjects as less than human and had become so detached that, without so much as a change in facial expression, they could kill a pleading victim as easily as they would crush a roach.

Initially, in 1939, Auschwitz was built and operated for the purpose of imprisoning and putting to work Polish and Soviet prisoners of war. By 1941, its mission was expanded to include extermination and some of the most heinous medical experiments known to man. Its vast reservoir of slave laborers was also used by chemical and pharmaceutical companies such as I. G. Farben, which at one time employed thirty thousand workers, including nine thousand camp inmates. On average, the Jewish inmates lasted about four months before falling ill and being sent to Birkenau for termination. The less fortunate become human guinea pigs.

Every medical experiment conducted on what were called subhumans was justified in the name of improving the life of Germany's citizens and military. Since so many German soldiers died of hypothermia on the eastern front, freezing and thawing experiments had to be done to find effective means of resuscitation. And because infections and diseases decreased an army's readiness and fighting capability, what better way to test new drugs than to inject subjects with disease or allow them to develop gangrene? Nazi doctors took their work seriously and they welcomed any opportunity to use human subjects rather than laboratory animals.

For the most part, early Nazi medical programs focused on negative eugenics. Beginning in 1939, the Nazi regime targeted disabled German nationals, euthanizing those deemed unworthy of life. Initially, that included infants and young children born with mental or physical disabilities, even common ones such as hearing and vision impairments. Adults were added to the target list later. A special unit known as Operation T4 sought out the victims, who were sometimes rounded up under the guise of getting treatment, and then shipped them to various euthanasia camps for the mercy killings. The following are two letters, one to the Reich minister of the interior, the other to the Reich minister of justice, discussing the practice of extermination of the insane:

September 5, 1940

Dear Reich Minister,

On July 19th I sent you a letter about the systematic extermination of lunatics, feebleminded, and epileptic persons. Since then this practice has reached tremendous proportions: recently the inmates of old-age homes have also been included. The basis for this practice seems to be that in an efficient nation there should be no room for weak and frail people. It is evident from the many reports which we are receiving that the people's feelings are being badly hurt by the measures ordered and that the feeling of legal insecurity is spreading which is regrettable from the point of view of national and state interest.

Dr. Wurm

Wuerttemberg Evangelical Provincial Church

September 6, 1940

Dear Reich Minister,

The measures being taken at present with mental patients of all kinds have caused a complete lack of confidence in justice among large groups of people. Without the consent of relatives and guardians, such patients are being transferred to different institutions. After a short time they are notified that the person concerned has died of some disease.

If the state really wants to carry out the extermination of these or at least of some mental patients, shouldn't a law be promulgated, which can be justified before the people—a law that would give everyone assurance of careful examination as to whether he is due to die or entitled to live and which would also give the relatives a chance to be heard, in a similar way, as provided by the law for the prevention of hereditarily affected progeny?

Chief of the Institution for the Feebleminded

The final step in Hitler's eugenics program included foreigners and non-German citizens believed to have inferior genes and therefore were seen as a threat to racial strength and purity. The exterminations were simply a way to prevent potentially bad genes from ever getting into the gene pool. The medical research program was designed to study genetic differences and to advance health and medicine using a seemingly endless supply of inferior human beings. There were virtually no limits to what Nazi doctors would do.

The bulk of Nazi medical research, however, centered on race and genetics. The principle focus of experiments was to improve the Nordic race through genetic manipulation and to determine the causes of genetic defects. Dr. Josef Mengele, medical commandant of Auschwitz from 1943 and known as the "Angel of Death," would sort through the men, women, and children as they arrived at the camp and decide who would be assigned to the experimental facilities. Twins, dwarfs, and those with unique physical characteristics were always chosen first. Mengele's principal interest in determining the genetic cause for the birth of twins was to establish the racialist theory about the superiority of the Nordic race.

After years of obscurity, Mengele had found the perfect human laboratory for his work in Auschwitz. Obsessed with the genetics of twins and dwarfs, he'd become almost fanatical in the selection process, insisting on being involved personally in choosing who would be sent to work camps, who would die in the gas chambers, and who would live to serve as a human subject. Selected twins had a special place in Auschwitz. Even the guards knew better than to tamper with Mengele's twins, lest they damage them before the research was done.

To the twins, it must have seemed odd to have such attention heaped upon them. They may even have believed that somehow God had spared them from the horrors they saw happening to everyone else around them. For while the crematoriums were operating day and night, there they were, eating better than anyone else, sitting in clean laboratories for hours at a time, and having every square inch of their bodies examined and measured as if they were something special.

Dr. Mengele himself often conducted the examinations, spending days examining the head, another day or so measuring limbs and studying bone structures. However, after the family histories and exhaustive physical examinations were done, it was time to get down to serious business.

The twins learned quickly that, although their uniqueness brought with it a little more time, it had, in fact, sealed their fate. A series of torturous experiments followed that made the twins wish they'd been culled for the gas chambers instead.

To create a race of blond-haired, blue-eyed individuals, according to some accounts (including Robert J. Lifton's *The Nazi Doctors*) Mengele experimented with various dyes, which he injected into the unanesthetized eyes of children, preferably twins. The excruciating procedure often caused injury and sometimes total blindness, at which time the children were exterminated. In some experiments, he sewed children together to simulate Siamese twins. In other experiments, he injected typhoid or tuberculosis to see how individuals of different races reacted to disease, or killed a set of healthy individuals simultaneously because he wanted to do autopsies on twins who'd died at precisely the same moment.

As painful as the eye, infection, and surgery experiments were, the respiratory and gastrointestinal procedures were worse. Tubes would be forced through an individual's nose and into his lungs for collection of fluids. As gas was pumped through the tubes, it triggered violent coughing to facilitate the collection process. If the lungs didn't tear or collapse, the victim was given a few days to recover, administered a two-liter enema, strapped to a bench table, and, without anesthesia, had his rectum distended for an intense and painful lower gastric examination. After tissue samples were taken from the kidneys, prostate, and testicles, the subject was taken to the dissection room, killed with a single injection of phenol or chloroform to the heart, and dissected for study of internal organs.

One of Mengele's pathologists, Dr. Miklos Nyiszli (whose account is admittedly controversial), described the specially constructed dissecting lab as an elaborate room for autopsies and pathological examinations of corpses. In his 1945 deposition before the Budapest Commission for the Welfare of Deported Hungarian Jews, Dr. Nyiszli recounted how he'd had to plead with his superior to spare the life of even a single individual, and he told of the procedure Mengele used to kill the twins whom he had collected:

> In the work room next to the dissecting room, fourteen Gypsy twins were waiting about midnight one night, guarded by SS men, and crying bitterly. Dr. Mengele didn't say a word to us,

> and prepared a 10 cc and 5 cc syringe. From a box he took evipan, and from another box he took chloroform, which was in 20 cc glass containers, and put these on the operating table. After that, the first twin was brought in, a fourteen-year-old girl. Dr. Mengele ordered me to undress the girl and put her on the dissecting table. Then he injected the evipan into her right arm intravenously. After the child had fallen asleep, he felt for the left ventricle of the heart and injected 10 cc of chloroform. After one little twitch the child was dead, whereupon Dr. Mengele had her taken into the morgue. In this manner, all fourteen twins were killed during the night.

Those who weren't killed by injection died on the operating table or as a result of intentional infections, or were sent to the gas chambers. After the poor creatures were prodded, shuffled, and then locked in large rooms that looked like communal showers, the gas canisters were dropped through a hole in the ceiling. Within seconds, when it became apparent that this was not a shower, the panic would begin. Outside the airtight doors, guards stood without emotion and listened to the screams of agony as the gas drifted upward to the victim's faces.

Sometimes it took ten minutes before the final whimpers stopped and everyone was dead. When the doors were unlocked, bodies could be seen piled one on top of another as if the strongest had tried to prolong their lives a few more seconds by knocking down the weakest and using them to reach for the gas-free layer of air near the ceiling. Urine and excrement were everywhere; on the floor, on the walls, and on the corpses, which were dragged from the room with special hook-tipped poles. In just about all cases, teeth were removed; in some cases limbs were chopped off to speed cremation or the heads sawed off before being taken to the examination and cataloging rooms.

The atrocities committed by Josef Mengele and other SS doctors were a culmination of a eugenics movement that had crossed the line from birth control to callous mass murder. The rationale for human experimentation was simple: If man would not allow nature to take its course and ensure that only the fit survive, then man himself had to make certain that evolution worked as it was intended. Had the Nazis not destroyed many of the documents, laboratories, and evidence before the Allied forces liberated the camps, we may have found that the extent of Germany's human medical experiments was far greater than we ever imagined.

Project Paperclip: Nazi Scientists in America

It was shortly after the war had ended in 1945 that Bill Donovan, head of the OSS (the wartime forerunner of today's CIA) and Allen Dulles, OSS head of intelligence, approached President Harry Truman with a plan that former President Franklin Roosevelt had previously rejected. Knowing that thousands of Nazi doctors and scientists had scattered like vermin across Europe, the men proposed to bring Nazi brain power to America rather than see it go to other countries, such as the Soviet Union. The proposal, initially known as Project Overcast, was approved by the joint chiefs of staff and signed off by the president. A memo initially released by the War Department never mentioned that the scientists were Nazis.

War Department
Bureau of Public Relations
Press Branch
Tel. RE 6500
Brs. 3425 and 4860

October 1, 1945

IMMEDIATE RELEASE

OUTSTANDING GERMAN SCIENTISTS
BEING BROUGHT TO U.S.

The Secretary of War has approved a project whereby certain outstanding German scientists and technicians are being brought to this country to ensure that we take full advantage of those significant developments which are deemed vital to our national security.

Interrogation and examination of documents, equipments

and facilities in the aggregate are but one means of exploiting German progress in science and technology. In order that this country may benefit fully from this resource a number of carefully selected scientists and technologists are being brought to the United States on a voluntary basis. These individuals have been chosen from those fields where German progress is of significant importance to us and in which these specialists have played a dominant role.

Throughout their temporary stay in the United States these German scientists and technical experts will be under the supervision of the War Department but will be utilized for appropriate military projects of the Army and Navy.

END

According to the plan, no fewer than one thousand Nazis were to be exfiltrated to the United States, given immunity from war crimes, and employed in government and civilian facilities. There was even debate about the quality of scientists being brought here, as illustrated by this memo:

UNITED STATES AIR FORCE
AIR UNIVERSITY HEADQUARTERS
USAF SCHOOL OF AVIATION MEDICINE

20 March 1951
Major General Harry G. Armstrong, USAF (MC)
The Surgeon General, USAF
Headquarters, USAF
Washington 25, D.C.

Dear General Armstrong:

In reference to your letter of 9 March, copy attached, concerning paperclip personnel, I wish to state that we are interested in obtaining first-class scientists and highly qualified technologists from Germany. The first group of paperclip personnel contained a number of scientists that have proved to be of real value to the Air Force. The weaker and less gifted ones have been culled to a considerable extent. The second group reporting here in 1949 were, in general, less competent than the original paperclip personnel, and a culling process will again be in order. A few vacancies currently exist, and more will occur within six months.

Informal information received here confirms your statement that the paperclip project may be revamped and revitalized in the near future. With this in mind, and in view of the statements made in the preceding paragraph, an effort is being made to obtain data on outstanding German scientists working in the medical sciences. In this connection, a copy of a letter written by the Director of Intelligence, Headquarters, USAF, is attached for your information.

The requirements of the Arctic Aero Medical Laboratory, and staffing of any future aeromedical installation at Eglin Field, should be kept in mind. An attempt will be made to locate paperclip personnel for these places if and when their requirements become known.

Sincerely,

Otis O. Benson, Jr.
Brigadier General, USAF (MC)
Commandant

The project's name was changed to Project Paperclip because the files of those selected to be brought to the States had paperclips attached to them as a clandestine signal to those familiar with the selection process. However, as a subsequent government memo illustrates, there were actually two classified projects:

Subject: Civilian Personnel Spaces to Accommodate the PAPERCLIP and PROJECT 63 Programs
Date: June 2, 1953

1. The Department of Defense has two classified projects, deemed of utmost importance, that result in the employment and exploitation of foreign scientists by the Department:

a. The first, PAPERCLIP, provides a means of obtaining services of foreign Specialists for specific assignments within the technical services of the Departments of Army, Navy, and Air Force. The primary function of this program is the utilization of the individual, the denial aspect being a highly desirable, although secondary feature. Such specialists sign a year's contract for a specific assignment prior to leaving their place of residence.

b. PROJECT 63 is primarily a denial program with utilization as a desirable feature. The aim of this program is to secure employment in the United States of certain preeminent German and Austrian specialists, thus denying their services to potential enemies. Such specialists sign a six-month Department of Defense contract which guarantees them an income until permanent employment is arranged with Department of Defense agencies or industry within the United States.

Hoyt S. Vandenberg
Air Force Chief of Staff

A large number of Nazis arriving in the United States were some of the world's vilest war criminals, but their backgrounds and expertise could be exploited in the West's struggle against communism. Researchers who'd used military prisoners like they were dogs, scientists working on biological warfare agents, and doctors who'd performed human medical experiments were actively sought out and given refuge. Two of the doctors that would have been prime recruits because of their work on human physiology were Sigmund Rascher and Herman Becker-Freyseng, whose experiments at Dachau were among the more gruesome.

Most of the subjects Dr. Rascher used were Jews, who had salt water forced into their stomachs through tubes or injected directly into their veins in an attempt to study how long pilots downed over the ocean could survive. During the course of the experiments, liver tissue was extracted without anesthesia by inserting long needles through the skin. All the test subjects died within a few weeks of the experiments.

In another study, Dr. Rascher had developed a special low-pressure chamber to test the effects of high altitudes on human physiology. The subjects chosen were Jews, Russian prisoners, and Polish resistance fighters. During the experiments, subjects were locked for thirty minutes without oxygen in the chamber, which simulated an altitude of seventy thousand feet. As scientists looked on and adjusted their instruments, the low pressure and lack of oxygen inside literally sucked the air from the men's lungs and brains. One observer testified that some men committed suicide by throwing themselves violently against the wall of the chamber rather than dying of suffocation or having their lungs explode.

According to one account, Dr. Rascher wrote, "Some experiments gave men such pressure in their heads that they would go mad and pull out their hair. They would tear at their heads and faces with their hands and scream in an effort to relieve the pressure on their eardrums." Dozens of unconscious subjects were then removed, drowned in vats of ice water, and immediately had their heads dissected so that scientists could examine the extent to which the blood vessels had ruptured from the pressure. So here was a Nazi scientist in America, allowed to live and work as if he'd done nothing out of the ordinary, even after the discovery of two letters he'd written to Heinrich Himmler.

April 5, 1942

Highly Esteemed Reich Leader:

Enclosed is an interim report on the low-pressure experiments so far conducted in the concentration camp of Dachau.

Only continuous experiments at altitudes higher than 10.5 km resulted in death. These experiments showed that breathing stopped after about thirty minutes, while in two cases the electrocardiographically charted action of the heart continued for another twenty minutes.

The third experiment of this type took such an extraordinary course that I called an SS physician of the camp as a witness, since I had worked on these experiments all by myself. It was a continuous experiment without oxygen at a height of 12 km conducted on a thirty-seven-year old Jew in good general condition. Breathing continued up to thirty minutes. After four minutes the experimental subject began to perspire and to wiggle his head, after five minutes cramps occurred, between six and ten minutes breathing increased in speed and the experimental subject became unconscious; from eleven to thirty minutes breathing slowed down to three breaths per minute, finally stopping altogether.

Autopsy report

One hour later after breathing had stopped, the spinal marrow was completely severed and the brain was removed. Thereupon the action of the auricle stopped for forty seconds. It then renewed its action, coming to a complete standstill eight minutes later. A heavy subarachnoid edema was found in the brain. In the veins and arteries of the brain a considerable quantity of air was discovered.

SS–Untersturmführer Sigmund Rascher

May 11, 1942

Highly Esteemed Reich Leader:

Enclosed I am forwarding a short summary on the principle experiments conducted up to date.

For the following experiments Jewish professional criminals who had committed race pollution were used. The question of the formation of embolism was investigated in ten cases. Some of the experimental subjects died during a continued high-altitude experiment; for instance, after one-half hour at the height of 12 km. After the skull had been opened under water an ample amount of air embolism was found in the brain vessels and, in part, free air in the brain ventricles.

To find out whether the severe psychic and physical effects, as mentioned under No. 3, are due to the formation of embolism, the following was done: After relative recuperation from such a parachute descending test had taken place, however, before regaining consciousness, some experimental subjects were kept

> under water until they died. When the skull and the cavities of the breast and of the abdomen had been opened under water, an enormous amount of air embolism was found in the vessels of the brain, the coronary vessels, and vessels of the liver and the intestines, etc.
>
> It was also proved by experiments that air embolism occurs in practically all vessels even when pure oxygen is being inhaled. One experimental subject was made to breathe pure oxygen for 2.5 hours before the experiment started. After six minutes at a height of 20 km, he died and at dissection also showed ample air embolism, as was the case in all other experiments.
>
> SS–Untersturmführer Sigmund Rascher

After experimenting with seawater injections and effects of low pressure, Dr. Rascher turned his attention to poison bullets. A partial transcript from the Nuremberg trials describing his work illustrates the callous approach he'd taken in dealing with his human subjects:

> On 11 September 1944, in the presence of SS Sturmbannführer Dr. Ding, Dr. Widmann, and the undersigned, experiments with aconite nitrate bullets were carried out on five persons who had been sentenced to death. The caliber of the bullets used was 7.65 millimeters, and they were filled with poison in crystal form. Each subject of the experiment received one shot in the upper part of the left thigh, while in a horizontal position. In the case of two persons, the bullets passed clean through the upper part of the thigh. Even later no effect from the poison could be seen. These two subjects were therefore rejected.
>
> The symptoms shown by the three condemned persons were surprisingly the same. At first, nothing special was noticeable. After twenty to twenty-five minutes, a disturbance of the motor nerves and a light flow of saliva began, but both stopped again. After forty to forty-four minutes, a strong flow of saliva appeared. The poisoned persons swallowed frequently; later the flow of saliva is so strong that it can no longer be controlled by swallowing. Foamy saliva flows from the mouth. Then a sensation of choking and vomiting starts.
>
> At the same time there was pronounced nausea. One of the

> poisoned persons tried in vain to vomit. In order to succeed he put four fingers of his hand, up to the main joint, right into his mouth. In spite of this, no vomiting occurred. His face became quite red.
>
> The faces of the other two subjects were already pale at an early stage. Other symptoms were the same. Later on the disturbances of the motor nerves increased so much that the persons threw themselves up and down, rolled their eyes, and made aimless movements with their hands and arms. At last the disturbance subsided, the pupils were enlarged to the maximum, the condemned lay still. Rectal cramps and loss of urine was observed in one of them. Death occurred 121, 123, and 129 minutes after they were shot.
>
> SS–Untersturmführer Sigmund Rascher

Dr. Rascher was by no means the most infamous Nazi scientist guaranteed hospitality under Project Paperclip. Theodore Benzinger, for example, ended up as a highly paid government researcher at Bethesda Naval Hospital in Maryland, where no one suspected that his experience in wound healing was honed on human subjects in various concentration camps. Dr. Eugene von Haagen, who'd spent years at the Natzweiler concentration camp infecting prisoners with biological agents and assorted diseases before autopsying those he'd killed, found himself working for a U.S. government germ weapons research program. And Dr. Walter Schreiber, the Nazi chief of medical science and director of some of the worst medical experiments conducted on inmates, was hired by the U.S. Air Force because of his valuable experience in epidemiology, preventive medicine, and his "peculiar" knowledge of public health.

In order to expand the pool of doctors and scientists, in 1947 the Joint Intelligence Objectives Agency (JIOA) initiated Project "National Interest," a program to add Eastern Europeans to the cadre of German and Austrian scientists. That year also happened to be when the CIA was formed as a separate intelligence agency. The sole standard for acceptance to the program was that the individual would contribute to U.S. national interests. At the time, keeping any prominent scientist from the Soviet Union was deemed vital for national interests and national security. So for the next few years, a steady stream of the world's refuse arrived at America's shores and integrated itself with the best of its new homeland. The following is a 1948 JIOA memo urging the facilitation of immigration and employment of German scientists by civil employers:

16 March 1948

Joint Intelligence
Objectives Agency
JIOA 902

MEMORANDUM FOR: Mr. John C. Green, Office of Technical Services, Department of Commerce

SUBJECT: Liaison between Joint Chiefs of Staff and the Department of Commerce in connection with the Employment of German Scientists

1. As you know this agency is responsible to the Joints Chiefs of Staff for liaison with the Department of Commerce in certain matters pertaining to the employment of German scientists and technicians in this country. Our contract with them has been with Mr. Hicks of your office. It has come to my attention that there is a possibility that the Office of Technical Services may be dissolved at the end of this fiscal year.

2. In order to permit advance coordination and planning it is requested that this Agency be informed, as far in advance as possible, of the appropriate division of the Department of Commerce for liaison in the matters listed below, in the event that your office is disestablished.

a. Sponsorship of the cases of specialists that are in this country under limited military custody who were brought here for civil exploitation or were transferred from military to civil exploitation after arrival, including such matters as insuring that employment is of benefit to the nation and that results of research and development are made available to the public; arranging for shipment of dependents and their reception on arrival; recommendation for and facilitation of immigration as agreed to in SWNCC 257/15; arranging for leaves of absence to Germany; arranging for further employment of specialists when contracts with original employer have expired, or return to Germany if no further employment is available; and approval of contracts between civil employers and specialists.

b. Cooperation with the military in placing specialists in private industry when further military employment is not available, and it is in the military interest to keep them employed in the United States.

c. Cooperation with industry in determining prospective availability of specialists employed by the military.

d. Arranging for allocation to the United States of any scientists or their families that may be brought to the United States under a program for entry not now in operation.

BOQUEST N. WEY
Captain, USN
Director

Entry into the United States was further facilitated by the CIA Act of 1949, which allowed U.S. entry "without regard to admissibility under any other laws" if the entry advanced national security. Many of the aliens were first sent to Canada or Mexico, after which they reentered the United States, where they were placed at major universities, such as the University of Texas, University of North Carolina, Washington University School of Medicine, and Boston University. If security was ever breached or if one of the war criminals was in danger of being discovered or arrested, he would be smuggled to a South American country such as Brazil or Argentina.

By 1973, Project Paperclip had ended. In all, about eight hundred doctors and scientists, some of them living out their lives in comfort and security, continued their work in the United States. The ultimate goal of Paperclip was to keep these men out of Soviet hands. To that end, the project was a success. But the price Americans paid to fight the new communist threat was to have the worst Nazi scientists and doctors living among them and not even know it.

Modern Population Control Programs

Sometime in the year 2000, the earth's human population reached six billion. However, at some point before that number was reached political leaders had already decided that the world was growing at a rate far greater than could be sustained or controlled. Poverty, starvation, social unrest, overpopulation, disease-infested slums, children who looked like skeletons dying in their mothers' arms, and a burgeoning rise in immigration was enough to stir into action those who'd finally had enough. Regardless of when the decision to act was made, the deciding was done by those who saw the rising tide of humanity as a direct threat to their security and vital national interests.

Following the war, the very term *eugenics* had become anathema. Because the Third Reich had clearly ruined it for the American eugenics movement with its extremism and the hideous nature of its human experiments, a more clandestine policy called "crypto-eugenics" was born. In a 1956 British eugenics resolution, the principles of eugenics would be vigorously preserved, but instead of calling openly for sterilization of the unfit, disabled, and poor, crypto-eugenics began operating under names like Planned Parenthood and the Population Council, whose aims were to prevent the birth of unwanted human beings. In America, the Eugenics Society changed its name to the Society for the Study of Social Biology, with its stated goals "to further the discussion, advancement, and dissemination of knowledge about biological and sociological forces which affect the structure and composition of human populations."

It was John D. Rockefeller III who launched the Population Council, a group advocating zero population growth and family planning in the Third World. Developed nations had already determined that global population growth would undermine the social and economic stability not only of Third World countries responsible for the explosive growth in the first place, but of literally every nation on earth. All they had to do was look at the numbers. While it took the human population until the early

1800s to reach one billion inhabitants, it took from then until 1920 to double that. Less than two generations later, two billion grew to an astounding three billion, and the realization that Malthus may have been right shook the political illuminati to their very core. Something had to be done to discourage or reverse the trend before the world literally exploded.

In 1957, intellectuals, scientists, and political leaders met secretly in Huntsville, Alabama, spurred by the words of President Dwight D. Eisenhower in that same year: "As a result of lowered infant mortality, longer lives, and the accelerating conquest of famine, there is underway a population explosion so incredibly great that in little more than another generation the population of the world is expected to double." An ominous tone permeated the gathering, with predictions of a total collapse of civilization and the extinction of the human race unless one or more of the following occurred: (1) a sudden reversal, by whatever means, in the number of humans; (2) a cutback in technologies that improved health and therefore increased life span or encouraged survival of the unfit; (3) the elimination of meat in the diets of certain populations; and (4) strict regulations and controls on human reproduction.

For the next decade, nothing seemed more urgent than reaching agreement on a plan for slowing the growth of the world's population. That urgency came to a head in 1968 when leaders of the newly formed Club of Rome, a global think tank headquartered in Hamburg, Germany, met to finalize a method to arrest a global pandemic out of control. Incredibly enough, the two-pronged attack involved lowering the birth rate through birth control methods, including sterilization, abortion, and hysterectomy, and increasing the death rate, which was obviously more problematic.

According to William Cooper, former member of U.S. Naval Intelligence and author of *Behold a Pale Horse*, Dr. Aurelio Peccei, a freemason who'd founded the club, proposed an unusual strategy, to say the least, for implementing the latter. His secret recommendation was to introduce a microbe into the general population that would target the immune system but that would be responsive to a vaccine available only to certain members of that population. A year later, coincidentally, a request was made (and granted) before a senate committee for funding to produce "a synthetic biological agent, an agent that does not naturally exist and for which no harmful immunity could have been acquired."

Whether Dr. Peccei's suggestion was taken seriously is doubtful, though some argue that a population explosion was considered as serious as any war that threatened national security. Adding to suspicions that the United States would accept any means of reducing populations were the comments of Thomas Ferguson, a Latin American officer for the U.S. State Department, who said, "Population is a political problem. . . . Once population is out of control it requires authoritarian government, even fascism, to reduce it. . . . The professionals aren't interested in lowering population for humanitarian reasons. . . . The quickest way to reduce population is through famine, like in Africa or through diseases like the Black Death. . . . We are letting people breed like flies without allowing for natural causes to keep population down. We raised the birth survival rates, extended lifespans by lowering death rates, and did nothing about lowering birth rates. That policy is finished."

There was almost a sense of desperation in those words, which seemed to make perfect sense to the likes of John Rockefeller and President Richard Nixon, who sought more research on birth control methods and family planning. In a 1969 message to Congress, Nixon joined a long list of American leaders who equated unrestrained population growth with a threat to the destiny of humankind:

> One of the most serious challenges to human destiny in the last third of this century will be the growth of the population. Whether man's response to that challenge will be a cause for pride or for despair in the year 2000 will depend very much on what we do today. If we now begin our work in an appropriate manner, and if we continue to devote a considerable amount of attention to this problem, then mankind will be able to surmount this challenge as it has surmounted so many during the long march of civilization.

The United States was becoming increasingly worried by reports from analysts around the world. To the south, Mexico and Latin America was seen as a seething hotbed of instability, the Middle East was a boiling pot, Africa's population was totally out of control, and Asia, it was predicted, would be the next great wave of humanity to overburden the earth's shrinking resources. While fear of these predictions enveloped Washington like a shroud, the U.S. State Department's Office of Population Affairs (OPA), established by Henry Kissinger in 1975, had already drafted the Carter administration's *Global 2000 Report*, a confidential document that outlined methods for worldwide population reduction. The sense that a war on population was needed to avert global disaster is evident in the dire predictions of another classified document, the National Security Study Memorandum 200 (NSSM-200), drafted a year before the OPA was even formed.

Believing that the world was ready to join in the effort to control birth, the United States AID (USAID) population office made a statement that sent shock waves through Third World nations. "Like a spring torrent after a long, cold winter, the United States has moved with crescendo strength during recent years to provide assistance for population and family planning throughout the developing world." The agency predicted that it was going to sterilize one-fourth of the world's women.

Though USAID's proclamation raised the specter of the United States's meddling, nations were privately embracing it. Throughout the world, the United States, either directly or indirectly, approved of eugenics or population-control programs. In most cases, the programs were implemented under the guise of simple birth control, but in reality many were coercive actions or laws that involved forced sterilization. It didn't take long after the initial draft of NSSM-200 came off the presses that its recommendations were put into practice.

During the 1970s, for example, Indira Gandhi's campaign of population control forced more than six million men to be sterilized; in Bangladesh and other Muslim countries, bribes were offered in exchange for permanent sterilization, and in some cases food and other aid was contingent on participation in forced birth control; and in Brazil, as many as 90 percent of women in some northeastern regions (where most of the population is dark skinned and of African descent) have been sterilized, claiming that they had never been told the procedure was permanent.

During the past decade, reports have surfaced that the WHO has been testing antipregnancy vaccines in Mexico and the Philippines. This wouldn't be startling news except for the fact that investigators claim that the women are allegedly receiving tetanus vaccines also containing human chorionic gonadotropin (HCG). Apparently, the purpose of the HCG is to stimulate antibodies to HCG to induce spontaneous abortion of subsequent pregnancies. Similar vaccines have been on the WHO's drawing board since the 1970s.

Much of the world's population control research has its origins and funding in the West. The reason, some say, is self-serving. While Western populations are shrinking, the rest of the world is expanding. In his new book, *The Death of the West*, Patrick Buchanan details some alarming statistics for population control advocates and eugenicists. According to Buchanan, Western nations have not only stopped reproducing; their populations have become so stagnant that not a single European country except for Albania (which is Muslim) will survive as we know it today. By 2050, says Buchanan, only 10 percent of the world will be of European descent. Even worse, the average age of a European will be 50 years. Buchanan's facts and figures are backed by UN studies, as well as a *London Times* analysis, which concludes that Europe's population decline will be the worst since the Black Plague of the Middle Ages.

A shocking look at demographics of the Third World corroborates Buchanan's claim that Europeans are a dying culture and that, because of socialism, which had eliminated the need for families, Europe as we know it will disappear. According to Buchanan, already sending shock waves through the eugenics community is the fact that the collective population of Europe's forty-seven nations will plunge from 728 million in 2000 to 556 million by 2050. By the same year, Germany's population will have decreased from 82 million to 59 million. Fifty-two percent of Italian women between the ages of sixteen and twenty-four plan to remain childless. Russia will lose twenty-two million people during the next fifteen years alone, and by the end of the century, people of British descent will be a minority in their own country.

Unless serious population controls are implemented in the Third World, eugenicists' worst fears will be realized. The nations and the people they have been targeting for nearly a century are on the threshold of a population and immigration explosion that some see as an irreversible tide of cataclysmic proportions. In light of the numbers and the nightmarish projections, there's no doubt that a secret war has already been declared.

In Mexico, Nicaragua, and the Philippines, millions of females were allegedly given tetanus vaccines laced with HCG that could cause miscarriages and sterilization. According to Dr. Alan Cantwell, author of *AIDS and the Doctors of Death*, "Suspicions were aroused when the tetanus vaccine was prescribed in the unusual dose of five multiple injections over a three-month period, and recommended only to women of reproductive age. When an unusual number of women experienced vaginal bleeding and miscarriages after the shots, a hormone additive was uncovered as the cause." Cantwell alleges that "the laced vaccine served as a covert contraceptive device" and that "20 percent of the WHO tetanus vaccines were contaminated with the hormone." The World Health Organization has denied the contentions concerning the tetanus vaccine.

One of the more sinister new developments in this war of population control is a sterilization agent known as quinacrine, a substance used during World War II as a drug for malaria. In its latest form, quinacrine is inserted into the uterus as a pellet that dissolves and releases the chemical that then travels to the fallopian tubes and causes formation of scar tissue. The scarring irreversibly blocks the tubes and prevents a released egg from being fertilized. Quinacrine, which has appropriate uses unrelated to birth control, is available in at least nineteen countries.

According to a 1993 article in the medical journal *Lancet*, thirty-two thousand Vietnamese women have been tested, despite warnings that quinacrine can cause nervous system disorders, hallucinations, toxic psychosis, a type of chemically induced insanity, uterine bleeding, and cancer. In one study of sixty women in Mexico, every woman treated reportedly experienced complications. In Malaysia and Chile, a significant number of women reported unusual bleeding and other side effects. Thus far, not a single country with the exception of Vietnam, which has chosen to accept quinacrine as a legitimate form of sterilization, has had problem-free results.

In 1993, the WHO declared that quinacrine should not be used because of its mutagenic properties. In 1998, the government of India banned it, as did the government of Chile. In the United States, quinacrine is approved as an antimalarial drug but not as a means of birth control. However, since the FDA permits approved drugs to be used for purposes other than those for which they were originally intended, that is, "off-label," physicians can prescribe it for other purposes. Currently, both the FDA and the WHO have recommended that more animal studies be done before any additional human trials are conducted.

The advantage of quinacrine, say population control advocates, is that it is permanent, inexpensive, and requires simple medical instruments that are currently used by clinics throughout the world. Never mind that the opportunity for abuse is high, as was acknowledged by the *International Journal of Gynecology and Obstetrics* in a May 29, 1989 article, which suggested that quinacrine sterilization could be done on a scale so massive that millions of women would be irreversibly sterilized each year.

Though the name *quinacrine* as a means of birth control may be new to most of us, we'd be mistaken to think that the technique is new. As early as the 1920s, Dr. Felix Mikulicz-Radecki, a German gynecologist, developed a method of scarring fallopian tubes by injecting carbon dioxide. For the next twenty years, his procedure was used on thousands of women, many of whom suffered severe complications and some of whom died of lung embolisms. The goal then, as now, was to sterilize as many women as possible at the lowest cost per unit.

The purpose of quinacrine, plain and simple, is destruction of a woman's reproductive organs. The fact that scar tissue is formed and that so many complications and side effects occur is evidence that the chemical is doing its job and doing it well. The distributors of quinacrine plan to continue supplying any country that wants it and any physician who requests it, including those in the United States.

Is there a eugenics angle to the question of quinacrine sterilization? There certainly seems to be, judging by the anti-immigrant tone of an interview by the distributor in a June 1998 *Wall Street Journal* article. "The explosion in human numbers," he says, "which after 2050 will come entirely from immigrants and the offspring of immigrants, will dominate our lives. There will be chaos and anarchy."

Evidence is mounting that quinacrine is a dangerous chemical. Despite that, more and more Third World nations and individuals see it as a cheap and effective way to reduce out-of-control populations. The West sees it as a principal line of defense, and some would like nothing better than to use it as an offensive weapon in a war on population growth. As long as the benefit of decreasing those future numbers outweighs the risk of some women becoming sick or dying, research and human testing will not only continue at full speed but will probably expand at levels that could make quinacrine the sterilization method of the next decade.

Quinacrine and other population control devices will also be used in conjunction with new tests developed to screen embryos for low intelligence. One such test, marketed by the British company Cytocell, analyzes DNA strands in each chromosome and predicts intelligence and learning difficulties. American and Spanish doctors have already used the test kit to identify retarded embryos and recommend selective abortion. The fear is that the test will ultimately be used by some to screen out "average" embryos or embryos with only slight defects.

The eugenics movement began as a blight on humankind, growing like a cancer and feeding on racism, nationalism, and paranoia to become a scourge that would infect the entire world. It was allowed to grow because few stood up against such evil until it was too late. Lest we forget how easy it would be for eugenics to once again rear its ugly head, we should remember the words of Martin Niemoeller, a Lutheran minister who lived in Nazi Germany during its darkest days.

> First they came for the Communists, but I was not a Communist so I did not speak out. Then they came for the Socialists and the Trade Unionists, but I was neither, so I did not speak out. Then they came for the Jews, but I was not a Jew so I did not speak out. And when they came for me, there was no one left to speak out for me.

4 HUMAN RADIATION EXPERIMENTS

They were known as atomic veterans, serving in exotic-sounding locations like Christmas Island and Bikini Atoll, just two of the many islands dotting the turquoise waters of the Pacific Ocean. When the men arrived, they thought they'd reached paradise. Even their upcoming mission of bomb damage assessment following the explosion of atomic weapons seemed like a fairly good trade-off considering where else they could have been deployed. Little did the men know that the tropical sun and pristine stretches of white sand were not the only things they were about to be exposed to.

From 1945 to 1962, the peace and tranquility of islands few people back in the States had ever heard of was broken regularly by violent detonations that scarred the fragile ecosystems and left in their wake poisoned craters the size of small villages. Following each detonation, the men took samples and measurements to assess the amount of fallout and the effectiveness of the blast. When some of the men fell sick, it became apparent that the cause was exposure to radiation.

Despite assurances by the military that any exposure was unintentional and insignificant, some of the men suspected all along that they'd been sent out as human guinea pigs to test the effects of radiation at various distances from ground zero. In some cases, personnel were ordered to locate themselves in areas where they would receive high doses of radioactivity. They were given no choice in the matter and were not told of the potential risks of exposure. The suspicions were always just that—suspicions—until the discovery of a 1951 document that confirmed their worst fears. The Joint Panel on the Medical Aspects of Atomic Warfare had apparently discussed the "bomb test–related experiments," and the panel's document had identified twenty-nine "specific problems" as "a legitimate basis for biomedical participation." So for seventeen years, the islands, with their beautiful shorelines and blue lagoons, were used not only for detonations and damage assessment but for human biomedical research as well.

One test, code-named BAKER, was conducted underwater in July 1946. As a fleet of ships reached the Bikini Atoll Pacific Proving Ground near the Marshall Islands, an atomic bomb with the same power as the one dropped over Nagasaki was suspended ninety feet beneath one of the ships anchored in the middle of the formation. When the signal was given, the ocean turned into a violent froth as the explosion tore through the water and sent radioactive spray across the entire fleet and every sailor on deck. So urgent was the need to conduct these kinds of tests and to understand radiological warfare that Dr. Joseph Hamilton, in a December 1946 army memorandum wrote:

> I strongly feel that the best protection that this nation can secure against the possibilities of radioactive agents being employed as a military tool by some foreign power is through evaluation and understanding of the full potentiality of such an agent.

While their buddies were being irradiated by salt spray on the beaches of places like Christmas Island, other soldiers were being targeted on the barren deserts of Utah, New Mexico and Nevada. Together with civilian populations, millions of unsuspecting victims, some as far away as New England, were unknowingly participating in a series of mass experiments. As scientists and military officials looked on, driven by a genuine desire to know as much as they could about this new power unleashed on the world, some of the most egregious radiation tests on human beings were taking place in the United States.

It was one thing to send soldiers into a contaminated field after the fact; it was another altogether to have nuclear fallout rain down on them as if they were nothing more than lab rats. These were the open-air ionization tests, designed to measure radioactive toxicity by inhalation and application to the body or clothing, its speed of action, its stability and danger, and how easily it penetrates protective devices. Today, surviving atomic veterans, sick from cancer and other illnesses, are still paying the price for what at the time was thought essential for national security.

Fear of a nuclear strike was foremost in the minds of Americans, who believed that if anyone were to do it first it would surely be the Russians.

The detonation of an atomic bomb by the Soviet Union was a watershed event for the United States. It marked the end of America's monopoly as the sole nuclear power and opened the door for an escalation in the research and development of bigger and better weapons of mass destruction. So for the next thirty years, the Department of Defense (DOD) nearly fell over itself designing every conceivable type of radiation experiment, study, program, and project involving human subjects. While the DOD oversaw the activities, it depended on a network of private companies and universities that supplied the technical skills and infrastructure necessary for such a complex and long-term mission.

The rationale behind human rather than animal studies, according to the DOD, was simple. Following World War II, the shock wave and thermal energy effects of a nuclear explosion were readily apparent, but the third effect—ionizing radiation—was completely unknown in the annals of modern warfare. It was evident that high levels of exposure led to imminent death; what was least known at the time were the long-term effects of a fatal or less than acute exposure. Since animal studies couldn't provide all of the data needed, authorization for human studies was given top priority.

As early as 1942, there was agreement about the need for human experimentation. Work on poorly understood fissionable material for the production of an atomic bomb was established under the Manhattan Project and later centered in Los Alamos. No one really had a handle on what exposure to this new "stuff" could do. So during an October 29, 1942 meeting of the Committee on Medical Research of the Office of Scientific Research and Development, the committee stated that "experiments on human beings were both desirable and necessary in certain types of medical research related to the war effort."

In January 1944, Glenn Seaborg, a Nobel Prize–winning chemist who codiscovered plutonium and worked on the Manhattan Project, warned that the hazards of plutonium might be so great that "a program to trace the course of plutonium in the body should begin as soon as possible." Officials knew that because of the extreme toxicity of the new materials there was cause for concern for two reasons:

1. If personnel were disabled or killed by exposure, keeping the project secret would be impossible and the bomb-making schedule would be jeopardized.

2. If excessive radioactivity were to escape and spread beyond the production facilities, it might be detected through adverse health effects and therefore compromise the project's secrecy as well.

As the years went by, it became apparent that little was still known about exposures, dosages, and effects of radiation on the human body. More had to be done, and quickly. During a May 23, 1950 Committee on Medical Research meeting, the committee members made the boldest statement yet: "The Committee on Medical Sciences endorses the view that it is essential to obtain all necessary scientific information concerning radiation doses and the effect on man by all means of biological experimentation, as promptly as possible, including if necessary human experiments under established principles of such experiments."

For its part, the U.S. Army recognized the need to determine radiation tolerances in humans. There was, after all, an abundance of evidence showing the extent of damage from long-term and high-dose exposure. So despite reservations concerning the difficulties and dangers of using human subjects, Assistant Secretary of the Army Archibald S. Alexander recommended that "consideration be given to establishing a significant experiment to validate the limits of human tolerance to radiation." Many of the answers derived from research at three major universities: University of California at Berkeley, University of Chicago, and University of Rochester. Some of the human subjects drank solutions of plutonium; others were injected with plutonium, polonium, or uranium; and still others inhaled radioactive substances. Results from the universities were used to establish exposure standards.

When the Korean War ended in July 1953, the drive to refine nuclear weapons continued, as did the need to learn as much as possible about how radiation was absorbed by the body and how best to remove it. The Atomic Energy Commission (AEC) conducted more than one hundred atmospheric tests in Nevada and at the Pacific Proving Grounds, but the most significant test, according to DOE records, occurred in February 1954 when a nuclear blast called BRAVO discharged enough radiation to contaminate seven thousand square miles. So great was the fallout that it reached inhabitants of the Marshall Islands and even descended on the crew of a Japanese fishing boat, the *Lucky Dragon*, a great distance away.

The frenzy of research activity involving radioactive substances had infected civilian doctors as well as military scientists. A great example of how even physicians got caught up in the human experiment mentality is found in a 1996 article by Jay Katz in the *Journal of the American Medical Association.* Describing arguments at Harvard Medical School in 1961 over whether or not the Nuremberg Code was even relevant when it came to medical research in the United States, Katz writes, "The medical research community found, and still finds, the stringency of the NC's first principle all too odious." Since the first principle of the Nuremberg Code requires that "voluntary consent of the person on whom the experiment is to be performed must be obtained," it's obvious that even in the hollowed halls of 1961 Harvard doctors felt it was overreaching to expect them to get consent to perform human experiments.

After the subjects have all died, the only memory of this chapter in history will be the documents that tell the real stories behind America's radiation secrets. Unfortunately, the DOD had very carefully written those documents in a way that would make it difficult to implicate the government directly in human research. As recently as 1995, the Advisory Committee on Human Radiation Experiments concluded that, "although there was a real possibility that human subject research had been conducted in conjunction with the bomb tests, the tests were not themselves experiments involving human subjects." So much for accountability by government to its own citizens.

Operation Crossroads: The Nuclear Legacy of Bikini Atoll and the Marshall Islands

Ten nautical miles east of Bikini Atoll, a fleet of one hundred and fifty ships took up safe positions and were told to stand by. The support fleet of more than forty-two thousand men, thirty-seven thousand of whom were navy personnel, had just left behind ninety abandoned vessels anchored in the middle of Bikini Lagoon. Among the target ships were older U.S. cruisers, destroyers, submarines, amphibious vessels, and captured German and Japanese ships. Aboard each of them, scientists had placed a variety of military equipment, radiation-recording devices, and some five thousand rats, pigs, and goats.

As the men took their positions, many watching from the decks, they could barely make out the B-29 flying high overhead and approaching the lagoon. Moments later, a twenty-three kiloton bomb exploded five hundred and twenty feet above the target fleet, creating a blast so powerful it sank five ships. Code-named ABLE, the July 1, 1946 test was the first of twenty-three such nuclear detonations around the Marshall Islands and the first of two in Bikini Lagoon.

Testifying before the Advisory Committee on Human Radiation Experiments on March 15, 1995, one of the crewmembers remembered being told that if they (the crew) wanted to see the detonation, they should look at a certain direction and listen to the countdown over the public address system. "We were not issued protective goggles," the witness said, "even though I later learned that a special directive to that effect had been issued to all the ships assigned to the test. When the countdown reached zero, I saw the flash on the horizon, and we immediately reversed course and headed back to Bikini. We entered the lagoon the next morning, anchored, and proceeded to work on damaged target ships.

"As before the test, when work was over, we stripped down and dove into the contaminated lagoon for our usual refreshing swim. Nobody told us that the lagoon water might be contaminated with alpha particles. Nobody told the young men who were sent aboard the target ships to retrieve the dead and dying experimental animals, cameras, and test equipment about the dangers of radiation. Why were some of the target ships decontaminated and declared Geiger sweet as opposed to Geiger sour when they were really not, so that the crews could go back aboard to resume their normal duties? We believe that this was experimental.

"Why did the surviving crew members of some of the support ships still refer to them as death ships? The bottom line is that they wanted to see how our ability to engage in combat operations during a nuclear war would be affected. They wanted to see how living aboard supposedly decontaminated target ships would affect our ability to perform our military duties, and we were also the victims of the insatiable curiosity of Dr. Strangeloves, like Dr. Edward Teller, who wanted to see what could be learned from the testing of bombs of many different designs and yields."

The second Bikini test occurred on July 25. Code-named BAKER and detonated underwater, this one sank eight vessels, damaged many more than did ABLE, and contaminated most of the ships with radioactive water spray and radioactive debris from the lagoon bottom. It took several weeks before decontamination teams could even enter the area or board the vessels for inspection. On August 10, 1946, the surviving target fleet was towed to the uncontaminated waters of Kwajelein Atoll, where stored ammunition aboard the ships was offloaded before the ships were destroyed and sunk.

Not long after the detonations, radiation detectors showed that the support fleet itself was being contaminated by radioactive marine life growing on the ships' hulls and by radioactive seawater flowing through the pipe systems. All contaminated ships sailed into navy shipyards, primarily at San Francisco, for decontamination, but according to DOD records this required "a great deal of experimentation and learning." It's not known or documented how many of the men aboard the ships developed cancer or other illnesses from exposure they couldn't avoid.

The purpose of the Bikini Atoll explosions was to test the effects of atomic weapons used for attack on naval vessels. The other part of the mission was to determine transport, uptake, and cycling of radiation through the ecosystem, to estimate doses and risk to living organisms, to evaluate methods for reducing radiation for people resettling the atolls, and to develop and maintain a database for future studies. In order to accomplish all that, more people needed to be displaced from the islands, and tests had to continue unabated.

Composed of five main islands and twenty-nine atolls, the Marshall Islands glisten like emeralds under more than seven hundred thousand square miles of Pacific sky. When President Harry Truman issued a directive in December 1945 to begin joint army and navy testing of nuclear weapons, the Marshall Islands were selected because of their obscurity and isolation in that part of the world known as Micronesia. Bikini Atoll, with its dry climate and northernmost location, was thought especially valuable for testing.

The order for the people of Bikini Atoll to pack up and leave their homeland was initially given in 1946. Not only did King Juda, the leader of 167 Bikinians, agree to the forced emigration, he stood before an assembly of his people and Commodore Ben Wyatt, the military governor of the Marshalls, and said, "If the United States government and the scientists of the world want to use our land and atoll for furthering development, which with God's blessing will result in kindness and benefit to all mankind, my people will be pleased to go elsewhere." Elsewhere was other islands and atolls, which, according to a history of their exodus by Jack Niedenthal, sometimes led to near starvation, families being forced to live in tents near military runways, numerous hardships caused by displacement and loss of skills, and contamination from the nuclear fallout that remains to this day.

For the next eight years, the test sites around the Marshall Islands were literally torn apart by nuclear explosions. The largest by far was BRAVO, which equaled fifteen million tons of TNT and produced fallout three to four times greater than was anticipated. Radiation from that single blast contaminated islands as far as three hundred miles downwind. In 1969, the U.S. government began clearing radioactive debris and initiated plans for the resettlement of Bikini Atoll, but unfortunately, when the families began moving back, radiological tests showed radiation levels higher than officials had expected, with the food and water much too contaminated with plutonium to be safe for human consumption.

Once again the inhabitants had to be moved to adjacent islands. Two decades later, in 1990, Congress appropriated ninety million dollars to clean up the water, topsoil, and vegetation, which is still laced with radioactive particles. No one is sure how many more years it will take before Bikini Atoll is livable again. An independent study by the U.S. Centers for Disease Control and Prevention (CDC) is also trying to determine how much of the thyroid disease currently seen in many Marshall Islanders is related to nuclear fallout from the atomic tests.

Unbeknownst to much of the world what had happened on Bikini Atoll in 1946, or how it would change forever the course of history, four days after the last nuclear test occurred the bikini swimsuit made its debut at a Parish fashion show. Named after one of the most beautiful regions in the world, it was ironic that so few people associated the term bikini with what had literally become a radioactive waste dump. Today, as the DOE continues to help in clean-up efforts, Bikini Atoll is a far cry from the legacy the U.S. government had in mind.

Ionizing Radiation Experiments

A disquieting silence permeated the air as the soldiers spread out and waited. Impatience and anxiety was visible in their sweat and on their faces as they lit cigarettes and stared nervously into the distance. Everywhere they looked they saw barren wasteland, a mix of desert sand, rock, and scattered brush that surrounded them and resembled something they might have seen in a science fiction movie. When the countdown began, they realized at once that this was no movie and that they were about to experience something they might never experience again.

Following a terrifying explosion that ripped through the silence and literally shook their innards, the men watched in amazement. The movement of loose dirt and rock felt like the earth had suddenly shifted around them. A thick plume then rumbled upward and mushroomed out against the blue sky as the troops hastily adjusted their goggles and waited in trenches only several miles away. A few of the men temporarily lost their vision, blinded by a flash fifty times brighter than daylight. But the routine "flashblindness" experiments, though sometimes causing permanent eye damage, were not as dangerous as what was to follow: a mass movement of troops from the relative safety of their trenches to ground zero in an effort to improve survival and military operations on a nuclear battlefield.

Like a wave of green ants, the soldiers emerged slowly onto the hot desert and moved in unison toward the cloud. Those blinded or dazed from the fireball were left behind. Others were selected for psychological examination to determine the effects of stress and the emotional impact of a nuclear detonation. The remaining troops conducted field exercises, maneuvering across the desert and through what looked like a brown and gray dust storm. Above them, a B-17 flew directly into the cloud, tracking its movement and analyzing how much was diffusing and how much was actually falling to the ground.

The men left nothing but the grit of rock beneath their boots and the distant warmth of the nuclear explosion. They marched carefully, almost in lockstep toward the unknown. In front of them, the cloud began to settle into something that must have seemed almost benign. In some cases, the troops were airlifted to the vicinity of ground zero. It was over, as far as they knew, though the invisible particles that reached deep inside them were already damaging their cells and, in one way or another, affecting them for the rest of their lives.

The approximately two hundred thousand personnel participating in what was called "Desert Rock" were there to observe explosions, test field operations, estimate target damage, evaluate fortifications and equipment, and simulate tactical maneuvers on a nuclear battlefield. One of the more critical goals was to measure and observe "physiological responses" to nuclear detonations. Back at the aircraft hangars, ground crews were exposed to contaminated surfaces and aircraft parts on planes that often flew into and tracked the movement of radioactive fallout for hours. Some of the pilots flew through the atomic clouds to measure how much radiation was being absorbed by their bodies.

At sea and off the coasts of Pacific islands, ionizing radiation experiments were doing similar damage. According to Richard Jenkins, a former navy radio operator writing in the May 27, 1991 issue of the *Los Angeles Times*, thirty nuclear bombs were detonated within a thirty-mile range in 1958 during Operation Hardtack. More than forty years later, Jenkins attributes his leukemia, kidney and liver disease, and chronic fatigue to radiation exposure. When he recalls what happened on board his ship, the destroyer *Mansfield*, he can't help thinking how a single day in 1958 had ruined the rest of his life.

There were three hundred men aboard the *Mansfield*, he remembers. Once it anchored off the Marshall Islands, he and his shipmates were issued protective sunglasses and radiation badges to wear around their necks. As the sailors stood on deck and observed the explosions, radiation drifted from fifteen miles away and engulfed the crew. Jenkins remembers looking down at the side of the ship closest to the blast and seeing the paint peeling back from the heat. Over the next several days, radiation badges turned red, a sign of exposure. At that point, the men were taken to special quarters, washed down, decontaminated, and issued new badges. In most cases, the health effects from this one mission would not show up until decades later.

But the men who'd witnessed the explosions weren't the only ones who eventually suffered from their effects. Civilians downwind often received higher doses. Even worse, a secret AEC document dated April 17, 1947 reveals that physicians were well aware of ionizing radiation hazards but ignored them. Under the title "Medical Experiments in Humans," the document reads: "It is desired that no document be released which refers to experiments with humans that might have an adverse effect on public opinion or result in legal suits. Documents covering such field work should be classified secret." As a result, downwinders were told by the Public Health Service that their cancers were caused by neurosis, and women with radiation sickness, hair loss, and burned skin were diagnosed as neurotic or having "housewife syndrome."

It wasn't simply the need to study the effects of ionizing radiation from nuclear weapons that triggered this kind of action. Since the DOD and the U.S. Air Force had been investigating the real possibility of developing a nuclear-powered aircraft known as NEPA (Nuclear Energy Propulsion for Airplanes), humans rather than animals had to be used to predict accurately what exposure to moderate doses of radiation would cause. Research programs involving both civilians and the military were established which, together with medical facilities such as the M.D. Anderson Cancer Center in Houston, Texas, studied the effects of radiation on cancer patients. In a January 1948 Advisory Committee on Biology and Medicine (ACBM) meeting, an exchange occurred between Colonel Cooney, head of the AEC Division of Military Application, Dr. Alan Gregg, chairman of the Advisory Committee on Biology and Medicine, Admiral Greaves, military radiation official, and Dr. Shields Warren, head of the Division of Biology and Medicine, in which the participants debated the propriety of using humans as subjects for radiation research.

"How much ionizing radiation can a healthy soldier take and still perform his duties?" asked Colonel Clooney. "I see no difference in subjecting men to this than I do to any other type of experimentation that has ever been carried on. Walter Reed killed some people. It was certainly the end result that was very wonderful." To which Dr. Warren responded, "In order to get a satisfactory answer to this problem in humans, I don't see how it is possible to have an answer that means anything, over and above what we already have in our animal data and our scattered human data, without going to tens of thousands of individuals." When the subject turned to the use of civilian prisoners, Dr. Gregg asked if that would fall in the category of cruel and unusual punishment. Admiral Greaves expressed the opinion that it would not, provided that the prisoners volunteer for that kind of work.

The debate over using human subjects raged within the DOD and the medical and scientific communities for nine years. Finally, in February 1953, Secretary of Defense Charles E. Wilson issued the Wilson Memorandum, which set the standard until mid-1974 for how each service would experiment with, treat, and protect human subjects.

Despite the safeguards, radiation detonation experiments were conducted until 1962, exposing thousands of military personnel and civilians to radioactive fallout. At the Dugway Proving Grounds in Utah and at Oak Ridge, Tennessee, the army also conducted at least seventeen radiological tests in which conventional weapons containing radiological agents were exploded to look at dispersal patterns and evaluate how much damage the radiation would do to crops, water supplies, and personnel.

Today, in states such as Nevada and Utah, where communities were downwind of the tests, the incidence of cancer is significantly higher than it is in other states. Even though the government has declassified many documents and is combing through thousands of files for hidden information on human radiation experiments, it admits that the records for many individuals who'd been exposed are inadequate or "inaccessible."

Atmospheric Iodine-131 Releases

Were it not for irate consumers taking their Kodak film back to stores and developing labs because it was fogged, no one would ever have known that a nuclear test in New Mexico had caused the spread of radiation to a small town in Indiana more than a thousand miles away. Yet that's exactly what happened.

In June 1946, a massive detonation in New Mexico called the Trinity Test spewed radioactive particles for miles into the desert atmosphere. The problem was that the fallout didn't stay where it was supposed to. Responding to photographers' complaints about fogged film, an Eastman Kodak scientist, who knew that corn husks from Indiana were used as packaging material for Kodak film, took some samples and discovered that the corn husks were contaminated with iodine-131 (I-131). After looking at wind and weather patterns and the coincidental timing of the Trinity Test, it didn't take much to conclude that the townspeople of Indiana were living in the middle of a radiation hot spot.

But the story doesn't end there. Six years later, the first nuclear detonation was unleashed in Nevada. After the radioactive fiasco of the New Mexico blast, would this one be any safer? Unfortunately, this time it was the children of Rochester, New York, more than sixteen hundred miles away, who woke up to play in and eat the freshly fallen snow without realizing that Geiger counter readings at the Kodak plant were registering twenty-five times above normal.

How did the AEC respond when confronted once again by Kodak? Without publicly acknowledging that individuals were exposed, it secretly assured the company that warnings would be issued in advance of any upcoming tests. Senator Arlen Specter, at a hearing on radioactive fallout, was outraged that "the government protected rolls of film, but not the lives of our kids." He thought it stunning that the government warned the entire photography industry and provided maps and forecasts of potential contamination but did not issue the same maps and forecasts to dairy farmers. "Why did they do that when they had all the information about hot spots and fallout," he questioned, "and yet they did not warn the people of this country about the dangers inherent in radioactive fallout, especially iodine-131?"

During the 1950s, more than ninety such above ground tests were conducted in Nevada alone, exposing one hundred sixty million Americans to large amounts of I-131. Children were particularly vulnerable because they drank large amounts of milk from cows that grazed in contaminated fields. Hot spots where radiation was especially high included areas as far away as New England and the Midwest, where children were found to have I-131 levels ten times higher than those of adults. Compounding the problem were other fallout ingredients such as cesium-137 and strontium-90, which has a half-life of nearly thirty years.

Studies by the National Cancer Institute (NCI) have shown that high exposure to I-131 can cause leukemia, bone cancer, thyroid cancer, and other illnesses. In fact, a 1997 NCI report claimed that millions of children had been exposed and that fallout from nuclear testing from 1951 to 1962 may have caused at least fifty thousand additional cases of thyroid cancer. Federal regulations today require that radiation from nuclear power plants be less than fifteen rads (radiation absorbed dose), but children in hot spots were receiving as much as 160 rads. Iowa Senator Tom Harkin put that number in perspective when he compared the 115 million curies of I-131 released in U.S. atmospheric tests to only 7.3 million curies released from the Chernobyl disaster in the former Soviet Union.

An NCI-funded study, eventually published in 1993, followed nearly twenty-five hundred children in Utah, Nevada, and Arizona who'd been exposed to nuclear fallout and were examined in the 1960s and then again in the 1980s. The study found a 3.5-fold increased risk of thyroid cancer in the children. Later, a Public Health Service study done on two southwestern Utah counties showed leukemia rates three times higher among children who were younger than nineteen living in those counties. The high rates of leukemia were subsequently found in northern Utah counties as well.

The amounts of radiation to which people had been exposed were incredible when one considers that a single chest X ray today is about 0.0025 rad. Only recently have we recognized how susceptible pregnant women are to even low-level radiation. According to the Nuclear Regulatory Commission (NRC), children exposed to as little as 2 rads in utero have an increased risk of developing cancer. In the workplace, pregnant women may not be exposed to more than 0.5 rad for fear of causing damage to the fetus. Five rads may be enough to trigger spontaneous miscarriages, while ten rads or more may cause mental retardation and eye abnormalities. No one knows how many children born to women exposed at the time of the tests developed cancer or other health problems, but some of the children in fallout hot spots were getting the equivalent of sixty-four thousand chest X rays.

Atmospheric releases weren't limited to nuclear explosions that sent radioactive clouds to neighboring states. There was real fear that foreign powers would resort to radiological weapons (dirty bombs) instead of atomic bombs against U.S. and British cities. The concern was great enough that radiation detectors were installed in Washington, New York, Chicago, and several other major cities. In order to test defenses, study contamination of land and structures, determine if nuclear materials could be detected at long distances from their source, and evaluate effectiveness of cleansing and detonation methods, intentional releases of radioactivity into the civilian population were necessary.

Triggering this extreme action was the Soviet Union's test of its first atomic bomb in August 1949. By adding radiological weapons to its nuclear arsenal, which would contaminate an area and deny its use by the enemy, the United States hoped to stay a step ahead of other nuclear powers. The officials at the AEC and DOD who made the decision to go ahead with the release not only kept it secret for national security reasons but never classified it as human experimentation because it was "not undertaken with the intent of testing the effects of the radiation in humans or designed to measure human exposure." That small technicality enabled the AEC to authorize on October 25, 1949 "Green Run," the code name for a massive release of I-131 into southeastern Washington and Oregon.

It was a typically cold and blustery day on December 2, 1949 when officials at one of the reactors at the Hanford Nuclear Reservation made final preparations for the test. The nuclear fuel stored at the world's first plutonium factory was normally stored for eighty-three to one hundred and one days to allow it to cool and decay before it was dissolved in nitric acid and the solution processed for extraction of plutonium. On this particular day, the fuel had only cooled for sixteen days and was referred to as "green" material, which resulted in a much greater release of I-131 than normal. However, there was a problem. Unusually inclement weather had already caused a one-week delay in the scheduled test. The new recommended release date, December 3 at 1 A.M., was changed to December 2 at 8 P.M. because of other weather concerns. No one could have predicted what happened next.

After the air scrubbers were intentionally deactivated for Green Run to maximize the release of radioactive material, four thousand curies of I-131 and seven thousand nine hundred curies of Xe-133 blew out of the reactor stacks. It was assumed that the air force's weather forecast was as reliable as possible and that the critical weather conditions necessary for the test to go forward were acceptable. But as air sampling began near dawn on December 3, pilots noticed that weather conditions were turning unstable, with several unpredictable changes occurring during the twelve-hour release.

Few were more stunned than the meteorologists who predicted optimal climatic events for the release. When wind speeds picked up unexpectedly, they decreased the strength of an existing temperature inversion, one of the key atmospheric conditions required for the test. Then, six hours into the release, more trouble came. A shift in wind direction affected the distribution of radiation and the amount that was deposited locally. As if that weren't enough, instead of the four thousand curies of I-131 and seven thousand nine hundred curies of Xe-133, somehow almost eight thousand curies of I-131 and twenty thousand curies of Xe-133 were released in an enormous swath forty miles wide and two hundred miles long over populated areas. Radioactivity along the radioactive pathway measured four hundred times the permissible levels thought safe for livestock. Inside the Hanford Reservation, collected animal specimens received thyroid radiation as much as eighty times the daily tolerance.

Taken aback by the mistake and the widespread contamination, officials agreed to suspend further planned releases. Until 1962, that is. For that entire year, Hanford intentionally released I-131 into the environment to study how radioactivity spread through the air, soil, and vegetation, as well as how it affected animals. But that wasn't all. The year-long tests were related to AEC biomedical studies of fallout and how I-131 spread and behaved in the atmosphere. Human volunteers were stationed in the expected path of the radioactive cloud and told to inhale the I-131 particles in order to measure thyroid uptake.

Other Hanford workers were recruited for a different kind of experiment: to drink milk from dairy cows that were fed I-131. The volunteers were given either a single dose or several daily doses and then monitored for a period of about a month. The precedent for feeding human subjects radioactive iodine was not new, but it was thought important enough to warrant separate research projects.

The Hanford nuclear site has been at the center of controversy since 1946, the year that General Electric assumed responsibility for producing plutonium. Hanford was also the main site for sampling and analyzing plutonium in bone, liver, and lung tissue obtained during autopsies from workers and residents who lived in areas where nuclear tests had been conducted. It was the first to conduct inhalation studies and to determine that plutonium caused lung tumors. Ironically, the tissue sampling program began the same year as Green Run, and today is known as the U.S. Transuranium Registry, administered by Washington State University.

What had once been the nation's maker of plutonium is now shut down, its resources devoted to environmental studies and the health effects of plutonium. There is still concern on the part of those who live or have lived near the site about long-term contamination and disease. Only time will tell how much long-term damage has been done to the residents in the name of national security.

Direct Administration of Iodine-131

It wasn't enough to study dispersion patterns and track the movement of radioactive clouds as they drifted out of Nevada and New Mexico. Scientists knew that I-131 concentrated in the thyroid, but they needed a body of information on exactly how it moved and behaved in the organs of humans, from embryos to adults. An entire series of studies was funded by the AEC to do just that.

One of the first such studies began at the University of Iowa in 1953 with aborted embryos from pregnant women who'd been given 100–200 microcuries of I-131. The results showed thyroid uptake at four weeks of development and demonstrated the ease at which radiation crossed the placental barrier. Newborn studies at the university came next. Twelve male and thirteen female infants, all younger than thirty-six hours and weighing between 5.5 and 8.5 pounds, received I-131 either orally or by intramuscular injection. Measurements were then taken at intervals of two to eight hours for three to four days to measure I-131 concentrations in the thyroid.

Other newborn studies were done at the University of Nebraska and the University of Tennessee. In Nebraska, twenty-eight normal, healthy infants from the nursery at the College of Medicine were given I-131 through a gastric tube and tested twenty-four hours later. The Tennessee study used healthy two- to three-day-old newborns to receive intravenous injections of I-131 at doses to the thyroid estimated at about 60 rads. In order to compare full-term infants with premature babies, in a study at Harper Hospital in Detroit oral I-131 was given to sixty-five infants weighing between 2.1 and 5.5 pounds.

Another critical issue as far as the AEC was concerned was the transport of I-131 through the air–vegetable–cow–milk–human food chain. From 1963 to 1968, the National Reactor Testing Station in Idaho conducted Controlled Environmental Radioiodine Tests (CERT), in which human subjects consumed milk from cows that had grazed on I-131-contaminated pastures. On twenty-three occasions, I-131 was released into the environment and allowed to disperse on plants and in water. Some volunteers were used in inhalation experiments, where they breathed in the contaminated air as I-131 was being released over a pasture. In other studies, AEC employees swallowed plastic capsules containing radioactive material or inhaled radioactive gas. Following the experiments, whole-body counts determined the level of I-131 uptake and its transport into and throughout the body. CERT was later expanded to include sulfur-35, chromium-51, potassium-42, cesium-134, and cerium-141.

Since it was known that the thyroid gland is involved in metabolism and temperature regulation, and because the military had become involved in Alaska, the Arctic, and other subfreezing areas, it also made sense to study the role of the thyroid in acclimatization to cold. The Ladd Army Airfield near Fairbanks, Alaska was a critical post against Japanese aggression, and it was later crucial in maintaining a large force to offset threats from the Soviet Union and Korea. The U.S. forces were faced with a new challenge: the effects of cold weather on human physiology. The thyroid was selected by the Arctic Aeromedical Laboratory because previous studies showed (1) increased thyroid activity in animals exposed to cold, (2) involvement of the thyroid in human acclimatization, and (3) elevated metabolic rates in native Alaskans.

One of the unique properties of iodine is that it tends to concentrate almost immediately in the thyroid. So what better population to use than native Alaskans, who had apparently adapted to arctic conditions? Beginning in August 1955, eighty-five healthy Eskimos and seventeen Athapascan Indians were recruited for the study. During the nearly two-year experiment, a total of two hundred doses of I-131 was administered and samples were taken of blood, thyroid tissue, urine, and saliva to track iodine movement and measure levels. Some of the subjects were used for the analysis of serum cholesterol and protein-bound I-131. Because of language barriers, no one ever explained to the natives exactly what was being done, and there was no follow-up to see if the test subjects suffered any subsequent health effects.

Iodine-131 was the only radioactive iodine tracer available during the 1950s, and there was concern that even small doses would be harmful. In its report, the Institute of Medicine wrote, "From an ethical perspective, the Committee concludes that aspects of the Arctic Aeromedical Laboratory study, especially the informed consent process, were flawed even by 1950s standards and thus the Alaska natives who participated and, to a lesser extent, the military subjects were wronged." Though many of the human volunteers were not properly informed about the tests and may have developed serious illnesses, what was done to the I-131 experimental groups was mild in comparison with what else was going on at the time.

Administration of Plutonium and Other Isotopes to Human Subjects

On January 8, 1947, a top secret memo was sent to the AEC that set the stage for a series of radiation experiments unlike any other. The memo, declassified in 1994, illustrates an early interest by the military to evaluate the effects of radioactive substances on human beings:

SECRET 707075

January 8, 1947

RHTG # 100,030
BOX # 603
United States Atomic Energy Commission
California Area
Berkeley, California
Attention: The Area Engineer

Subject: ADMINISTRATION OF RADIOACTIVE SUBSTANCES TO HUMAN SUBJECTS

1. Reference is made to Progress Report P-770 under contract no. W-7405-Eng-48-A dated November 26, 1946 from the Berkeley Area.

2. The first paragraph of this report indicates that certain radioactive substances are being prepared for intravenous administration to human subjects as a part of the work of the contract.

3. Until the Atomic Energy Commission is able to consider sponsoring this type of experimentation authorization cannot be given for the use of radioactive materials in human subjects under this contract. It is suggested, however, that if the physicians at the University of California wish to administer radioactive isotopes they may make application to the isotopes branch of the Research Division of the Atomic Energy Commission for the purchase of such isotopes.

4. If it is felt by the contractor that the program involving experimentation with human subjects will be of ultimate interest or value to the Atomic Energy Commission it is suggested that such program be prepared for submittal to the Commission for their approval.

ATOMIC ENERGY COMMISSION

E. E. KIRKPATRICK
Colonel, Corps of Engineers
Acting Manager, Field Operations

Actually, the human experiments had been underway for almost two decades before the memo was sent. Between 1931 and 1933, patients at the Elgin State Hospital in Elgin, Illinois, were injected with radium-266 as an experimental therapy for mental illness. In 1944, laboratory workers at the University of Chicago and the University of Minnesota injected subjects with phosphorus-32 to study the metabolism of hemoglobin. A year later, on April 10, 1945, the unthinkable happened. A car accident victim, surrounded by medical staff of the U.S. Manhattan Engineer District in Oak Ridge, Tennessee, became the first person to be injected with plutonium.

Though the Manhattan Project scientists realized how hazardous plutonium was, the need for information at a time when plutonium was being produced for atomic bombs was critical. That same year, patients at four different hospitals (in Rochester, New York; Chicago, Illinois; and San Francisco, California) were administered various amounts of plutonium and their excreta collected and analyzed to establish mathematical equations for plutonium excretion rates. The patients who happened to die during the course of the experiments were autopsied to examine how the plutonium had dispersed through their organs.

By the time the 1946 memo was issued, studies had not only been well established but were becoming more common and increasingly dangerous. In May 1946, for example, six male employees of a metallurgical laboratory in Chicago were given a water solution containing plutonium-239 to drink. The purpose was to determine how plutonium is absorbed in the gastrointestinal tract. At the Argonne National Laboratories in Chicago, male and female subjects were injected intravenously with arsenic-76 to provide information about the uptake, distribution, and excretion of arsenic. Other human arsenic studies were done at the departments of medicine and surgery at the University of Chicago, with subjects receiving one to four arsenic-76 injections and having tissue biopsies taken before and after each administration.

The University of Rochester was an especially important hub for the Manhattan Project and became a major center for biomedical research and human radiation experiments. Human metabolism studies from 1945 through 1947 involved at least five different radioactive substances: plutonium, uranium, polonium, radioactive lead, and radium. Subjects received injections or were given oral doses in tap water and then monitored for health effects. Blood, urine, and feces samples were taken to determine absorption, excretion, and activity of the substances in the body.

During one period from August 1946 to January 1947, four men and two women, ranging in age from twenty-four to sixty-one, were injected with highly enriched uranium (uranium-234 and uranium-235) in amounts from 6.4 to 70.9 micrograms per kilogram body weight. The purpose was to measure how much uranium could be tolerated before kidney tissue toxicity set in. When the subject who received 70.9 micrograms developed toxicity, ammonium chloride had to be administered to induce acidosis (a decrease in alkalinity relative to acidity in bodily fluids). He or she was then given a second uranium injection at a dose of 56.6 micrograms per kilogram and continued with the experiment.

According to DOE declassified documents, these kinds of experiments began in the early 1930s and continued for decades at some of the most prestigious institutions in America, including the Mayo Clinic in Rochester, Minnesota. There was optimism among physicians and scientists that radionuclides could be used to combat various diseases. In many cases, the individuals were hospital patients and knew that they were being injected with or asked to drink radioactive substances. However, during the last decade evidence has surfaced linking the U.S. government to secret programs conducted long after the dangers of radioactivity were known, in which healthy people, including civilians, were intentionally irradiated with some of the most dangerous substances known to man. One of the blatant loopholes that allowed this to happen was that the AEC had no requirement that private researchers obtain consent from subjects!

Even as atmospheric detonations and exposures were ongoing, government researchers were turning their attention to more surreptitious uses of atomic energy. The outcry over nuclear fallout had not yet surfaced, and it was thought that as long as communities downwind had no idea they were being bombarded with radioactive fallout or didn't know the dangers, human experiments could simply go on. By the 1960s alone, there had been more than a half million shipments of radioisotopes to physicians and research scientists doing radiation experiments. Although many human subjects were secretly radiated under the guise of improving medical treatment or offering cures for disease, Stephen Klaidman of the

President's Advisory Committee on Human Radiation Experiments criticized the practices of earlier human experiments and said, "We have no idea that the subjects of these experiments were not terminally ill, not suffering from cancer, and may not even have been chronically ill. We were doing experiments of unknown risk on people who potentially had a full, long life ahead of them."

In the 1960s and 1970s, for instance, the AEC funded various medical experiments, including a University of Washington study that looked at the effects of radiation on testicular function in prison inmates. The study was really a sterilization experiment. Initially, 232 healthy men were recruited, sixty of whom were irradiated with acute doses of X rays ranging from 7.5 to 400 rads directly to the testes, with the rest of the men serving as controls. After tissue samples were taken, the results showed that 75 rads destroyed existing sperm cells and 100 to 400 rads produced temporary sterility. Recall that a single chest X ray today is only about 0.0025 rad. The prisoners who were subsequently released were never followed to see if they produced defective offspring as a result of irradiation, but testimonies by wives indicate that they might have. One prisoner's wife, Rosalie Jones of West Jordan, Utah, told congressmen that she personally knew of four babies born to men who were the subjects of these experiments and of the four, three had died of birth defects, including her infant.

A similar study from August 1963 to May 1971 used inmates at the Oregon State Prison in Salem. In this study, sixty-seven healthy men ranging in age from twenty-four to fifty-two were irradiated with doses from 8 to 640 rads at least once. Six of the men also received a second irradiation, one received three irradiations, and one received a series of eleven weekly irradiations. The typical payment was five dollars per month, as well as twenty-five dollars for each biopsy performed on the testicles, which showed that even a single testicular dose of 600 rads caused disruption of testicular function.

By the 1960s, a host of radionuclides were widely available but were not sufficiently understood and tested to permit proper calibration of analytical equipment. At Idaho's National Reactor Testing Station, in 1974 renamed the Idaho National Engineering Laboratory, the AEC initiated a set of experiments to determine how much could be swallowed or inhaled. From 1965 to 1972, the radionuclides tested were Ar-41, K-42, Mn-54,

Co-60, Zn-65, Kr-85, Zr-95, Nb-95, Ru-106, Ag-110, I-131, Cs-132, Xe-133, Cs-137, and Ce-144. One of the main purposes of the seven-year study was to develop and evaluate new whole-body counting equipment and to calibrate that equipment based on the results of human tests.

Although there were always ulterior motives behind the radiation projects, many researchers saw the human experiments as valuable to society and humanity. At the Fernald School in Massachusetts, for example, children were fed radioactive oatmeal so that nutritionists could study how preservatives moved through the body blocking absorption of vitamins and minerals. Patients scheduled for limb amputation had large amounts of plutonium injected into the affected limb so that researchers could see how the plutonium dispersed throughout the tissue. It was believed that all of these medical and scientific studies served the ultimate good of humanity.

As one of the workers at the Lawrence Livermore Laboratories said when asked about its plutonium experiments, "We were always on the lookout for somebody who had some kind of terminal disease who was going to undergo an amputation. These things were not done to make people sick or miserable . . . they were done to gain potentially valuable information. . . . It doesn't bother me to talk about the plutonium injectees because of the value of the information they provided." It's easy to see from statements like this how establishing and accepting human experimentation would lead to a scientific culture in which the sick and dying are seen as nothing more than guinea pigs.

Radioactive Tobacco: How Much Did Scientists Know?

Did researchers know that radioactive substances such as uranium and polonium caused lung, trachea, and bronchial cancer? If they didn't at first, they certainly should have made the connection when they observed lung cancer rates in smokers soar significantly after 1950. To some the link was obvious because tobacco fields at the time were increasingly being fertilized with chemicals that contained radioactive substances, and smoking a pack and a half of cigarettes per day was becoming equivalent to having almost one thousand chest X rays a year. To illustrate the strong correlation, consider the following cancer rates from 1950 to 2000 taken from CDC cancer mortality statistics:

DEATH RATES FROM CANCER OF THE LUNG, TRACHEA, AND BRONCHI PER 100,000

		1950	1960	1970	1980	1990	2000
MALE							
	White	21.9	38.2	57.7	70.4	73.5	78.7
	Black	15.7	37.9	55.1	93.3	107.7	106.5
FEMALE							
	White	4.9	5.6	11.1	21.1	32.1	42.2
	Black	3.8	5.6	11.7	21.5	32.0	41.2

Tobacco corporation memos recently made public show that industry executives and scientists were well aware of the radioactive cigarette problem and even developed methods to deal with it (U.S. Patent No. 4,194,514 describes removing radioactive tobacco components with a dilute acid solution). In retrospect, perhaps executive denials of a link between smoking and cancer were a way to keep the radioactive aspect out of the public forum. After all, what would make smokers think about quitting more than being told that they're inhaling not only carcinogens but *radioactive* carcinogens? And what would outrage consumers more than discovering that they are actually being sold products the company has known all along are radioactive? A 1961 Philip Morris memo clearly shows that the company knew early on about strontium-90:

> Mr. Gonzalo Segura in our Radiochemistry Laboratory says he can readily measure 5 to 10 microcuries of strontium-90 so that we are in a position to monitor our products if this becomes nec-

> essary. I am asking him to make exploratory determinations just to find out what order of magnitude we are dealing with in tobacco. If it turns out that we are well on the safe side, I suggest that we defer any further measurements until there is reason to believe we need to be concerned about the problem.

The problem that tobacco companies faced but wanted desperately to keep from the public was that more and more studies were showing that phosphate fertilizers contain natural radioactive particles that are transformed into radon, which then decays to radioactive polonium (^{210}Po) and lead (^{210}Pb). During the growing season, the radon is taken up by the root system and accumulates in the leaves. With years of regular fertilizer applications, the radioactivity increases and the plants absorb even higher levels of radiation.

Executive silence was one thing, but why didn't research scientists speak up for the public good? The answer, quite simply, is money and careers. I'd lived in Winston-Salem, North Carolina for fifteen years and still know many R. J. Reynolds Tobacco Company employees, researchers, and scientists at Wake Forest University who'd received grant money from R. J. Reynolds. The loyalty, which doesn't really surprise me given the financial incentives, is something to behold. To this day, scientists with doctoral degrees who'd worked at Reynolds for decades refuse to admit that smoking causes cancer. But become their friend, talk to them a little more, and with a smile and a wink they'll tell you that life in retirement is great and they owe it all to R. J. Reynolds.

The one person who certainly wasn't afraid to rock the boat was former Surgeon General C. Everett Koop, who went so far as to claim that 90 percent of all tobacco-related cancer is caused by radioactivity. An internal memo from Liggett & Myers Tobacco Company has proven once and for all that company executives had full knowledge of the radiation risk but chose to ignore it (Appendix XII).

During the famous tobacco trials that led to a landmark settlement against major U.S. tobacco companies, virtually nothing was reported to the public about radioactive cigarettes. Were it not for disclosure of secret internal memos and documents, few outside the scientific community would know that lighting a cigarette triggers a volatile reaction of radioactive particles more dangerous than nicotine. Company executives knew it all along, as did scientists who sometimes tried in vain to warn them. Thankfully, buried in mounds of paperwork so voluminous that teams of lawyers had a hard time sifting through them, the true link between tobacco and lung cancer may have been uncovered and, with increased public awareness, may someday prove to be the final nail in tobacco's coffin.

Other Human Radiation Experiments

The 1950s marked the beginning of an unusual relationship. For obvious reasons, the DOD needed data on the physiological effects of radiation, while cancer researchers wanted to study total-body irradiation (TBI) and partial-body irradiation (PBI) as treatment approaches to different types of cancer. Funding opportunities grew significantly, and researchers at institutions throughout the United States were increasingly becoming a conduit for information the DOD needed for its nuclear database.

The first experiments took place from 1951 to 1956 at the University of Texas M. D. Anderson Cancer Center in Houston. However, cancer treatment was the last thing the U.S. Air Force School of Aviation Medicine had in mind when it issued contract AF-18-600-926. The motive behind the funding was the real possibility that the U.S. Air Force would develop a nuclear-powered aircraft and, therefore, needed to know the potential adverse health effects of ionizing radiation on flight crews. Researchers at M. D. Anderson were happy to serve a dual function: to investigate TBI as a treatment for cancer and to determine how radiation would affect the simple and complex mental and psychomotor tasks performed by pilots.

A total of 263 male subjects aged nineteen to seventy-six and in various stages of cancer participated in the five-year study. The psychomotor experiments evaluated ionizing radiation on motor skills needed to operate aircraft. After receiving either a single dose or repeated doses of radiation, the men were subjected to three perceptual motor tasks to measure how much radiation deteriorated their skills. During the course of the study, investigators noticed a variety of radiation sicknesses after exposure to 125 roentgens, including nausea, vomiting, and bone marrow depression. At 200 roentgens, 10 percent of the subjects developed complications that required management.

At the same time, Baylor University College of Medicine began a twelve-year DOD study funded by the Armed Forces Special Weapons

Project (contracts DA-49-007-MD-302 and DA-49-007-MD-28) and the Defense Atomic Support Agency (contract DA-49-146-X2-032). During the first few years of the project, from 1952 to 1956, the goal was simple: to identify the therapeutic effects of TBI as a cancer treatment and to determine the effects of low-dose radiation over time. By 1956, the focus of the research was expanded to include the cumulative effects of radiation over time and to answer the DOD's questions about continuous radiation exposure in the event of a nuclear war.

Baylor scientists recruited a total of 112 cancer patients and divided them into three groups: Group I received single exposures of 25 to 250 roentgens, group II received exposures of 25 to 545 roentgens from two to sixty-three days, and group III received repeated exposures of 170 to 500 roentgens over several months or, in some cases, years. The data were to be compiled by the military and used to gain information about how individuals would be affected following a nuclear attack.

At the Sloan-Kettering Institute for Cancer Research in New York, Dr. J. J. Nickson received five DOD grants from the Armed Forces Special Weapons Project (contracts DA-49-007-MD-533, DA-49-007-MD-669, DA-49-007-MD-910, DA-49-007-MD-1022, and DA-49-146-XZ-037). Thirty-four subjects ranging in age from nineteen to sixty-three were used. The projects, which ran from June 1954 to January 1964 and were monitored by the Office of the Army Surgeon General, looked at the effects of TBI as a cancer treatment and on physiological parameters useful to the military, such as urine excretion, blood composition, and brain function.

Also involved in TBI research and receiving grants from the Research and Development Division of the Office of the Army Surgeon General (DASA) was the University of Cincinnati College of Medicine (contracts DA-49-146-XZ-029, DA-49-146-XZ-315, and DASA-01-69-C-0131). From January 1, 1960 to March 1972, the university used new, deep-penetrating TBI technology to study the biological and clinical features of injuries caused by radiation, something in which the DOD was very much interested. Halfway through the study, the DOD also wanted to determine the effects of radiation on intelligence and on psychological responses to radiation exposure.

Of all the university DOD projects, the University of Cincinnati's was the most controversial because the majority of research subjects were indigent African Americans with low IQs. All of the patients received single or multiple exposures of 16 to 300 rads and showed a decrease in intellectual functioning immediately after the exposures. Records show that the men often experienced nose and ear bleeds, intense pain, and vomiting. Because of its nature, and allegations that consent forms were inadequate, the project came under intense scrutiny, and a January 1972 critical report called for cancellation of the project.

According to the DOD Report on Human Radiation Experiments, TBI continues to this day. The Advisory Committee on Human Radiation Experiments states that "since the 1980s, TBI has again been used to treat certain widely disseminated, radioresistant carcinomas at doses as high as 1,575 rads in conjunction with effective bone marrow transplantation, which became routinely available in the late 1970s."

In all, more than 430 experiments were conducted on more than sixteen thousand men, women, and children exposed to radiation in one way or another. If we include witnesses, military personnel involved in nuclear detonations, and those who unwittingly participated directly or indirectly, the number is in the hundreds of thousands. Add the "downwinders" and there are literally millions.

Injections, inhalations, oral doses, TBI, and PBI were all carried out for the purpose of finding out what radioactive substances will do in the atmosphere, in the environment, and in the human body. Admittedly, many of the experiments have led to breakthroughs in nuclear medicine, cancer treatments, and the establishment of occupational standards. Yet some were heinous in their nature, exposing retarded and institutionalized children, pregnant women, fetuses, newborns, psychiatric patients, and prisoners who did not fully understand what they were getting into.

For more than seventy years, human beings have been used as radioactive guinea pigs. The problem is not so much that it was done as it is the fact that is was often done with duplicity and secrecy. Were people downwind ever told they were drinking radioactive milk or eating contaminated vegetables? Were they offered treatment after the link was made between radiation and cancer? Did cancer patients think they were getting new and innovative treatments only to become sicker and die sooner because the military needed to know exactly what radiation would do to them? It's difficult to say how many individuals have become sick or died as a result of intentional exposure because many government documents are "born classified." In other words, they are classified from the very start and can only be declassified by "specially qualified personnel" who review each page of each document. Since the DOE has 3.2 million cubic feet of human radiation experiment records alone, many who had been affected will be long gone before the job is complete.

A major problem with government documents that include experiments done with military personnel is that individual medical records may not be included, and even when included usually do not contain information about the studies the individuals participated in. There were never standardized guidelines imposed by either the DOD or the Veterans' Administration to include a copy of the informed consent form or research proposal in the medical records of exposed human subjects. When medical records *do* contain relevant information, they are often difficult to obtain and, at times, have been illegally removed from veterans' files. An added hurdle is that veterans' service records are still stored in thousands of locations, which makes them easy to lose, misplace, or to purposely discard without anyone else knowing about their existence.

We now know that "informed consent" was not always informed. In some cases, individuals weren't even classified as experimental subjects if the tests weren't specifically designed to involve human subjects. If individuals happened to have gotten in the way, so much the better. And if the U.S. government has lied about and kept hidden from the public the dangers of something as serious as radiation, what else might they be keeping from us? Hopefully, as more documents are uncovered and declassified, the outcry will prevent these kinds of activities from happening again.

5 THE CIA AND HUMAN EXPERIMENTS

In Salem, Massachusetts in 1692, a young girl, her blond hair tied neatly in a bow against the back of her head, stood silently before three robed men. Outside, the blazing sun bore down like a hot stone on a crowd of spectators straining to see and hear the strange goings-on inside. The men and women seated almost reverently on the wooden courtroom benches leaned forward in anticipation when one of the judges folded his hands and asked a simple question: "Are you a witch?"

The fourteen-year-old in front of him had already spent a month with town doctors trying to come up with a physical answer for the hallucinations, delusions, and strange behavior she'd been exhibiting. It was baffling to say the least. Bouts of normalcy were followed by spontaneous outbursts and convulsive fits, lucid conversations by childlike gibberish. But the final diagnosis of possession was no doubt confirmed when the girl, in one of those fits, claimed that the devil was speaking to her.

Once again the judge looked down from his bench and asked, more emphatically this time, "Are you a witch?" Instead of answering, the girl began to tremble. Suddenly, as if she'd been overcome by an invisible force, her trembling progressed to twitching and then spasms. Hallucinating, she looked past the men and screamed at the terrifying images she saw behind them. Her face red and swollen, the pain in her extremities causing her to scratch and claw at her skin as if truly possessed by the devil, she was restrained and immediately carted away to Gallows Hill, a barren field on the outskirts of Salem Village, for hanging.

Judgment was typically swift, as was the grim execution. Many accused individuals languished in prison; most of the convicted were hanged, though one elderly man was crushed to death by heavy stones. The young girl in this case was led to the gallows, her hands bound tightly behind her. As a hood was placed over her head, she twisted and contorted violently and then cried out in an incoherent garble, another telltale sign to those watching the poor creature that Satan had surely taken hold of her.

Two and a half centuries after her death, the Central Intelligence Agency (CIA) initiated a program of human experimentation, mind control, and behavior modification based in part on a fungus that had likely been the cause of the young girl's mental illness. One of the principal mind-altering drugs with which the CIA began experimenting was lysergic acid diethylamide, or LSD. When it was first isolated, the original source of LSD was the ergot of *Claviceps purpurea*, a fungus that grows mainly on rye wheat. The term "ergot" refers to the resting stage of the fungus, which contains compounds used in various pharmaceutical drugs and toxins that can be fatal if eaten.

Although it has only recently been theorized that three hundred years of burning innocent people at the stake in Europe and America for witchcraft may have been attributed to ergot poisoning, scientists had known for some time what ergot could do to one's mind. On August 13, 1951, for example, doctors in Pont St. Esprit, France were inundated with patients complaining of intense lower abdominal pain. By the next day, the town's hospital was filled with patients who literally had to be tied to their beds or restrained in straitjackets to prevent their escaping the hospital and running through the streets screaming like madmen, hallucinating, or convulsing in epileptic-type seizures. A desperate search of the victims' homes turned up a common denominator: all had eaten bread they'd bought from the same baker. Analysis of the bread found that it was tainted with alkaloids belonging to rye ergot.

A year and half later, an accident occurred that would trigger two decades of secret CIA research. Dr. Albert Hoffman, the Swiss chemist who'd first isolated LSD from ergot in 1938 as a possible headache treatment but put it aside because it had no effect on his lab animals, accidentally absorbed some through his skin and had to go home because of dizziness that made him feel as if he were drunk. In his description of what happened after that, Hoffman said, "As I lay in a dazed condition with my eyes closed, I experienced daylight as specially bright. There surged up from me an uninterrupted stream of fantastic images of extraordinary plasticity and vividness and accompanied by an intense, kaleidoscopic-like play of colors."

The reports must have sent the CIA into a frenzy, especially when they found out how little it took to produce the effect. When Dr. Hoffman tried to replicate the experiment with two hundred and fifty micrograms and literally had an out-of-body experience, it had to have been a happy day indeed for CIA officials looking to expand their arsenal of clandestine weapons. With this, they figured, they could add behavior modification and mind-altering chemicals to their portfolio of interrogation drugs, hypnosis, and truth serums.

Although the CIA was established as a foreign intelligence–gathering service, it didn't take long for the agency to grow beyond the gathering and spying business. With backing and cooperation from other agencies such as the Department of Defense (DOD), it initiated a variety of classified programs with an immunity that stuns us today. We still don't know the extent of human experimentation because former CIA director Richard Helms had many of the documents destroyed on his watch. Interestingly, not a single individual could be found who remembered any of the details about any of the experiments, including researchers from the eighty-six universities and institutions involved. Known primarily for intelligence gathering, the CIA's darker side includes some of the most controversial human experiments of the twentieth century.

A Brief History of the CIA

Since the founding of this country, there has always been spying. As the world became exceedingly more complex, however, President Franklin D. Roosevelt decided that an organized agency was needed to collect and analyze information from a myriad of sources. The job of laying out the plans fell on William Donovan, who established the Office of Strategy Services (OSS) in June 1942. By October 1945, it became apparent that the OSS was not enough and that a centralized intelligence service was needed to operate in a postwar environment. As Donovan put it, the United States had to have an organization that would "procure intelligence both by overt and covert methods and will at the same provide intelligence guidance, determine national intelligence objectives, and correlate the intelligence material collected by all government agencies." On September 18, 1947, the National Security Council (NSC) and the CIA were established.

The go-ahead for the CIA to initiate projects involving human experimentation originated with the National Security Act of 1947. Besides charging the CIA with coordinating intelligence activities that affected national security, the Act directed it "to perform such other duties and functions as related to intelligence as the NSC might direct." During the 1940s, everything was deemed a threat to national security and, therefore, a no-holds-barred attitude was justified if it meant security for the nation. However, it was the next piece of legislation that truly allowed the CIA to become independent enough to do whatever it saw fit to do.

In 1949, the Central Intelligence Agency Act was passed and subsequently replaced the 1947 National Security Act. This significant change exempted the agency from many of the financial limitations and restraints it had previously been under. The importance of this was that the CIA could now receive funding through the transfer of money from budgets of other departments, such as Defense, without restrictions and disclosures. In other words, if a program was to remain secret, the DOD or the State Department could divert part of their funds to the CIA and be protected by law from revealing where the money was going and for what. The exemption was written into the Act to protect intelligence sources, officials, methods, and organizations.

As the CIA evolved and grew more independent, it was also expanding. In January 1952, the CIA's intelligence functions were grouped under the Directorate of Intelligence (DI) and eventually divided into seven main components the most important of which were: Office of Research and Reports (ORR), which gathered economic and political intelligence; Office of Scientific Intelligence (OSI), which conducted various weapons-related experiments and engaged in secret projects; Office of Current Intelligence (OCI), which gathered information about current political activities; Office of Operations (OO), which collected information about various covert activities; and Office of Collection and Dissemination (OCD), which collected and disseminated intelligence.

One of America's greatest fears was that countries hostile to the United States would use an array of chemical and biological agents against it. Defensive measures weren't good enough and soon became secondary to the goal of using offensive drugs to obtain information, gain control of enemy agents, and apply coercive interrogation techniques. Some of these techniques are exposed in a 1963 CIA interrogation manual that is almost impossible to access but that I was able to obtain. When the research programs to alter human behavior first started in the late 1940s, they included only willing human subjects. However, they soon expanded to include unwitting nonvolunteers in order to test the effects of chemical and biological agents on individuals unaware that they had received a drug.

The intention of early CIA projects wasn't only to kill but to debilitate and control. Drugs and methods of deception were among the earliest experiments conducted. Testimony by Admiral Stansfield Turner, then director of the CIA, illustrates this as he discusses the use of magician's art and the spook business, how to surreptitiously administer material to someone, how to distract someone's attention while doing something else, how to covertly communicate without having others know it, and so forth. When Senator Huddleston asked, "Was there any evidence that there were other motives that the Agency might be looking for drugs that could be applied for other purposes, such as debilitating or even killing another person?" Admiral Turner replied, "Yes, I think there is. I have not seen in this series of documentation evidence of desire to kill, but I think the project turned its character from a defensive to an offensive one as it went along, and there certainly was an intention here to develop drugs that could be of use."

The 1960s marked a new era in science and technology. In response, the deputy directorate for science and technology (DDS&T) was created in 1963, its functions including to this day research, development, operations, data collection, and analysis. One of the most important aspects of DDS&T's operations is that it relies on expertise and advice from outside the agency. As such, projects using human subjects have involved not only government researchers but private and public universities and laboratory facilities as well. The CIA's current organizational chart is much different from its 1953 version.

Today the CIA reports regularly to the Senate Select Committee on Intelligence and the House Permanent Select Committee on Intelligence, as required by the Intelligence Oversight Act of 1980. Moreover, the agency reports regularly to defense subcommittees of both houses of Congress, the Senate Foreign Relations Committee, the House Committee on Foreign Affairs, and the armed services committees of both houses of Congress. Despite that, there is still a shroud of secrecy when it comes to certain aspects of CIA activities, and no one is absolutely certain that human experimentation has not continued to some degree.

Psychochemicals and Mind Control Projects

Interest in mind control and human behavior is centuries old. Potions, herbal medicines that alter behavior, drug-induced personality disorders, and hypnosis as a means of controlling human beings have always fascinated scientists. However, since the discovery of the mind–body connection and the development of drugs that can actually change perception, thought processing, and brain patterns, this fascination has quickly given way to the real possibility that psychochemicals may be used as weapons. The CIA's interest was heightened by what it knew was active Soviet pursuit of parapsychology, telepathy, extrasensory perception, and "bioinformation," a method of obtaining information through normal senses as well as through other than normal senses.

The U.S. government's research on psychochemicals can be traced to the late 1940s and early 1950s, when testing of various drugs on volunteers, as well as on unwitting human subjects, began. Initial experiments were designed to determine the potential effects of chemical or biological agents against individuals unaware that they had received a drug. For this reason it was imperative that nonvolunteers be used and that drugs be administered without an individual's knowledge or consent. To ensure secrecy, few individuals, even inside the CIA, had knowledge of a program's existence. Likewise, presidents and members of Congress were never informed at the time the research was being conducted. Those who needed to be involved to a lesser degree in terms of funding, overseeing contracts, and the like, no doubt were kept in the dark for the sake of "plausible deniability." If ever asked, one could simply deny any knowledge without fear of committing perjury or of compromising vital national interests.

The rationale behind the secrecy of these initial programs was rather obvious. First, there was a matter of national security; and issues of national security, even today, are often clouded in secrecy. The public, it was agreed, did not need to know that the Soviet Union was already years ahead of us in research involving the effects of drugs on human beings. Second, there was the matter of human experimentation. The very idea of using humans as research subjects is abhorrent to most people. According to the CIA inspector general at the time, "the knowledge that the Agency is engaging in unethical and illicit activities would have serious repercussions in political and diplomatic circles and would be detrimental to the accomplishment of its missions." This would have been especially true if the programs resulted in tragic consequences, which they did.

Besides concern over progress made by the Soviet Union and other communist bloc countries, stunning events were taking place during the Korean War that took U.S. leaders by surprise. Prisoners of war (POWs) who were refusing to come home and being used as propaganda agents made us believe that they'd been brainwashed. We suspected that POWs were being subjected to drugs that induced them to talk or that punished them, and there was an urgency to develop ways to detect chemical and biological agent use and to implement effective countermeasures.

The fear and paranoia about the Soviet Union, China, Korea, and other communist regimes led to the establishment of projects designed in response to a threat we thought could be as dangerous as nuclear weapons. After all, the odds of a country launching a nuclear attack was far less than those of a country using psychochemicals against us. So, from 1947 until the mid-1970s, the U.S. government engaged in a series of programs and projects unlike any in American history.

Based on intelligence gathering during World War II, there existed a series of reports documenting "amazing results" in the use of truth drugs by the Soviet Union. In response, shortly following the end of World War II, the U.S. Navy began Project Chatter in 1947. This top secret project, which lasted six years, focused on the identification and testing of drugs for use in interrogations and in the recruitment of agents, and included experiments on both animals and human subjects. Two of the drugs, scopolamine and mescarine, were tested on humans as possible truth serums.

Three years after Project Chatter began, the CIA decided that it, too, needed to get involved in biological and chemical agent research. Project Bluebird was approved by the director of the CIA in 1950 and later headed up by the Office of Scientific Intelligence (OSI), focused on five things:

1. Developing a way to prevent the extraction of information from CIA agents.
2. Developing the means to control individuals through special interrogation techniques.

3. Developing ways to enhance memory.
4. Establishing defensive means for preventing hostile control of agency personnel.
5. Developing offensive uses of unconventional techniques, including hypnosis and drugs.

Project Bluebird was renamed Project Artichoke in August 1951 and included both in-house experiments on interrogation techniques and overseas experiments utilizing LSD, sodium pentothal, and hypnosis. The project was initiated in response to concerns that drugs like LSD would be employed against U.S. personnel. Overall responsibility for the project was transferred from OSI to the Inspection and Security Office, which is currently the Office of Security. While the CIA maintains that Artichoke ended in the fall of 1953, evidence such as excerpts of a declassified memo suggests that "special interrogation techniques" continued long after that (Appendix XI).

In April 1953, Project MKDELTA was established to study biochemicals in clandestine operations and to use them for harassment, discrediting, and disabling purposes. At the same time, MKNAOMI began, its purpose to stockpile incapacitating and lethal materials, to develop gadgetry for the disseminations of these materials, and to test the effects of certain drugs on animals and humans.

During MKNAOMI, the CIA had a close relationship with the Special Operations Division (SOD) of the Army Biological Laboratory at Fort Detrick. From 1952 to 1970, when MKNAOMI was terminated, research focused on the development of two different types of suicide pills and a successful operation using biological weapons materials such as *Staphylococcus* enterotoxin. A major goal was to replace the standard cyanide pill issued to agents in hazardous situations and U-2 pilots for suicide purposes in the event of capture.

There were also other activities of peculiar CIA interest. One development specifically associated with the CIA was the "microbioinoculator," an extremely small dart device that could be fired through clothing to penetrate the skin and inoculate the target without the perception of being hit and without being detected during autopsy. The special darts were coated with biological agents that could remain potent for weeks or months. Much of the work was done on human incapacitation. By the late 1960s, a stockpile of biological agents and toxins was maintained on a regular basis. The supply included food poisons, infectious viruses, lethal botulinum toxin, paralytic shellfish toxin, snake venom, and *Microsporum gypseum*, which produces severe skin disease.

Human beings weren't the only targets. It was thought that, in certain cases, attacking crops would be even more effective than attacking humans directly. In a 1967 memo to the SOD, showing how close it was to implementing the research, the CIA states, "Three methods and systems for carrying out a covert attack against crops and causing severe crop loss have been developed and evaluated under field conditions. This was accomplished in anticipation of a requirement which was later developed but was subsequently scrubbed just prior to putting into action."

There is no official record to date of MKNAOMI being used in actual operations, although discussions with those involved in the project indicate that hand launchers with darts loaded with dog incapacitant were delivered for use in Southeast Asia. In one report, North Vietnamese embassy dogs ate meat treated with dog incapacitant so that the embassy could be penetrated. According to a 1975 declassified memo, "while no direct connections to assassination planning have been found, there are some disturbing similarities between the agents being investigated at Fort Detrick and some of the reported schemes."

Project QKHILLTOP was the cryptonym for a 1954 project to study Chinese Communist Party brainwashing methods and to develop interrogation techniques. It's believed that most of the early studies were conducted by the Cornell University Medical School human ecology study programs. The Society for the Investigation of Human Ecology, later the Human Ecology Fund, was an outgrowth of QKHILLTOP.

Project MKULTRA: The CIA's Program of Research for Behavior Modification

In 1952, a proposal was submitted to the director of central intelligence outlining funding mechanisms for highly sensitive CIA research and development projects that would study the use of biological and chemical materials in altering human behavior. Part of the rationale for the special funding was the extreme sensitivity of the projects. In the proposal, Richard Helms, assistant deputy director for plans, outlined the following two-pronged objective:

1. To develop an offensive capability in the covert use of biological and chemical materials, including the production of various physiological conditions which could support clandestine operations.
2. To develop a comprehensive capability in the field of covert chemical and biological warfare that would give us knowledge of the enemy's theoretical potential, thus enabling us to defend ourselves against a foe who might not be as restrained in the use of these techniques as we are.

On April 13, 1953, MKULTRA was established for the express purpose of researching and developing chemical, biological, and radiological materials to be used in clandestine operations and capable of controlling or modifying human behavior. During its ten-year life, MKULTRA projects also included radiation, electroshock, harassment substances, and paramilitary devices and materials. In a proposal describing MKULTRA, Richard Helms wrote, "We intend to investigate the development of a chemical material which causes a reversible, nontoxic aberrant mental state, the specific nature of which can be reasonably well predicted for each individual. This material could potentially aid in discrediting individuals, eliciting information, and implanting suggestions and other forms of mental control."

The human behavior/psychology component consisted of three phases: (1) a search for materials to be tested, (2) laboratory experiments with human subjects in various institutions, and (3) application of MKULTRA materials in real-life situations (a code word for the use of unwitting nonvolunteers). Although many of the records were ordered destroyed by Dr. Sidney Gottlieb, chief of the Technical Services Division, enough documentation was inadvertently saved to expose the program. The manner in which the documents were uncovered is an example of how a single individual with the dedication to do a job thoroughly can make a profound difference.

During the Church Committee investigations in 1975, a search was launched for MKULTRA-related materials, which had been sent to the Retired Records Center outside of Washington, DC. The committee members searched active and retired records of all branches of the CIA thought to have been even remotely associated with MKULTRA. What they didn't know was that some of the documents, for reasons still unknown, had mistakenly been forwarded to the Budget and Fiscal Section, where they were filed away in retired records. This was a complete departure from normal protocol. Senate investigators would have never thought to look in those files and, fortunately, neither did the individuals who'd been ordered to locate and destroy any MKULTRA documents they could.

The dedicated employee given the assignment of digging out MKULTRA information left no stone unturned. As such, he also probed the retired files of the Budget and Fiscal Section and uncovered a treasure trove of documents missed in previous searches by people trying to get rid of evidence. The mistaken route the boxes of paperwork had taken kept them from getting destroyed, and what little was subsequently discovered about MKULTRA revealed a massive program of human experimentation that spanned ten years.

Based on declassified records, there were 149 MKULTRA subprojects, many of them involved with research into behavior modification, hypnosis, drug effects, psychotherapy, polygraph studies, truth serums, pathogens and toxins in human tissue, knockout drops, and testing or administering drugs surreptitiously. Some of these studies were conducted on unwitting subjects. There were also thirty-three other subprojects concerning intelligence activities. Together, these CIA activities are grouped into the following fifteen categories:

1. Research into the effects of behavioral drugs and/or alcohol:

    (a) Seventeen subprojects probably not involving human testing.

    (b) Fourteen subprojects definitely involving tests on human volunteers.

    (c) Nineteen subprojects probably including tests on human volunteers. While not known, some of these subprojects may have included tests on unwitting subjects as well; 6 subprojects involving tests on unwitting subjects as well.

2. Research on hypnosis: eight subprojects, including two involving hypnosis and drugs in combination.
3. Acquisition of chemicals or drugs: seven subprojects.
4. Aspects of magicians' art useful in covert operations, such as surreptitious delivery of drug-related materials: four subprojects.

5. Studies of human behavior, sleep research, and behavioral changes during psychotherapy: nine subprojects.
6. Library searches and attendance at seminars and international conferences on behavioral modification: six subprojects.
7. Motivational studies, studies of defectors, assessment, and training techniques: twenty-three subprojects.
8. Polygraph research: three subprojects.
9. Funding mechanisms for MKULTRA external research activities: three subprojects.
10. Research on drugs, toxins, and biologicals in human tissue; provision of exotic pathogens and the capability to incorporate them in effective delivery systems: six subprojects.
11. Activities whose objectives cannot be determined from available documentation: three subprojects.
12. Subprojects involving funding support for unspecified activities connected with the U.S. Army's Special Operations Division at Fort Detrick, Maryland. Under CIA's Project MKNAOMI, the army assisted the CIA in developing, testing, and maintaining biological agents and delivery systems for use against humans as well as against animals and crops. The objectives of these subprojects cannot be identified from the recovered material beyond the fact that the money was to be used where normal funding channels would require more written or oral justification than appeared desirable for security reasons or where operational considerations dictated short lead times for purchases. About eleven thousand dollars was involved during this period 1953–1960: three subprojects.
13. Single subprojects in such areas as effects of electroshock, harassment techniques for offensive use, analysis of extrasensory perception, gas-propelled sprays and aerosols, and four subprojects involving crop and material sabotage.
14. One or two subprojects on each of the following:
    "Blood grouping" research, controlling the activity of animals, energy storage and transfer in organic systems; and stimulus and response in biological systems.
15. Three subprojects canceled before any work was done on them having to do with laboratory drug screening, research on brain concussion, and research on biologically active materials to be tested through the skin on human volunteers.

One of the first MKULTRA studies conducted was at the National Institute of Mental Health Addiction Research Center in Lexington, Kentucky. At the time, it was working hand in hand with the CIA to test and develop new, mind-altering drugs. Young patients, usually drug addicts serving various sentences for drug violations, were offered a chance to volunteer as guinea pigs in exchange for the drug of their addiction. Naturally, the CIA got inundated with eager volunteers jumping at this wonderful opportunity to get free drugs while they were in prison. Each was given a physical examination, administered one of eight hundred or so hallucinogenic drugs, and observed for a few days. They were then given heroin, morphine, or anything else they wanted as payment for their participation.

The more controversial research projects by far were those that involved the administration of LSD. After receiving reports that the Soviet Union was engaged in research to produce LSD, the CIA's greatest fear was that it would be used against the United States. As one agent said, "The drugs would be useful in order to gain control of bodies whether they were willing or not." In response to the perceived threat, programs were conceived and implemented at all social levels—high, low, native American, foreign—in order to test how a variety of individuals would react to certain drugs.

In the United States, many LSD experiments were conducted at the Army Chemical Warfare Laboratories in Edgewood, Maryland: The army was especially interested in how LSD sprayed over a battlefield would disorient an enemy and force surrender. The program, EA1729, included ninety-five human subjects who participated in three structured experiments:

1. LSD was administered surreptitiously at a simulated social reception to volunteer subjects who were unaware of the purpose or nature of the tests in which they were participating.
2. LSD was administered to volunteers who were subsequently polygraphed.
3. LSD was administered to volunteers who were then confined to "isolation chambers."

There were no attempts to secure approval for the most controversial of the programs. In fact, the surreptitious use of LSD continued until

1963, not only because of its potential as an interrogation drug but also as a way to publicly humiliate officials and world leaders. Dr. Gottlieb, who directed MKULTRA through its most sinister days, said, "Giving LSD to high officials would be a relatively simple matter and could have a significant effect at key meetings, speeches, etc." MKULTRA programs involved physicians, toxicologists, and other specialists in mental, narcotics, and general hospitals, as well as prisons.

From the beginning, it was assumed that there would be dangers; and CIA officials at the outset readied themselves for casualties in this war for national security. However, it was on November 18, 1953 that the tragedy of these experiments was first realized. On that day, ten scientists from the CIA and Fort Detrick gathered for a review and analysis conference at a secluded cabin in Deep Creek Lake, Maryland, where they agreed that unsuspecting human subjects would be needed to verify the effects and potency of LSD. Dr. Robert Lashbrook, one of the CIA officers, poured seventy micrograms of LSD into a bottle of Cointreau to be served after dinner the following evening. After final preparations and deliberations, the plan was set to administer the drug to anyone who unwittingly took a drink of the liqueur.

The following evening, several conference participants, including Dr. Frank Olson, a civilian employee of the army working for the SOD, raised their glasses and had no idea that they were about to take a swallow of LSD. Twenty minutes later, Dr. Gottlieb, observing the group's increasingly boisterous behavior, told the victims what they had just done. Dr. Olson felt especially edgy. That night he couldn't sleep, and when he came down from his room the next morning he was completely fatigued and unable to concentrate.

Shortly after the experiment, Olson became paranoid, his behavior more erratic. Colonel Vincent Ruwet, Olson's immediate supervisor, noticed that Olson "appeared to be agitated at breakfast" but that he "did not consider this to be abnormal under the circumstances." Within days, however, Olson sank into a depression that bordered on such severe despair that Lashbrook, the man who'd laced the liqueur with LSD, recommended immediate medical treatment. The next afternoon, Olson, Ruwet, and Lashbrook flew to New York to meet with Dr. Harold Abramson, an expert on LSD who'd been cleared by the CIA. It was agreed at the meeting that Dr. Olson would need further psychiatric care.

Seriously depressed and afraid to face his family, Dr. Olson spent the next several days in New York before flying back to Washington to spend Thanksgiving with his family and explain the events to his wife. However, since the evening he'd been given the LSD, his mind was not the same. A paralyzing fear had suddenly gripped him at the airport, and he told Ruwet that he simply could not face his family. After some deliberation, the three decided on a new plan. Olson and Lashbrook would return to New York for more psychiatric consultations while Ruwet went on to Frederick, Maryland to explain things to Mrs. Olson.

Returning to New York, Olson met with Dr. Abramson the next morning, who suggested that Olson be placed under regular psychiatric care closer to home. Not able to get a flight out of New York that day, Olson and Lashbrook remained behind and checked into the Statler Hotel in Manhattan. Everything seemed fine as Dr. Olson settled in and thought about the last few days. He watched television, had a few martinis at the hotel lounge, ate dinner and went to bed. Observers at the hotel that evening didn't notice anything out of the ordinary. Lashbrook himself said that Olson "was cheerful and appeared to enjoy the entertainment." He added that Olson "appeared no longer particularly depressed, and almost the Dr. Olson I knew prior to the experiment."

But at 2:30 A.M., a loud crash of glass broke the silence of the hotel room, waking Lashbrook from a deep sleep. In horror he saw that the blinds had been torn from their holders and the window completely shattered. He ran to the opening and, looking out, couldn't believe his eyes. Ten floors below, the body of Dr. Olson could be seen lying in a pool of blood, his secret dying with him until it was eventually uncovered decades later.

Despite setbacks like the Dr. Olson incident, the government forged ahead anyway. LSD, which is colorless, odorless, tasteless, easily dissolvable, and required in such small amounts as to be easily slipped into food and drinks, had too much going for it. Its purpose soon shifted from a passive means of coercing someone to talk to an offensive psychological weapon used to force information from an unwilling source by disrupting brain waves, confusing thought patterns, altering behavior, and breaking down resistance.

Collaborating with the Federal Bureau of Narcotics, the CIA intensified its research, testing LSD surreptitiously on patrons in San Francisco and New York bars in Operation Midnight Climax. In their book, *Acid Dreams: The CIA, LSD, and the Sixties Rebellion*, Martin Lee and Bruce

Shlain tell of drug-addicted prostitutes being hired by the CIA to pick up men and bring them back to CIA bordellos where they would be enticed into drinking alcohol laced with LSD. As the men succumbed to the drug, CIA agents sat in the next room, observing them through two-way mirrors. Other agents were running similar operations at the same time but with more unsavory characters. Unwitting individuals living in safehouses in San Francisco and New York were also given LSD. Some of the test subjects were drug addicts and prostitutes who at times became violently ill and had to be hospitalized. The CIA even experimented with aerosolized LSD, as evidenced in a U.S. Senate hearing in which Senator Edward Kennedy questioned David Rhodes, a former CIA employee, about the use of "safe houses" where unsuspecting citizens would be taken and given LSD.

"And the first trip you made to California with Mr. Pasternak was to understand the different ways of delivering LSD to unsuspecting citizens, is that correct?" Kennedy asked. Rhodes replied, "That is correct," and added, "We were testing a particular device, to determine if LSD could be given in small quantities via an aerosol delivery." When questioned about how individuals were selected, Rhodes said, "We lined up people that we thought we could invite to a party." He no doubt turned some heads when he admitted, "At the party the intent was that we would be able to spray the aerosol, which as I understood it, had a sufficiently small quantity, or the amount that could be ingested would be sufficiently small, so that you would need practiced people to observe any differences in behavior of people, but to see if it could be delivered in that fashion."

We also know from CIA reports that some human subjects chosen were sexual psychopaths at state hospitals or mental patients. From 1954 through 1959, for instance, 142 criminally insane individuals were used for experiments with straight interrogation, hypnosis, hypnosis in conjunction with LSD, and LSD with interrogation. In addition, experiments were performed with knockout or "K" drugs on unsuspecting hospital patients who at the same time were being given experimental pain killers.

In one case, Harold Blauer, a patient undergoing testing at the New York Psychiatric Institute, was purposely given an intravenous injection of a synthetic mescarine derivative. This was during the same period that the institute was under contract with the U.S. Army Chemical Corps, which was also heavily involved in LSD and mescarine research. The chief officer of the Army Chemical Corps, General William Creasy, had even made the statement that psychochemicals like LSD would be the weapons of the future. Mr. Blauer's death from circulatory collapse and heart failure while at the hospital was immediately attributed to mescarine, and the hospital denied that he was ever used as a test subject.

Apparently the death of Dr. Olson, who may be the most famous casualty of LSD research, did nothing to deter the CIA from continuing its use of unwitting subjects. When Dr. Gottlieb was questioned by senators about that and about the lack of experimental safeguards, his responses were not very reassuring. When it was his turn, Senator Chafee, pursuing the fact that the CIA dismissed Olson's death as an unfortunate incident, asked, "You still had unwitting subjects, so as best as you can recall, despite the concern that was shown over the death of Mr. Olson and the fact that you got medical testimony in which the whole subject of the tie-in between LSD and Mr. Olson's death was discussed—despite all of that, things went on just as in the past as far as unwitting subjects were concerned? The decision was, don't change anything?" Dr. Gottlieb answered, "Well, the best I can respond to that, that seems to be the case."

Chiming in, Senator Schweiker brought up the experiments that may have been conducted at a hospital research facility that the CIA helped to finance. His concern was that one-sixth of the total space in the new hospital wing was available to the chemical division of TSS, agency sponsorship of sensitive research projects was completely deniable, professional cover was provided for up to three biochemical employees of the chemical division, and human patients and volunteers were available for experiments. "Why would you go to such trouble and expense to arrange all that," Schweiker asked, "if you weren't planning to experiment on people in the hospital?" Admiral Turner agreed without hesitation, saying, "Those were clearly the intentions."

When the subject of cancer patients was brought up, Senator Schweiker pursued his interrogation even more vigorously. "You do acknowledge in your statement, and it is clear from other documents, that these kinds of experiments were at some point being done somewhere. My question is, is there any indication that cancer patients were experimented with in this wing?" After Admiral Turner admitted that they were, Senator Schweiker brought up the most controversial MKULTRA subject of the day, involving techniques to cause brain concussions and amnesia by using weapons or sound waves to strike individuals without leaving any clear physical marks. "The other question I had relates to the development of something called the perfect concussion," Schweiker said. "A series of experiments toward that end were described in the CIA documents. I wonder if you would just tell us what your understanding of perfect concussion is?"

A bit flustered, Admiral Turner, at first claiming that project No. 54 was never carried out, had to relent when confronted with evidence that for at least a year or two, investigations were ongoing to produce concussions with special blackjacks, sound waves, and other methods as described in backup materials. "I will double-check that and furnish the information for the record," Turner responded. By the end of the day, one thing was clear: Despite the continued attempt at secrecy, enough had been learned to convince the public that human experimentation was a routine part of CIA operations.

The Sleep Room: CIA Brainwashing Experiments in Canada

In 1957, a young girl was led to a small room at Montreal's Allan Memorial Institute, her arms and legs strapped to a bed and electrodes from a Page-Russell electroconvulsive therapy machine attached to her head. When the signal was given, a switch was thrown, causing her frail body to stiffen then convulse uncontrollably from electroshock forty times more intense than was considered safe. Over the next few weeks, she was awakened three times a day and subjected to these multiple shocking sessions known as "depatterning."

Another patient, this one an older man, was kept in a drug-induced state for several months while being forced to listen to an audiotaped message twelve to sixteen hours per day. The purpose of this experiment, as well as the electroshock treatment, was to see how long it would take before repeated messages, physical shock, drugs, or a combination of all three would destroy a person's personality, clear the brain, induce changes in behavior and memory, and promote a new and healthier personality. In both instances, the medical abuse had been perpetrated by Dr. Ewen Cameron, a renowned psychiatrist paid by the CIA to study the potential of depatterning as a brainwashing tool.

In October 1988, the U.S. Justice Department quietly paid nine Canadian citizens an out-of-court settlement of seven hundred and fifty thousand dollars rather than risk further investigations into what had happened some thirty years earlier. The Canadians, more than three hundred of them at the time, had been recruited by the CIA and the Canadian government for the experiments because of what the U.S. government saw shortly after prisoners began returning home from the Korean War. Intelligence officers who'd debriefed the men learned that they'd been coercively interrogated, brainwashed, mentally tortured, and physically and emotionally abused. To the surprise of U.S. intelligence, some of the soldiers even expressed sympathy for their captors. This mind control had to be countered; and since CIA experts had heard of Dr. Cameron's "psychic driving" techniques being compared to coerced interrogation and brainwashing, they solicited a grant application from Cameron to fund his work. In a matter of months, experiments had begun at Allan Memorial Institute to develop ways to cleanse the brain.

The CIA had known about Dr. Cameron's methods since 1956, the year he'd published an article in the *American Journal of Psychiatry* entitled "Psychic Driving." In the article, Cameron described his brainwashing technique as a two-stage process. In the first stage, patients were depatterned or forced into a vegetative state through electroshock, drugs, and sensory deprivation. Once this state was achieved, the patient was ready for psychic driving, which consisted of weeks or even months spent listening to tape loops through headphones, football-like helmets, or speakers. By the time treatment was over, patients were basically programmed to think and function in a certain way.

Before that, Cameron worked at Brandon Hospital, conducting bizarre experiments on schizophrenics. In a 1950 article published in the *British Journal of Physical Medicine*, he described how he'd exposed patients to red light (chosen because it is the color of blood) from fifteen 200-watt lamps. The patients were forced to lie in the bright light for eight hours per day, some being treated in this manner for as long as eight months. Other Cameron experiments included heat treatments, in which patients' body temperature was raised to 103 degrees Fahrenheit, and insulin coma therapy, in which patients were given large insulin injections and went into a coma for up to five hours. In these latter experiments, some of the patients were given injections every day for up to two months. All this must have been just what CIA operatives were looking for, and Dr. Cameron seemed like a perfect candidate for recruitment to MKULTRA.

The problem was that Dr. Cameron used the CIA national security directive as an excuse to expand his already inhumane research and carry on some of the most hideous experiments since World War II. Operation Knockout had become a part of MKULTRA, subjecting unwitting mental patients to modern torture and stripping them of their memories and identities. In *The Search for the Manchurian Candidate*, author John Marks wrote, "The frequent screams of the patients (usually women) that echoed through the hospital did not deter Cameron or most of his associates in their attempts to depattern their subjects completely. Other hospital patients report being petrified by the 'sleep rooms,' where the treatment took place, and they would usually creep down the opposite side of the hall."

As part of his human experimentation, Dr. Cameron used combinations of electric shock so intense as to sometimes cause apnea (cessation of breathing), hallucinogenic drugs, barbiturates, and such prolonged bouts of induced sleep as to leave his patients in a stupor for months at a time. When they awakened, patients found themselves dazed and confused, incontinent, and barely able to function. The experiments ended in 1964. When Dr. Cameron died in 1967, the CIA was guaranteed that their secrets would remain secret a little while longer.

CIA Mind Control Experiments on Children

Reports have recently surfaced showing that children were used as experimental subjects in secret mind control projects. Much of the testimony heard before the President's Committee on Radiation on March 15, 1995 has been difficult to uncover, probably because it includes sworn statements and shocking tales by individuals claiming to have been experimented on as children. Former human subjects tell of being brainwashed by the CIA, psychologically tortured, hypnotized, given drugs, threatened with death, and even raped. Jon Rappoport, an investigative reporter and author of *AIDS, Inc.* claims to have congressional documents proving that children from Mexico and South America were used not only as human guinea pigs but also as sex agents for the purpose of blackmailing politicians, businessmen, and educators, and selected because they were considered expendable.

The following are actual statements taken from the March 15, 1995 public meeting of the Advisory Committee on Human Radiation Experiments.

Statement of Dr. Valerie Wolf

In listening to the testimony today, it all sounds really familiar. I am here to talk about a possible link between radiation and mind-control experimentation that began in the late 1940s.

The main reason that mind-control research is being mentioned is because people are alleging that they were exposed as children to mind-control radiation drugs and chemical experimentation, which were administered by the same doctors who are known to have been involved in conducting both radiation and mind-control research.

Written documentation has been provided revealing the names of people and the names of research projects in statements from people across the country. It is also important to understand that mind-control techniques and follow-ups into adulthood may have been used to intimidate these particular research subjects into not talking about their victimization in government research.

As a therapist for the past twenty-two years, I have specialized in treating victims and perpetrators of trauma and their families. When word got out that I was appearing at this hearing, nearly forty therapists across the country, and I had about a week and a half to prepare, contacted me to talk about clients who had reported being subjects in radiation and mind-control experiments.

The consistency of people's stories about the purpose of the mind-control and pain-induction techniques, such as electric shock, use of hallucinogens, sensory deprivation, hypnosis, dislocation of limbs and sexual abuse, is remarkable. There is almost nothing published on this aspect of mind-control used with children, and these clients come from all over the country, having had no contact with each other. What was startling was that therapists reported many of these clients were also physically ill with autoimmune problems, thyroid problems, multiple sclerosis, and other muscle and connective tissue diseases as well as mysterious ailments for which a diagnosis cannot be found.

While somatization disorder is commonly found in these clients, many of the clients who have been involved in the human experimentation with the government have multiple medically documented physical ailments, and I was really shocked today to hear one of the speakers talk about the cysts and the teeth breaking off, because I have a client that that's happening to.

Many people are afraid to tell their doctors their histories as mind-control subjects for fear of being considered crazy. These clients have named some of the same people. . . . [Dr. Wolf then identified a series of individuals.]

It needs to be made clear that people have remembered these names and events spontaneously with free recall and without the use of any memory-retrievable techniques, such as hypnosis. As much as possible, we have tried to verify the memories with family members, records and experts in the field.

Many attempts have been made through Freedom of Information Act filings to gain access to the mind-control research documentation. These requests have generally been slowed down or denied, although some information has been obtained, which suggests that at least some of the information supplied by these clients is true. It is important that we obtain all of the information contained in the CIA and military files to verify or deny our clients' memories. Although many of the files for MKULTRA may have been destroyed, whatever is left, along with the files for other projects, such as Bluebird and Artichoke, to name only two, contain valuable information.

Furthermore, if, as the evidence suggests, some of these people were used in radiation experiments, there might be information in the mind-control experiment file on radiation experiments. We need this information to help in the rehabilitation and treatment of many people who have severe psychological and medical problems, which interfere with their social, emotional and financial well-being.

Finally, I urge you to recommend an investigation into these matters. Although there was a commission on mind-control, it did not include experiments on children because most of them were too young or still involved in the research in the late 1970s to come forward. The only way to end the harassment and suffering of these people is to make public what has happened to them in the mind-control experiments. Please recommend that there be an investigation and that the files be opened on the mind-control experiments as they related to children.

Statement of Christina DeNicola

Good afternoon. I'm Christina DeNicola, born July 1962. I was a subject in radiation as well as mind-control and drug experiments performed by a man I knew as [a doctor is named by the witness at this point and will be referred to herein as Dr. B]. I was a subject from 1966 to 1976. Dr. B performed radiation experiments on me in 1970, focusing on my neck, throat, and chest in 1972, focusing on my chest and my uterus in 1975.

Each time I became dizzy, nauseous, and threw up. All these experiments were performed on me in conjunction with mind-control techniques and drugs in Tucson, Arizona. Dr. B was using me mostly as a mind-control subject from 1966 to 1973. His objective was to gain control of my mind and train me to be a spy assassin. The first significant memory took place at Kansas City University in 1966. [A relative] took me there by plane when my mom was out of town. I was in what looked like a laboratory, and there seemed to be other children. I was strapped down naked, spread-eagle on a table, on my back.

Dr. B had electrodes on my body, including my head. He used what looked like an overhead projector and repeatedly said he was burning different images into my brain while a red light flashed aimed at my forehead. In between each sequence, he used electric shock on my body and told me to go deeper and deeper, while repeating each image would go deeper into my brain, and I would do whatever he told me to do.

I felt drugged because he had given me a shot before he started the procedure. When it was over, he gave me another shot. The next thing I remember, I was with my grandparents again in Tucson, Arizona. I was four years old. You can see from this experiment that Dr. B used trauma, drugs, post-hypnotic suggestion, and more trauma in an effort to gain total control of my mind. He used me in radiation experiments, both for the purposes of determining the effects of radiation on various parts of my body and to terrorize me as an additional trauma in the mind-control experiments. The rest of the experiments took place in Tucson, Arizona, out in the desert. I was taught how to pick locks, be secretive, use my photographic memory, and a technique to withhold information by repeating numbers to myself.

Dr. B moved on to wanting me to kill dolls that looked like real children. I stabbed a doll with a spear once after being severely traumatized, but the next time, I refused. He used many pain induction techniques, but as I got older, I resisted more and more. He often tied me down in a cage, which was near his office. Between 1972 and 1976, he and his assistants were sometimes careless and left the cage unlocked. Whenever physically possible, I snuck into his office and found files with reports and memos addressed to CIA and military personnel. Included in these files were project, subproject, subject, and experiment names with some code numbers for radiation and mind control experiments, which I have submitted in your written documentation.

I was caught twice, and Dr. B ruthlessly used electric shock, drugs, spun me on a table, put shots in my stomach and my back, dislocated my joints, and hypnotic techniques to make me feel crazy and suicidal. Because of my rebellion and growing lack of cooperation, they gave up on me as a spy assassin. Consequently, the last two years, 1974 to 1976, Dr. B used various mind control techniques to reverse the spy assassin messages to self-destruct and death messages.

His purpose. He wanted me dead, and I have struggled to stay alive all of my adult life. I believe it is by the grace of God that I am still alive. These horrible experiments have profoundly affected my life. I developed multiple personality disorder because Dr. B's goal was to split my mind into as many parts as possible so he could control me totally. He failed. But I've had to endure years of constant physical, mental, and emotional pain even to this day. I've been in therapy consistently for twelve years, and it wasn't until I found my current therapist two and a half years ago, who had knowledge of the mind control experiments, that I finally have been able to make real progress and begin to heal.

In closing, I ask that you keep in mind that the memories I have described are but a glimpse of the countless others that took place over the ten years between 1966 and 1976, that they weren't just radiation but mind control and drug experiments as well. I have included more detailed information of what I remember in your written documentation. Please help us by recommending an investigation and making the information available so that therapists and other mental health professionals can help more people like myself. I know I can get better. I am getting better, and I know others can, too, with the proper help. Please help us in an effort to prevent these heinous acts from continuing in the future.

Statement of Claudia Mullen

Good afternoon. Between the years of 1957 and 1974, I became a pawn in the government's game, whose ultimate goal was mind control and to create the perfect spy, all through the use of chemicals, radiation, drugs, hypnosis, electric shock, isolation in tubs of water, sleep deprivation, brainwashing, verbal, physical, emotional, and sexual abuse.

I was exploited unwittingly for nearly three decades of my life, and the only explanations given to me were that "the end justifies the means" and "I was serving my country in their bold effort to fight Communism." I can only summarize my circumstances by saying they took an already abused seven-year-old child and compounded my suffering beyond belief. The saddest part is I know for a fact that I was not alone. There were countless other children in my same situation, and there was no one to help us until now.

I've already submitted as much information as possible, including conversations overheard of the agencies responsible. I'm able to report all this to you in such detail because of my photographic memory and the arrogance of the doctors—the arrogance of the people involved. They were certain they would always control my mind. Although the process of recalling these atrocities is not an easy one, nor is it without some danger to myself and my family, I feel the risk is worth taking.

[The witness names a doctor with a name similar to that of Dr. B] who claimed to have received fifty million dollars from . . . as part of a TSD or technical science division of the CIA, once described to [another doctor] that "children were used as subjects because they were more fun to work with and cheaper, too." They needed lower profile subjects than soldiers or government people.

So, only young willing females would do. Besides, he said, "I like scaring them. They and the agency think I'm a god, creating experiments for whatever deviant purposes Sid and [a third doctor] could think up." Sid being Dr. Sidney Gottlieb, . . . In 1958, I was to be tested, they told me, by some important doctors . . . and I was instructed to cooperate. I was told not to look at anyone's faces, and not—try hard not to ignore—to try hard to ignore any names as this was a very secret project, but I was told that all these things would help me forget.

Naturally, as most children do, I did the opposite, and I remembered as much as I could. . . . Then I was told by Sid Gottlieb that "I was ripe for the big A" meaning Artichoke. By the time I left to go home, just like every time from then on, I would remember only whatever explanations [another doctor is named] gave me for the odd bruises, needle marks, burns on my head, fingers, and even the genital soreness. I had no reason to believe otherwise. They had already begun to control my mind.

The next year, I was sent to a lodge in Maryland . . . to learn how to sexually please men. I was taught how to coerce them into talking about themselves, and it was [various doctors and officials named], who were all planning on filming as many high government agency officials and heads of academic institutions and foundations as possible, so that later, when the funding for mind control and radiation started to dwindle, projects would continue.

I was used to entrap many unwitting men, including themselves, all with the use of a hidden camera. I was only nine years old when this sexual humiliation began. I overheard conversations about a part of the agency. . . . Once a crude remark was made by Dr. Gottlieb about a certain possible leak over New Orleans involving a large group of retarded children who were being given massive doses of radiation. He asked why [another individual] was so worried about a few retarded kids; after all, they would be the least likely to spill the beans.

Another time, I heard. . . . the director of the scientific office . . . . state that, "In order to keep more funding coming from different sources for radiation and mind control projects, he suggested stepping up the amounts of stressors used and also the blackmail portion of the experiments." He said it needed to be done faster and to get rid of the subjects or they were asking for us to come back later and haunt them with our remembrances.

There's much more I could tell you about government-sponsored research, including project names, cell project numbers, people involved, facilities used, tests and other forms of pain induction, but I think I've given more than enough information to recommend further investigation of all the mind control projects, especially as they involve so much abuse of the radiation. I would love nothing more than to say that I had dreamed the whole thing up and need just to forget it, but that would be a tragic mistake. It would also be a lie.

All these atrocities did occur to me and to countless other children, and all under the guise of defending our country. It is because of the cumulative effects of exposure to radiation, chemicals, drugs, pain, and subsequent mental and physical distress that I've been robbed of the ability to work and even to bear any children of my own.

It is blatantly obvious that none of this was needed nor should it ever have been allowed to take place at all, and the only means we have to seek out the awful truth and bring it to light is by opening whatever files remain on all the projects and through another presidential commission on mind control. I believe that every citizen of this nation has the right to know just what is fact and what is fiction. It is our greatest protection against the possibility of this ever happening again. In conclusion, I can offer you no more than what I've given you today, the truth, and I thank you for your time.

Statement of Suzzanne Starr

A whole part of my life just came together. This is phenomenal. Here I am, living in a remote area of New Mexico, and I start remembering this really bizarre stuff. Then I go back and I find the place where it happened, a place I never thought I had been in my life, and by gosh, it looks just like my recall of it, and now I sit here today, and I hear from people I have never met, never seen. They have been through the same thing I'm experiencing.

I don't have the names, but you know one thing that just shocks me is through all of my work, I keep coming up with this darned Delta code, Delta 5133867. Until today, I didn't know what that was. It's an experimentation code. I kept wondering, why do I write Delta 5133867. What's an alpha code? What's a beta code? Those are things that this nation needs to find out for the sake of our future, and really and truly, without mistake, for the sake of the salvation of our planet.

I'm just shocked. I'm surprised. I am a survivor of secret experimentation conducted by our government on healthy children. I recalled and began to recall these incidences two years ago. I have been working for weeks to overcome the terror program so that I could be here and speak to you with dignity today.

I know I survived my childhood for this moment. These horrid secrets undermine the core of our society. They exist only out of the power of evil. As long as atrocity [*sic*] to human beings, particularly children, go unbelieved, they can continue. I have come to realize from my awakening that reality is a dimension beyond human beings' ability to conceive the truth. When the truth comes to the light and is believed, there is an incredible healing for ourselves and our nation. That is my hope.

I was born in 1949. We were very poor. I lived in the mountains of Colorado. Both of my parents have died of cancer. All but two of my aunts and uncles have either died of cancer or have cancer. As a child, my parents were victims of a mind control organization that permitted me to be inducted into experimentation. I have early recollections of people coming to my house. My father was picked up on a false arrest for a ticket, parking ticket, and put in jail. They came to my house, and they tortured me, and they held my mother until she signed a paper.

I believe and I know that if she had not signed that paper, I would not be here today. I believe that her signing this paper is related to me being brought into these experiments. Either she signed or I died. I believe . . . [a] physician, who was retired from the military, got children from the mountains of Colorado for the experiments. He was the only doctor I ever saw until I was twenty years old. The first memory I have of environmental deprivation was in the basement of this doctor's office.

His office adjoined a meeting hall that was used for satanic rites. I was astounded when I returned to this city not that long ago, two years ago, and discovered that his office and the adjoining chambers and the subchambers in that city were exactly as I had remembered it. Of course, I would remember my doctor's office, but I had no knowledge prior to my return and my investigation of the subchambers and of the secret things about his office.

The incidences I have recalled happened to me between ages of three and twelve years old. I was taken to a college campus in the summer. We were kept in a locked dorm and taken to the experiment by way of underground tunnels. I provided the name of that institution in my narrative. I believe you have my narrative. I don't want to say that here in public. One day, there was a lot of confusion, and a door was left open, and I slipped out. I went across the campus and entered into another dorm. I heard some people yelling. I wandered down the hall. I was a very type of inquisitive kind of a slip-out child, and when I went into the room and looked around the corner where the people were yelling, there was a high official from the United States military. There was the man that the people in the program called the Nazi doctor. They called him a Nazi. I don't know who this man is. I believe I could recognize a picture if I was given the opportunity, and there was one of the technicians at the head of the program.

I was caught and taken into electroshock sessions, something was put up my nose, and I passed out. In recovering this incident, I had convulsions, which I have. I'm not a seizure person, but when I am recalling these incidences, frequently I go into a convulsive type of episode. It's not grand mal. It's just extreme shaking. A year and a half ago, on an investigative trip, my husband and I returned to that campus. I was amazed to find it exactly as I had recalled it. The two buildings where we were used for the experiments had been torn down in 1968, but the dorm that I wandered into was exactly as I remembered it.

I recall being in a classroom with other children. We were all in institution pajamas. We were told that we were chosen to help serve our country. A careful record of the procedures was kept. The technicians were highly trained professionals. They were just doing their job. We were not to be angry at them. An American flag hung in the room. The experiments are discussed in more detail in my narrative. One of the doctors, who supervised the experiments, was called the Nazi when he was out of the room. The experiments involved environmental deprivation, to the point of forced psychotic states, and you know why I remember the forced psychotic states that had a great impact on me because I realized something. After they put me in that little cell and treated me like a dog and kept me there until I went into psychotic states, they gave me electroshock and told me we returned you to sanity, so we can take your sanity away, if you ever speak, and I'm speaking today, and I'm not going to lose my sanity.

The experiments also included extreme sensation on the brain, spin programming, breeding of children, and injections. I was given frequent electroshock and mind control sessions with the threat of death or insanity if I ever spoke, and through my recollection and these years that I have struggled for my freedom and the phrase that says thank God I'm free at least means a lot to me. Through these times, I have fought self-destructive programmed messages to kill myself, and I know what a programmed message is, and I don't act on them. I know the difference.

Obviously they misjudged my spirit and my desire to be free. The experiment I wish to speak about involved radiation. I was strapped face down, straddled on a device like a chair that curved my spine in a haunch. Needles were put in three places in my spine: my coccyx, my midspine, and the base of my skull. To the right, there was a device with five orifices. Five IV tubes came out and joined into one, with controls for the amount of fluid and frequency. This tube was connected to the needles at the base of my skull.

I was given a timed injection at my coccyx. The technician had a monitor; I believe it was a Geiger counter. They checked my head with it. There would be timed releases—released injections through the IV into the needle at the base of my skull repeatedly, which was monitored. When the injections went into my brain, it felt like ice spreading throughout my skull. It was agonizing. I had cuffs on my upper arms and things on my fingers, I believe for vital signs. Wires were connected to my head simulator to an EEG. Often, they would say get some fluid. They did something to the needles in my middle spine. I believe they were testing my spinal fluid.

Sometimes something happened to the cuffs on my arms that caused horrible pain. Readings were taken again. The procedure was being taught to someone. I believe—I believe that's what was happening. They talked as if I was unconscious and not even human. I recall it was explained that the injections were referred to as "trace" but enough to make this kid's head light up like a Christmas tree. They thought this was funny. They kept making jokes about my head glowing. They sat me up and put a tube in my nose. I could feel something horrible in the front of my brain, and I blacked out.

In another experiment, when I—they thought I was dead, they took me out of the chair, and the technician looked at me, and he said, "It looks like we lost this one. Well, there's plenty more where she came from. If she's brain dead, we can institutionalize her and use her for further experiments. If she's dead, we will arrange an accident as is procedure with her family."

Another experiment involved inserting air into my uterus and expanding the abdominal cavity with air. This experiment was torturous. Measurements were taken periodically. X rays of my uterus and fallopian tubes were taken by injecting radioactive dye. I know that this is a salpinghistiogram. I had to have this done during fertility testing when my husband and I were trying to conceive a child. Fertility testing was so traumatic that I had to stop trying. I have never had a normal pregnancy or been able to conceive a child. However, I do remember at the age of twelve having an induced pregnancy. My baby boy was taken for the experiments. That is the only child I have ever had, unless there are other abortions that I'm not aware of.

I am willing to discuss my experiences in more detail, if any of you wish to. I have suffered all my life because of this. My life has completely changed now because of my recovery. Five years ago, I began my quest for truth. I didn't perceive how much I was suffering until finally my symptoms diminished. I have recovered these incidences with the help of a caring professional. He has been careful to maintain a neutral position and does not hypnotize or lead me or influence me in any way, and he said he will attest to that.

Once early in my healing, I spoke to a man who helps people deprogram from mind control groups. He told me freedom is in the struggle. The good Lord knows I have struggled to be free. I am thankful that I started working on healing my body in my thirties. The past five years, I have healed my mind and spirit. Now I am strong enough to speak the truth.

There's one more thing I didn't mention. During the many times, there were forced rapes. I wanted to say one thing—a memory I've had all my life. I always knew about, I always wondered what it was. When I was four or five years old, I used to lie in the bedroom when my sisters went to school in the morning, and I played Nazi concentration camp, and I would be the Jewish princess, and they would be experimenting on me and military people would come and rape me, and I held up to it all because I was such a brave girl. I think I was a very brave girl. I really do.

I always wondered, why did a four-year-old fantasize that she was being experimented on? Why did she think that people were raping her? Now I know why. Because it was the truth. I wish to thank the people at the task force for helping me trust enough to testify. I would never have trusted a government project without their support. I also wish to thank President Clinton for appointing this commission, and each of you especially for having the courage and the integrity to listen to us, the survivors of America's most horrid secret.

I am deeply committed to exposing this most horrid secret. Of course, I am terrified of repercussions, but if one of you hears us today, if one of you takes action, if someone in this room takes action, even if it's ten years from now, this can change. I am terrified of repercussions, but I will not purchase peace at the price of my silence. If life's so dear or peace so sweet as to be purchased at the price of chains of slavery, forbid it, almighty God. I know not what course others may take, but as for me, give me liberty or give me death, and I imagine you all know who said that. My hero when I was a little girl, Patrick Henry. I do not choose death; I choose freedom, freedom to speak the truth. Thank you.

Project Third Chance and Derby Hat

The CIA's eagerness to use human subjects for LSD experiments didn't stop with civilians. In its search for volunteers, it found willing participants in the ranks of the U.S. Army. The testing program was divided into three phases: In the first phase, LSD was administered to more than one thousand American soldiers who volunteered to be subjects in chemical warfare experiments. It's not known if they were ever told that the chemical was LSD. In the second phase, Material Testing Program EA1729, ninety-five volunteers were given LSD in clinical experiments designed to evaluate potential intelligence uses of the drug. Experiments included physical and psychological techniques to exploit a subject's mental state and to maximize stress. In the third phase, code-named Third Chance and Derby Hat, unwitting nonvolunteers were interrogated after receiving LSD without their consent or knowledge as part of operational field tests.

The main difference between these and the MKULTRA LSD tests is that Third Chance and Derby Hat also involved administration of LSD to unwitting subjects in Europe and the Far East. The first field tests—Third Chance—were conducted in Europe by an army Special Purpose Team (SPT) from May to August 1961 and involved eleven separate interrogations of ten subjects. None of the subjects were volunteers and none were aware that they were to receive LSD.

Before Third Chance even began, there was concern that it had not been coordinated with the CIA and FBI. General Willems the army assistant chief of staff for intelligence, was quoted as saying, "If this project is going to be worth anything, LSD should be used on higher types of non-U.S. subjects or staffers." This despite the conclusion of a 1959 study raising warning flags about the dangers of LSD: The view has been expressed that EA1729 is a potentially dangerous drug whose pharmaceutical actions are not fully understood, and there has been cited the possibility of the continuance of a chemically induced psychosis in chronic form, particularly if a latent schizophrenic were a subject, with consequent claim or representation against the U.S. government.

The second series of field tests—Derby Hat—was conducted by an army SPT in the Far East from August to November 1962. Seven subjects were interrogated, all of whom were foreign nationals either suspected of dealing in narcotics or implicated in foreign intelligence operations. The purpose of this second set of experiments was to collect additional data on the utility of LSD in field interrogations and to evaluate any different effects the drugs might have on "Orientals." In one case, a suspected Asian agent was given six micrograms of LSD per kilogram of body weight and became comatose.

Ironically, greater care was taken to protect foreign nationals abroad than to ensure the safety of American subjects at home. According to Senate testimony, medical examinations were often performed prior to giving individuals LSD; and during LSD interrogations, local physicians, who had no idea what the individuals had been given, were on call in the event that something happened. In fact, both the Office of Security and the Office of Medical Services stated that LSD "should not be administered unless preceded by a medical examination, and should be administered only by or in the presence of a physician who had studied it and its effects." No such criteria were extended to unwitting American subjects receiving LSD.

The CIA after MKULTRA

In 1966, just two years after MKULTRA was terminated, U.S. intelligence was at it again. Funding for a new, six-year program, MKSEARCH, was initiated to develop, test, and evaluate (1) capabilities in the covert use of biological, chemical, and radioactive materials and (2) techniques for producing predictable changes in both human behavior and physiology through the use of drugs. In essence, MKSEARCH represented a continuation of a limited number of the ULTRA projects.

Even less is known about MKSEARCH then about MKULTRA, principally because it was primarily a CIA project. Both were part of a larger envelope that included a DOD program but not DOD responsibility for those particular subcomponents.

In 1967, the Office of Research and Development and the Edgewood Arsenal Research Laboratories undertook a new program for identifying and characterizing drugs that could influence human behavior. Edgewood was chosen because it had the facilities for a full range of laboratory and human clinical testing. The phased program consisted of acquisition of drugs and chemical compounds believed to have effects on human behavior, and testing and evaluation of these materials through laboratory procedures and toxicological studies. Compounds considered promising based on animal tests were then to be evaluated clinically with human subjects at Edgewood. These compounds would then be analyzed structurally as a basis for identifying and synthesizing possible new and even better derivatives.

The program was divided into two projects. Project OFTEN, which was part of an ongoing DOD program, dealt mainly with testing the toxicology, transmission, and behavioral effects of drugs in animals but ultimately in human subjects. Project CHICKWIT was designed to gather information on new drug developments in Europe and Asia and to acquire samples for testing and analysis.

From its inception in 1953 until it ended thirteen years later, MKULTRA and its projects were shrouded in absolute secrecy. It was a time in U.S. history when the threat to national security was thought to be endangered by rogue nations such as Korea and superpowers such as the Soviet Union. Human experimentation was justified in the name of vital national interests and rationalized as a means of protecting society and maintaining the American way of life. It was not surprising, then, that the U.S. military and agencies such as the CIA were allowed free reign to conduct research and implement programs that would guarantee security.

But if history teaches one anything, it is that even the most guarded secrets are often exposed, sometimes purely by chance. So it was with MKULTRA. Nearly a decade after it was approved by the DCI, MKULTRA began to unravel. A wide-ranging inspector general survey of the Technical Services Division in the spring of 1963 stumbled across the top secret program and, after some tenacious digging, learned of horrific research and experiments involving the surreptitious administration of LSD to unwitting, nonvoluntary human subjects. That revelation started an investigation that essentially put a stop to all research. Following a detailed report by the inspector general, MKULTRA was finally terminated in 1964.

However, the end of MKULTRA did not put a stop to related programs or to subsequent research deemed necessary to maintain national security. It was, after all, the mission of security agencies such as the CIA to protect U.S. vital interests at all costs. So for the next several decades, the military, along with the CIA and private corporations, engaged in ongoing research to one extent or another in a variety of areas. According to hearings before the U.S. Senate, the various projects were turned on and turned off in a never-ending web, especially the most recent ones dating from the late 1960s to the early 1970s.

Throughout the 1970s and 1980s, biological and chemical agent research continued, although emphasis was placed on defensive measures. A cancer virus research program was established at Fort Detrick. In addition, efforts continued in the development of weapons that could target specific ethnic groups on the basis of differences in their genetic makeup. These programs, some of which still exist today, are the focus of upcoming chapters.

6 SILENT CONSPIRATORS: THE GOVERNMENT-INDUSTRY CONNECTION, FROM ASPARTAME TO AZT

In 1969, Dr. Herbert L. Ley, former commissioner of the U.S. Food and Drug Administration (FDA), made a statement that must have shocked even the most vocal critics of the oft-criticized government agency:

> The thing that bugs me is that people think the Food and Drug Administration is protecting them—it isn't. What the FDA is doing and what the public thinks it's doing are as different as night and day.

Not one of the more congenial or play-by-the-book commissioners (he only lasted a year), Dr. Ley told it like it was. His main objection, as he saw it, was the increased influence that money and power had on government agencies such as the FDA, the centers for Disease Control and Prevention (CDC), the National Institutes of Health (NIH), and the National Cancer Institute (NCI) whose business it is to protect people first. Shortly before his departure from the FDA, Dr. Ley proclaimed, "Unless there is a major change in the drug industry emphasis on sales over safety, the industry as we know it today may well be buried within the next several years in a grave it has helped dig—inch by inch, overpromotion by overpromotion, bad drug by bad drug."

Powerful words, to say the least, considering who his target was. But even today, Dr. Ley's words reverberate throughout the world, where people are beginning to suspect that perhaps agencies such as the FDA are not always what they seem to be. For example, in a May 2001 article appearing in the renowned medical journal *The Lancet*, the editor wrote, "The FDA, which safeguards the health of 274 million people and regulates over one trillion dollars worth of products, was compromised by funding from the drug industry and pressure from Congress." Other editors and experts have been equally critical of the NIH, whose mission is to sponsor research that leads to better health; the CDC, the lead agency for protecting public health and safety of people through disease prevention and control; and the NCI, part of the NIH that was established in 1937 for cancer research and training.

In an ideal world, all scientists, especially scientists working for government agencies, would be sharing their work, describing their experiments in detail, and verifying results in order for other scientists to be able to replicate what they've done and prevent harm to human subjects. After all, criticism and scrutiny are an integral part of the process and ensure that truth prevails above all else. Most scientists I've known play by those rules and are more than open to scrutiny. But when researchers hide behind a veil of secrecy, when federal workers delete data or censor critical information, they not only hurt science and those who do honest research but jeopardize the lives of people directly affected by research. Often these cases are kept hidden because a revelation might uncover fraud or collusion.

Through the years, agencies such as the FDA and NIH have often been in competition. Because they have different missions and budgets, each has not always known what the other was doing, nor did they care. In some instances, collusion between two or more agencies, under pressure from Congress, has resulted in decisions based on politics rather than science. For the most part, federal agencies have done what they are supposed to do, with dedicated career employees ensuring that safety issues are addressed. In some cases, however, their actions have bordered on the criminal, sacrificing lives for jobs and careers, and placing the interests of corporations above the people they are supposed to be protecting.

Indeed, the biggest problem with the system can be financial incentives, with some scientists rewarded like entrepreneurs to develop whatever they're paid to develop. This works amazingly well when the goal is to find cures or to synthesize new and useful products, and that should never be tampered with. It doesn't work so well, and can fail miserably, when the goal is to uncover product defects or do anything else that would threaten company profits or stock prices, especially if the product is in late-stage development or already on the market. This is precisely why industry hires its own researchers, designs its own experiments, and then develops relationships with federal agencies that rely on the honesty and integrity of the research to make its recommendations. In the majority of cases, the system works just fine. However, any irregularity along the way is multiplied throughout the process and can be a prescription for disaster.

According to an investigative article in *USA Today*, 54 percent of the time experts hired by the government have a direct financial relationship with the drug company whose product they are hired to evaluate; and even though federal law prohibits the FDA from using experts with such conflicts of interest, since 1998 the FDA has waved this restriction more than eight hundred times! Upon further investigation, *USA Today* discovered that at more than one hundred meetings dealing with the fate of a specific drug, 33 percent of the experts who had influence over the approval process had financial interests with the drug company. This would not have been such a big deal were it not for an April 15, 1998 *Journal of the American Medical Association* article showing that fatal adverse reactions to FDA-approved drugs were between the fourth and sixth leading cause of death in the United States, with nearly one hundred thousand people killed each year by drugs that the FDA and its panels of "conflict of interest experts" have said were safe and effective for people to use.

To illustrate how a federal agency could easily be influenced, I'll begin with the story of aspartame (NutraSweet) because this sweetener's approval was the most controversial and political in the FDA's history. Hopefully, what happened between the U.S. government and the manufacturer of aspartame is not as common as some have feared.

The Aspartame Story: How the FDA's Approval Process Broke Down

In his daily Pentagon briefings on the war in Afghanistan, Secretary of Defense Donald Rumsfeld was as charming as he was brilliant. The

"Rummy Show," as newspaper reporters called the anticipated briefings, had gotten higher ratings during the height of the war than just about any show in its timeslot. Few people watching the humorous bantering between Rumsfeld, a former congressman and chief of staff under President Gerald Ford, and the press corps knew that one of the main architects of the successful Afghan campaign had been the chairman and CEO of G. D. Searle Pharmaceuticals (now owned by Monsanto) at a time when the company's artificial sweetener NutraSweet was in the throes of FDA approval.

The history of aspartame is a fascinating one. Discovered in 1965 when Dr. James Schlatter, a Searle researcher, accidentally licked remnants of a new antiulcer drug from his fingers and was startled by its very sweet taste, aspartame flew onto the front burner for preliminary FDA studies. Some early tests on primates, however, proved troubling. When Dr. Harry Waisman, a biochemist at the University of Wisconsin's Joseph P. Kennedy, Jr. Memorial Laboratory, gave the new chemical to seven infant monkeys, one died in less than a year and five others had grand mal seizures. These significant effects were not reported to the FDA when Searle filed the initial approval application. Other studies proved not much better, but only the industry studies with "single dose" tests of aspartame showing no effects were published and disseminated to the public. And though brain tumors and seizures found in some nonindustry studies were just some of the unexpected side effects, the research continued unabated, winding its way into the final stages of clinical trials.

It was especially important for the industry to fill the void and corner the gold mine artificial sweetener market at all costs, since cyclamate was banned in 1970 for causing cancer in mice. That year, in a memo dated December 28, 1970, Mr. Helling, a Searle company executive, wrote, "The basic philosophy of our approach to Food and Drug should be to try to get them to say 'yes,' . . . even if we have to throw some [questions] in that have no significance to us other than putting them into a 'yes'-saying habit. We must create an affirmative atmosphere in our dealing with [the FDA]. It would also help if we can get them to get the people involved to do us any sort of favor, as this would also bring them into a subconscious spirit of participation."

So less than a decade after the startling discovery, aspartame was FDA approved for limited use as a dry foods additive in 1974, but with one glitch. Dr. John Olney, a neuroscientist at Washington University in St. Louis, began a careful review of the research data and found studies showing brain tumors. He was especially concerned about the possible effects of aspartame on children and fetuses. In fact, Olney pointed to a large increase in brain tumor incidence about three years following aspartame's introduction into the market (although industry spokesmen point out that the increase began prior to aspartame's introduction). Olney immediately reported his findings to the FDA, citing grave concerns for public safety and uncovering previous studies showing that aspartic acid, one of the main components of aspartame, may have caused microscopic holes in the brains of rats.

While not saying that aspartame had been established to be unsafe, Olney and his colleagues have repeatedly called for more studies. Specifically, both Dr. Olney and Dr. Richard Wurtman, a researcher at the Massachusetts Institute of Technology who testified in the aspartame hearings, do not contend that aspartame should be banned, but urge placebo-controlled studies to determine the cause of alleged adverse reactions in some consumers.

In contrast, the FDA and Deputy Commissioner David Friedman concluded that Olney's hypothesis was "not a convincing line of evidence" and that brain tumors were in no way correlated to aspartame.

In fact, one of the opposing theories proposes that because hundreds of millions of people consume products with aspartame, there may be a certain small percent who simply have allergic reactions, just as some would have to any product that is otherwise safe for most people. Taking into account the percentage of individuals allegedly affected compared with the total number of individuals exposed, some experts claim that the adverse symptoms purportedly attributed to aspartame are not a significant part of the overall consumer population.

At a U.S. congressional hearing, Olney stated, "Scientists informed G. D. Searle that aspartic acid caused holes in the brains of mice. G. D. Searle did not inform the FDA of this study until after aspartame's approval. None of the tests submitted by G. D. Searle to the FDA contradicted these findings." In response to the charges, the FDA put a hold on aspartame and began an internal review of Searle's research and facilities. It subsequently found such serious flaws with the experiments and the product that it feared a rash of lawsuits against Searle should the safety issues be realized. The statement by Dr. Adrian Gross, an FDA toxicologist and task force member, speaks volumes about what investigators thought of Searle's research:

> At the heart of FDA's regulatory process is its ability to rely upon the integrity of the basic safety data submitted by sponsors of regulated products. Our investigation clearly demonstrates that, in the case of the Searle Company, we have no basis for such reliance now. . . .
>
> "We have noted that Searle has not submitted all the facts of experiments to FDA, retaining unto itself the unpermitted option of filtering, interpreting, and not submitting information which we would consider material to the safety evaluation of the product. Finally, we have found instances of irrelevant or unproductive animal research where experiments have been poorly conceived, carelessly executed, or inaccurately analyzed or reported. Some of our findings suggest an attitude of disregard for FDA's mission of protection of the public health by selectively reporting the results of studies in a manner which allay the concerns or questions of an FDA reviewer.

On April 8, 1976, aspartame became the focus of a Senate hearing chaired by Senator Edward Kennedy. The exchange between Kennedy and the FDA's then commissioner, Dr. Alexander Schmidt, was a preview of what the FDA task force had found and was about to report. "Is this the first time, to your knowledge, that such a problem has been uncovered of this magnitude by the Food and Drug Administration?" Kennedy asked. "We have never before conducted such an examination as we did at Searle," Dr. Schmidt responded. "From time to time, we have been aware of isolated problems, but we were not aware of the extent of the problem in one pharmaceutical house." Senator Kennedy could only shake his head and say, "The extensive nature of the almost unbelievable range of abuses discovered by the FDA on several major Searle products is profoundly disturbing."

The two-year FDA review culminated in the 1977 Bressler Report, which contains a number of scathing allegations that set the stage for an FDA–Searle showdown. Besides finding gross deficiencies in thirteen studies submitted to the FDA, which Searle claimed showed no genetic damage, and fifteen "missing" rat fetuses in a toxicity test, here are just some of the actual quotes contained in the report:

> (1) "In some cases, original data could be recorded in several areas, making it difficult, and sometimes impossible to determine which was actually the original. This was a particular problem in dealing with dates of deaths, as some conflicted on the 'source' documents. Many of the responsible individuals involved with the study, including stability testing of DKP, are no longer employed by Searle. Dr. K. S. Rao, Study Monitor, the only individual who could have possibly answered some questions, had left Searle. He was contacted, but permission for an interview was refused by his attorney. Due to the absence of various individuals it was not always possible to accurately determine methods used in some analyses and operations carried out in conducting this study. In a number of areas, including chemistry, statistics, diet preparation and feeding, it was necessary to use assumptions, or information supplied by current employees who were not involved with the study."
>
> (2) "Observation records indicated that animal A23LM was alive at week 88, dead from week 92 through week 104, alive at week 108, and dead at week 112."
>
> (3) "Analytical records A-9129 for DKP lot 5R showed an assay of 1000%. Examination of laboratory notebooks showed that eleven (11) samples had been analyzed from this lot, and the analytical record only reflected an average of the last three of these. The other assays (not reported) ranged from 87.93% to 114.83%."
>
> (4) "Ninety-eight of the 196 animals that died during the study were fixed in toto and autopsied at some later date, in some cases more than one year later. A total of 20 animals were excluded from the study due to excessive autolysis. Of these, 17 had been fixed in toto and autopsied at a later date."
>
> (5) "Records for approximately 30 animals showed substantial differences between gross observations on pathology sheets, when compared with the gross observations on pathology sheets submitted to FDA."

(6) "Excising masses (tumors) from live animals, in some cases without histological examination of the masses, in other words without reporting them to the FDA. Searle's representatives, when caught and questioned about these actions, stated that these masses were in the head and neck areas and prevented the animals from feeding. Also, failure to report to the FDA all internal tumors present in the experimental rats, e.g. polyps in the uterus, ovary neoplasms as well as other lesions."

(7) "Laboratory records of one sort or another for all assays reported in the submission were obtained. In some cases data sheets were noted with results of assays carried out at treatment days not indicated in the protocol or protocol amendment. For example, serum cholesterol determinations were done at days 796 and 798 (terminal bleeding) but not included in the submission to FDA. Because the submission to FDA (Vol. 1 p. 286) reported a significant decrease in serum cholesterol that was more perceptible towards the end of the study, and may have been related to compound administration, the omitted data is of some importance. No data was seen for two assays (serum insulin and serum ornithine carbamyl transferase), which were called for in an amendment to the protocol. Original data was not always available for authentication of results or examination of procedures for conversion of raw data into the calculated values submitted to FDA."

(8)"A total of 49 disparities were noted between statistical computations reported by Searle in the submission and those calculated by FDA. The disparities are constituted by the values for 6 means, 23 standard errors, and 20 significant differences."

(9)"Presenting information to FDA in a manner likely to obscure problems, such as editing the report of a consulting pathologist, reporting one pathology report while failing to submit, or make reference to another more adverse pathology report on the same slide."

Dr. Adrian Gross was stunned by what he'd uncovered. "They lied and they didn't submit the real nature of their observations because had they done that, it is more than likely that a great number of their studies would have been rejected simply for adequacy," he said. Dr. Philip Brodsky, head of the FDA task force, was equally blunt. "I'd never seen anything as bad as G. D. Searle's studies," he said, to which Dr. Alexander Schmidt added, "[The studies] were incredibly sloppy science. What we discovered was reprehensible." Even the National Soft Drink Association (NSDA) initially had grave misgivings about adding aspartame to beverages. In a draft thirty-page report, the NSDA condemned the additive and called for more extensive studies because of their fear that it would lead to major health problems (Appendix XIX). Ultimately, the NSDA changed its position, and the report was not submitted to the FDA. Here are some of the NSDA's original concerns about aspartame according to the unsubmitted draft:

> Collectively, the extensive deficiencies in the stability studies conducted by Searle to demonstrate that APM and its degradation products are safe in soft drinks intended to be sold in the United States, render those studies inadequate and unreliable. It is not possible on the basis of these studies to conclude that the petitioner has demonstrated that, notwithstanding its inherent instability, APM is safe for use in soft drinks.
>
> The concern of the commentator, Dr. Richard J. Wurtman, Professor of Neuroendocrine Regulation at the Massachusetts Institute of Technology, was that increased brain levels of Phe and Tyr are likely to affect the synthesis of certain neurotransmitters—substances vital to the regulation of brain function—and that changes in the levels of neurotransmitters could in turn cause adverse physiological effects (by, for example, modifying the function of the autonomic nervous system) and/or behavioral effects.
>
> For these reasons, Searle has not met its burden of demonstrating to a reasonable certainty that the unlimited use of aspartame, especially in combination with carbohydrates, will not adversely affect human health. The questions posed by Dr. Wurtman are significant because of the seriousness of the potential effects (e.g., changes in blood pressure) and because of aspartame's anticipated widespread use—use that includes consumption by potentially vulnerable sub-groups, such as children,

> pregnant women, and hypertensives. Dr. Wurtman's concerns are shared by other distinguished scientists expert in this field (affidavits attached). It is Searle's legal burden to submit data sufficient to resolve the concerns.
>
> Aspartame has been demonstrated to inhibit the carbohydrate-induced-synthesis of the neurotransmitter serotonin (Wurtman affidavit). Serotonin blunts the sensation of craving carbohydrates and thus is part of the body's feedback system that helps limit consumption of carbohydrates to appropriate levels. Its inhibition by aspartame could lead to the anomalous result of a diet product causing increased consumption of carbohydrates.

Hundreds of millions of research dollars had been spent over a dozen or so years to bring aspartame to market as a food additive. So it was quite a shock to Searle when the Public Board of Inquiry voted unanimously on September 30, 1980 to withdraw FDA approval and reject the use of aspartame until further studies were done to ensure that the product would not cause brain tumors. The following is the decision that sent Searle executives scrambling:

> DEPARTMENT OF HEALTH AND HUMAN SERVICES
> FOOD AND DRUG ADMINISTRATION
> (Docket No. 75P-0355)
> ASPARTAME
> DECISION OF THE PUBLIC BOARD OF INQUIRY
>
> Skip to page 49: V. ISSUE NUMBER 3
>
> a) Should aspartame be allowed for use in foods, or, instead should approval of aspartame be withdrawn?
>
> b) If aspartame is allowed for use in foods, i.e., if its approval is not withdrawn, what conditions of use and labeling and label statements should be required, if any? 44 Fed. Reg. 31717
>
> On the basis of the conclusion concerning Issue Number 2, the Board concludes that approval of aspartame for use in foods should be withheld at least until the question concerning its pos-

sible oncogenic potential has been resolved by further experiments. The Board has not been presented with proof of reasonable certainty that aspartame is safe for use as a food additive under its intended conditions of use.

The foregoing constitutes the Board's findings of fact and conclusions of law.

Therefore, it is ORDERED that:

1. Approval of the food additive petition for aspartame (FAP 3A2885) is hereby withdrawn.

2. The stay of the effectiveness of the regulation for aspartame, 21 CFR 172.804, is hereby vacated and the regulation revoked.

3. Pursuant to 21 CFR 12.125, exceptions to this Initial Decision must be received by the Hearing Clerk within 30 days; replies to exceptions must be received by the Hearing Clerk not more than 20 days thereafter. In the absence of the timely filing of exceptions, or of a review notice by the Commissioner under 21 CFR 12.125(f), this Initial Decision will become the Final Decision of the Commissioner upon the expiration of the date for filing for appeal or review and shall be effective upon publication of a notice to that effect in the Federal Register.

Dated this 30th day of September, 1980

Signed: Walle J. H. Hauta, M.D., Ph.D. Chairman
Peter W. Lampert, M.D. member
Vernon R. Young, Ph.D member

ASPARTAME PUBLIC BOARD OF INQUIRY

The board's concern was certainly not unfounded or frivolous, since some preliminary research had not ruled out a potential link between aspartame and a number of disorders. This was not a product free from questions, and surely not one that the public would have embraced so quickly had it known what aspartame is composed of and what happens chemically once it's swallowed and broken down in the body.

Essentially, aspartame is a combination of three molecular compounds: aspartic acid, phenylalanine, and methanol, also known as wood alcohol, which is used to bind the first two. When ingested and then absorbed into the body, methanol tends to accumulate in the liver and the nervous system and is converted to highly toxic formaldehyde (a class A carcinogen and the preservative once used as embalming fluid) and formic acid (the chemical used commercially in ant poison and paint stripper). A by-product of aspartame metabolism is diketopiperazine (DKP), which some studies have shown to cause brain tumors (Figure 6.1). Methanol itself is highly toxic and in larger doses destroys nerve tissue. In pregnant women, it crosses the placenta and blood-brain barrier and may affect the developing brain and nervous system.

Certainly, many foods that are generally considered safe for human consumption contain trace amounts of chemicals that, if given in large amounts, could be dangerous or even lethal. The question with any substance is: How much is too much? Some experts, who note that many of these same breakdown products can be found in natural foods, don't believe there is a danger to humans. In fact, many leading researchers and public health organizations think that aspartame is harmless.

According to both the unfiled draft NSDA report and the Environmental Protection Agency (EPA) report during the approval process, one of the major concerns was the possible release of free methanol from aspartame when it is heated to above 86 degrees Fahrenheit (30 degrees Celsius). The theory was put forth that cans of diet soda and other products stored in hot warehouses could liberate the methanol and make it that much easier for it to break down further into formaldehyde and formic acid. One of the theories now swirling around Gulf War illness is that Desert Storm troops drank diet sodas that had been sitting in the hot Arabian sun for as long as eight weeks. One Danish study allegedly showed a significant increase in the health problems reported by female Gulf War veterans following a significant increase in diet soda consumption. Despite this one study, the origins of Gulf War illness remain a mystery, and it cannot be said that any direct link has been established between the ailment and aspartame or any other hypothesized cause.

To illustrate how significant the EPA thought the issues were and what they thought of the FDA approval process, here is a direct quote from the EPA's own preapproval report on aspartame:

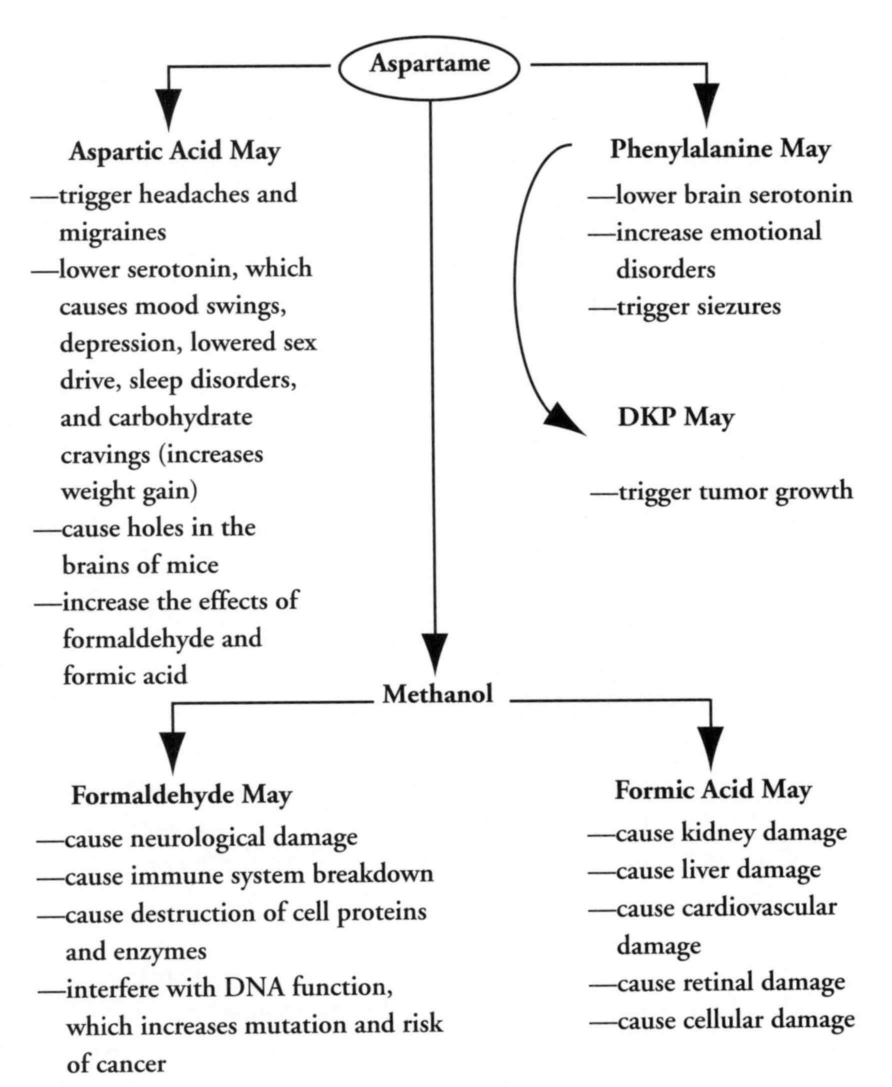

Figure 6.1. Aspartame's chemical pathway.

Certainly, many foods that are generally considered safe for human consumption contain trace amounts of chemicals that, if given in large amounts, could be dangerous or even lethal. The question with any substance is: How much is too much? Some experts, who note that many of these same breakdown products can be found in natural foods, don't believe there is a danger to humans. In fact, many leading researchers and public health organizations think that aspartame is harmless.

> Through our efforts, we have uncovered serious deficiencies in Searle's operations and practices which undermine the basis for reliance on Searle's integrity in conducting high quality animal research to accurately determine or characterize the toxic potential of its products.
>
> Searle has not met the above criteria on a number of occasions and in a number of ways. We have noted that Searle has not submitted *all* of the facts of experiments to FDA, retaining unto itself the unpermitted option of filtering, interpreting, and not submitting information which we would consider material to the safety evaluation of the product. Some of our findings suggest an attitude of disregard for FDA's mission of protection of the public health by selectively reporting the results of studies in a manner which allays the concerns or questions of an FDA reviewer. Finally, we have found instances of irrelevant or unproductive animal research where experiments have been poorly conceived, carelessly executed, or inaccurately analyzed or reported. While a single discrepancy, error, or inconsistency in any given study may not be significant in and of itself, the cumulative findings of problems within and across the studies we investigated reveal a pattern of conduct which compromises the scientific integrity of the studies. We have attempted to analyze and characterize the problems and to determine why they are so pervasive in the studies we investigated.

Despite all of these concerns, on January 21, 1981 Searle reapplied to the FDA for approval. That date happened to be the day after President Ronald Reagan was inaugurated and Donald Rumsfeld became part of the Reagan transition team. According to U.S. Senate records, Patty Wood-Allott, a former Searle salesperson, told senators that Rumsfeld said, "If necessary he would call in all his markers and that no matter what, he would see to it that aspartame would be approved that year." Also, the NSDA suddenly reversed itself and was now behind the additive 100 percent.

On January 25, 1981, less than a week after the inauguration, Dr. Jere Goyan, the FDA commissioner appointed by Jimmy Carter, was unexpectedly suspended and a new commissioner, Dr. Arthur Hayes, a Department of Defense (DOD) contract researcher, appointed in his place. Not more than six months later, in one of his first major decisions as commissioner, Hayes approved aspartame for use in dry foods on July 18, 1981. In November 1983, the same month that Hayes left the FDA to join Searle's outside public relations firm as senior medical advisor, the FDA further approved the use of aspartame in soft drinks. It was one of the most contested and controversial approval processes in FDA history. Today, the chemical is distributed worldwide in more than one hundred countries and can be found in more than nine thousand consumer products.

Since that approval, aspartame has been blamed, correctly or incorrectly, for countless health problems. A 1988 epidemiology survey appearing in *Journal of Applied Nutrition* reported that 551 people claimed to have suffered acute and chronic toxicity effects from aspartame. Even before the survey, study after study had been submitted to the FDA with little or no action, even though scientists had documented more than seven thousand alleged toxicity reactions. Considering that only a fraction of the reactions from any substance typically get reported to the FDA, the actual number of claims may be much higher. In fairness, many distinguished experts and organizations have refuted critics' claims that these reactions are caused by aspartame, and they contend that the evidence shows that the substance is safe. Long-term studies are currently underway that may provide the last word.

There is a divergence when one compares studies done by nonindustry scientists with those working for the industry. According to Dr. Ralph Walton, chairman of the Center for Behavioral Medicine and professor of clinical psychiatry at Northeastern Ohio University College of Medicine, 92 percent of nonindustry-sponsored studies found problems with aspartame in comparison with 0 percent of industry-sponsored studies!

To his credit, Senator Howard Metzenbaum called for hearings to investigate but was repeatedly blocked from trying to uncover what he thought had been a serious FDA mistake. A letter from Senator Metzenbaum to Senator Orrin Hatch of Utah illustrates Metzenbaum's frustration at what seemed like an attempt to squelch the investigation (Appendix XIV). In his letter, Metzenbaum raised health concerns of nine different scientists along with new and significant data on aspartame's effects on brain chemistry. "There have been many reports of seizures, headaches, mood alterations, etc., associated with NutraSweet," he wrote. "Dr. Coulombe's research, as well as the other research cited in my report, raises new health concerns which have not been resolved. We need to hold hearings on aspartame—which is being used by over one hundred million Americans. With an issue that is critical to the health of half the American population, how can you in good conscience say 'no'?" he added.

The showdown came to a head on October 30, 1987, when the senior science advisor from the EPA's pesticide division wrote Senator Metzenbaum a letter detailing the concerns relating to aspartame. He followed up with another letter urging the FDA to reconsider its approval based on alleged flaws in Searle's research and flagrant errors in the FDA's approval process (Appendix XV).

The following week, a Senate hearing was finally underway. In his opening statement, Senator Metzenbaum noted that the FDA had received close to four thousand complaints, ranging from alleged seizures to mood alterations, and that some medical journals had warned of neurological and behavioral effects. It didn't take long for the stunning revelations to take the committee members by surprise. Jim Turner, an attorney for the Nutrition Institute, was one of the first to testify, claiming that the FDA began referring individuals complaining about NutraSweet symptoms to the AIDS hotline, where their complaints would be noted, collected, and filed away. The CDC, according to Turner, also reported getting complaints from six hundred and fifty people whose symptoms purportedly stopped when they ceased using products with aspartame and then suddenly returned when they either accidentally or purposely used it again.

So what about the studies that had shown no adverse effects from aspartame? Were they reliable? And how much can the public really trust studies conducted by company researchers or scientists receiving funds from the very companies that would be affected by the research results? To answer that, let's examine two of the more controversial health cases: the effects, if any, of aspartame on Parkinson's disease patients and the effects, if any, of aspartame on epileptic seizures. First the Parkinson's disease case:

Following FDA approval there were contentions from some consumer organizations and physicians about aspartame's use worsening the symptoms of Parkinson's disease, yet an industry-sponsored study widely used to promote aspartame showed no adverse effects. Critics questioned the study on three grounds. First, in the industry study the subjects received aspartame for only one day rather than a more realistic, real-world time frame, such as weeks or months. Second, aspartame was administered in capsules, which, according to the critics, could decrease absorption into the body. Third, the dose given was 60–80 percent less than the FDA deemed an acceptable daily intake. In other words, the conclusions drawn from a defining aspartame study were based on a single dose given on a single day containing so little aspartame that no positive results could possibly have been achieved.

There has also been ongoing research into the relationship, if any, between aspartame and seizures. According to the Department of Health and Human Services, almost 10 percent of the claims of aspartame toxicity reactions sent to the FDA reportedly involve seizures and convulsions. Even the U.S. Navy's magazine, *Navy Physiology*, in a 1992 article titled *Aspartame Alert*, warned that aspartame might make pilots more susceptible to seizures and vertigo. Furthermore, in one double-blind study of aspartame in children, the researchers reported that a single dose of forty milligrams per kilogram of body weight increased the amount of seizure time by 40 percent. The researchers hypothesized that the intensified seizures could be due to methanol produced as a result of aspartame breakdown.

Contrasting nonindustry results, several industry studies have concluded that aspartame does not cause seizures. Critics have said that in some of these studies: (1) Ninety percent of the test subjects in certain industry studies were taking antiseizure medication at the time of the experiments compared to none in the independent studies; (2) as in the Parkinson's study, aspartame was given in capsules, which may decrease its absorption; (3) a single dose was given on which to base the results; (4) aspartame was given with meals, which could slow down absorption of the breakdown products; and (5) in the animal trials, rodents were given dosages sixty times less than humans would have received. The critics contend that since humans are much more sensitive to methanol toxicity than all other mammals, the researchers should have taken further steps to adjust for differences between rodent and human metabolism.

Dr. Jacqueline Verrett, a toxicologist and former member of an FDA investigative team from the Bureau of Foods, claimed that she was told in no uncertain terms, after questioning some of Searle's DKP studies, not to be concerned with or to comment on the validity of the studies. This, she was told, would be carried out at a "higher" level.

Later, at a 1987 Senate hearing, Dr. Verrett testified that the subsequent review "discarded or ignored the problems and the deficiencies outlined in the team report and concluded that, even in toto, all of these problems were insufficient to render the study invalid." As a scientist who saw first-hand the kind of research that the FDA had approved, she said, "It is unthinkable that any reputable toxicologist giving a completely objective evaluation of this data resulting from such a study could conclude anything other than that the study was uninterpretable and worthless and should be repeated. This is especially important for an additive such as aspartame, which is intended for and is now being used in such a widespread and uncontrolled fashion." Dr. Verrett concluded by saying that the entire aspartame DKP experiment she reviewed should have been discarded.

What Searle presumably feared was a repeption of the cyclamate experience. The pot of gold, if the health concerns proved true, would vanish in a heartbeat of public outcry. But nearly thirty years after aspartame's initial approval, questions still swirl around the FDA's approval process. During those three decades, aspartame has received the support of many responsible scientific and public health organizations who continue to regard it as safe. Research is underway that may finally resolve any lingering controversies attributable to the FDA's handling of the approval process. For example, a three-year study at King's College is currently reexamining the safety of aspartame and whether it has any effect on different cell types.

In the case of aspartame and the initial FDA approval process, there seemed to be a disconnect between the independent process of evaluating products to ensure public safety and the FDA's desire to get a product to market. No doubt one of the main reasons that so much pressure is brought to bear for approval of products once they get past stage II FDA trials is the enormous financial loss that could result in both company revenue and its stock price if a substance does not reach the market. In 1996, however, FDA Commissioner David Kessler signed a blanket FDA approval for use of aspartame in virtually all foods and beverages. Today, NutraSweet Co. sells more than one billion dollars worth of aspartame annually.

$100 Million and Counting: How Drugs Get To Market

Started in 1862 as a single chemist in the U.S. Department of Agriculture, the FDA today is part of the Department of Health and Human Services, with a staff of more than nine thousand and a budget of 1.3 billion dollars. In 1906, with the passage of the Federal Food and Drugs Act, the FDA became responsible for evaluating the safety and effectiveness of both domestic and imported foods and drugs. The system's checks and balances, independent investigators, and rules and regulations are there to make sure that nothing slips through the cracks. In most cases, the system works just fine, but when it does not, the consequences can be deadly.

In order to protect consumers and ensure that marketed drugs are safe and effective, the FDA has in place a series of preclinical and clinical trials that weigh risks, test optimal dosage, establish proper duration of treatment, and determine what group of patients would most benefit from treatment. In most cases it takes at least twelve years for a drug to make its way from preclinical research to FDA approval. However, a fast-track system is in place to speed up trials and FDA approval for drugs deemed critical in saving many lives (HIV drugs, for example).

The expense of bringing a new drug to market is very high. Therefore, it's important that ineffective or dangerous drugs be weeded out early. By the time a drug gets the go-ahead to enter Phase III clinical trials, there's a reasonably good assumption that it has proven itself to be safe and effective. However, we know that that's not always the case. About 45 percent of drugs entering Phase II trials go on to Phase III, and 85 percent of those entering Phase III actually complete the last stage. In the end, only about 40 percent of drugs beginning Phase II complete Phase III, but of those that finish, there's more than a 70 percent chance of FDA approval.

With so much time, effort, and money invested in bringing a drug to market, it's no wonder that sometimes products trump people. Add to that the pressure brought to bear on scientists who depend on industry funding for their livelihood and you have the potential for misconduct. Lotronex may be one example of a drug that was brought to the market too early.

Approved by the FDA in February 2000, GlaxoSmithKline's bowel drug, Lotronex, had to be withdrawn nine months later after five patients died. According to Richard Horton, editor of *The Lancet*, the FDA knew about studies done during the preapproval process showing serious side effects. However, even though safety concerns were raised, there were no further discussions and no call for more studies, even after an independent review found serious flaws and recommended that more experiments be done. Based on his investigations, Horton alleges, "The FDA is not only compromised because it receives so much funding from industry, but because it comes under incredible congressional pressure to be favorable to industry. This has led to deaths." He adds, "It is an impossible conflict for safety issues to be overseen by a center that receives funding from industry to review and approve new drugs."

Is the FDA the nation's lead protector against bad products or has it become a servant of industry? Over the last few decades, enough evidence has surfaced to raise doubts and question just how independent the approval process really is. The fact that dangerous and sometimes deadly products have been knowingly approved, despite mounds of evidence that should have raised warning flags and called for additional studies, is proof enough that the FDA has made its share of mistakes, been unduly influenced, and has not always had clean hands. An agency that basically relies on the honor system, where drug manufacturers do all their own testing and then present their data to the FDA for evaluation, the FDA has shown itself prone to political pressure and has been portrayed as a vital cog in the wheel of a very complex government–industry connection. In addition, since the FDA cannot legally serve scientists with subpoenas, it has no real power to investigate fraud.

But if the FDA's American industry connections are not enough to worry about, something else is. From the moment the ink dried on the North American Free Trade Agreement (NAFTA) and the General Agreement on Tariffs and Trade (GATT) treaties, a well-coordinated and often secret plan was set into motion that, if implemented, will change the entire structure of America's health care system. With the blessings of the FDA, a shadowy international organization known as the "Codex Alimentarius Commission" is rewriting the rules, taking governing power away from U.S. agencies, and could literally regulate the health supplement industry out of existence. So, as Americans decide each day what over-the-counter medication or health products they should buy to improve their health, a behind-the-scenes international government–industrial complex is gradually taking steps to secure as much power as it can and ultimately do the deciding for them.

Codex Alimentarius, WTO, and the FDA: The New Health Nazis

Imagine walking into a neighborhood Wal-Mart, wheeling your shopping cart to the pharmacy section, and seeing a team of clerks removing all the vitamins and herbal supplements from the shelves. Since one of the reasons you drove to Wal-Mart in the first place was to buy a bottle of multivitamins, you ask the clerk with a blank expression on his face what the deal is. The young man shrugs his shoulders and tells you all he knows is that the store is no longer allowed to sell any vitamins or dietary supplements.

Sound too ridiculous to be true? Not if a secretive organization known as Codex Alimentarius has anything to do with it. For while most Americans who worry about going to work each day have never even heard of this organization, its influence and impact on American citizens will be profound. Consider this: Under both NAFTA and GATT, to which the United States is a signatory, all laws, including food and drug laws, are to be "harmonized" to international standards. In other words, the United States will be forced to regulate and restrict all dietary supplements based on the global standards of other nations. NAFTA, despite the rhetoric about free trade and increased prosperity, is very much about conforming to a world standard despite whether anyone agrees with it or not. The reason these international treaties will have such power over supplements is that both of them contain within their articles sanitary (health) and phytosanitary (health of plants) agreements.

Still not convinced? Then consider this: Right now, in Norway and Germany, the entire health food industry is being regulated by drug companies that are in charge of selling vitamins as prescription drugs for inflated prices. What we take for granted in this country is a criminal violation in these two nations, where it is illegal to sell vitamin C at dosages above two hundred milligrams (which also happens to be the new minimal daily requirement). In Canada, it is no longer legal to sell tryptophan and carnitine, which used to be available as health supplements for fourteen dollars a bottle, but which are now being sold as prescriptions for up to two hundred dollars a bottle. Norway's giant drug company, Schering-Plough, has control over echinacea, gingko, and many other herbal products, which must be approved by the government-controlled pharmacy.

In Spain, it is illegal to use the word "natural" on any label, even if the product truly is natural, or to describe the benefits of any dietary supplement. Even more shocking is that in much of Europe, selling simple herbs as foods carries the same criminal penalty as selling illegal drugs. Nations around the world have been busy harmonizing their laws to conform to these kinds of World Trade Organization (WTO) regulations or are on track to do so within the next few years. When most of the GATT signatories are on board, Americans will simply have to step in line or face the economic consequences.

Why haven't most of us heard of the Codex Alimentarius Commission, which is the force behind the new health world order? It could be that the architects of this diabolical scheme want us kept in the dark while rules are being implemented step by step until everyone is on board and conforming to a binding set of global regulations. The architects, of course, are international drug and pharmaceutical companies that would be the main beneficiaries of the new health rules. According to a little publicized FDA document, the United States is seriously considering changing its own laws to conform to global rules. Fortunately, a few U.S. congressional representatives had gotten a whiff of this madness and were concerned enough to hold hearings.

I've always found it a bit suspicious when groups need to meet in secret and then classify or try to hide their documents and meeting agendas. The Codex Alimentarius Commission, which began when the World Health Organization (WHO) and the Food and Agricultural Organization (FAO) were authorized by the United Nations to develop a universal food code, meets every two years in either Geneva or Rome to discuss, among other things, their health agenda, and to set future policies for food and nutrition standards. The majority of attendants are delegates from drug firms and officials from the nations' government agencies. In 1996, the commission met in Germany to create global trade rules for health supplements. In 1998, it was a principle advocate of the "Green Paper" (EEC6565), a report calling for the classification of any herb that affects physiological function in any way as a drug to be sold "exclusively" as a drug. Anyone caught selling herbs as foods would be charged as a drug dealer and subject to the same penalties as one who sells narcotics. Its current proposed guidelines include:

1. *Limits on vitamins, minerals, and dietary supplements.* No supplement could be sold for therapeutic or preventive use, and any dietary supplement that exceeds dosage levels set by Codex could only be sold through and by pharmaceutical companies. Even products such as garlic and peppermint would be classified as drugs.
2. *Limits on potencies and combinations.* Higher potency supplements (such as vitamin C) would be regulated by drug and pharmaceutical companies, and supplements without an RDA (Recommended Daily Allowance) would be classified as drugs and made illegal. Any new product would be banned unless tested and approved by Codex.
3. *Allowing genetically engineered foods on the market without having to be labeled as genetically engineered.* In this way, no one would be able to discriminate against companies producing genetic products.
4. *Transfer of regulatory powers from agencies such as the FDA to international agencies such as Codex Alimentarius.*

The real goal of Codex, and the reason it has become so secretive, is its harmonization agenda. The concept of foreign nations dictating to the FDA what kind of health and dietary supplements Americans are allowed to buy, based on foreign rules and regulations, seems almost shocking. Yet that's exactly what is proposed and, according to the GATT treaty, required under penalty of heavy trade sanctions. The United States has already lost seven trade disputes in this area.

According to John Hammell, founder of International Advocates for Health Freedom, "If Codex Alimentarius has its way, herbs, vitamins, minerals, homeopathic remedies, amino acids, and other natural remedies you have taken for granted most of your life will be gone. The name of the game for Codex is to shift all remedies into the prescription category so they can be controlled exclusively by the medical monopolies and its bosses, the major pharmaceutical firms."

For the majority of Americans, a silent conspiracy to regulate how they choose to improve their health is happening right under their noses. The reason, thought noble at first by individuals believing their governments were looking out for their well-being, may not be so noble after all.

Upon closer investigation, the real reason may be as simple as the cost of what's in their next bottle of vitamins.

Cro-Magnons, Heart Disease, and Vitamin C: What the FDA and CDC Are Not Telling Us

Huddled in a dank cave somewhere in northern Europe, a prehistoric family gathered around a fire and watched as one of the males, festering sores all over his body and blood running from his nose, took a few labored breaths and slumped over. Within minutes his breathing stopped. The family, who had seen this before, dragged the wasted body outside and prepared it for burial. Tomorrow, another member of the tribe would die a similar death and get a similar burial. In both cases, there had been gradual destruction of blood vessels that led to each man literally bleeding to death.

Time passed slowly for the tribe, whose existence depended on food gathering, hunting, and reproducing, and not much more. Winters came and went. One by one, generation by generation, the weak died and the strong survived, passing their genes on to their offspring until one day the strange bleedings stopped and a new disease took its place. The new disease was atherosclerosis, and now, thanks to some astounding research and studies, we may know what the two have in common. Scientists at the CDC and FDA know as well, but they've not seemed too anxious to spread the word about a possible cure for what has become the leading cause of death in the United States.

In my example of the prehistoric family, the individuals died of scurvy (caused by vitamin C deficiency) because human beings, along with only three other mammals—guinea pigs, gorillas, and fruit bats—do not naturally produce ascorbic acid, or vitamin C. Early on during evolution, scurvy was a much more sinister disease, causing a marked breakdown of collagen, the main structural protein in the body and a principal component of blood vessels, and leading to cardiovascular deterioration and intense bleeding from holes in the arteries. As more of these individuals died, the survivors who had the ability to repair their blood vessels in the absence of ascorbic acid were left to inherit the earth. Thus, our ancestors, through mutations and adaptations, have given us a molecular defense mechanism that makes up for our vitamin C shortcoming and repairs cracked blood vessels. That molecular defense is low-density lipoprotein or LDL, the lipoprotein largely responsible for heart disease throughout the world.

What exactly do scientists know that many in the medical establishment and pharmaceutical industry don't want the rest of us to know? And why did the rate of mortality from coronary heart disease take a staggering 30–40 percent downturn during the 1970s? Was it better drugs and treatment, as we've been told, or did something else happen in 1970 that triggered the sudden drop, as illustrated in the graph that follows:

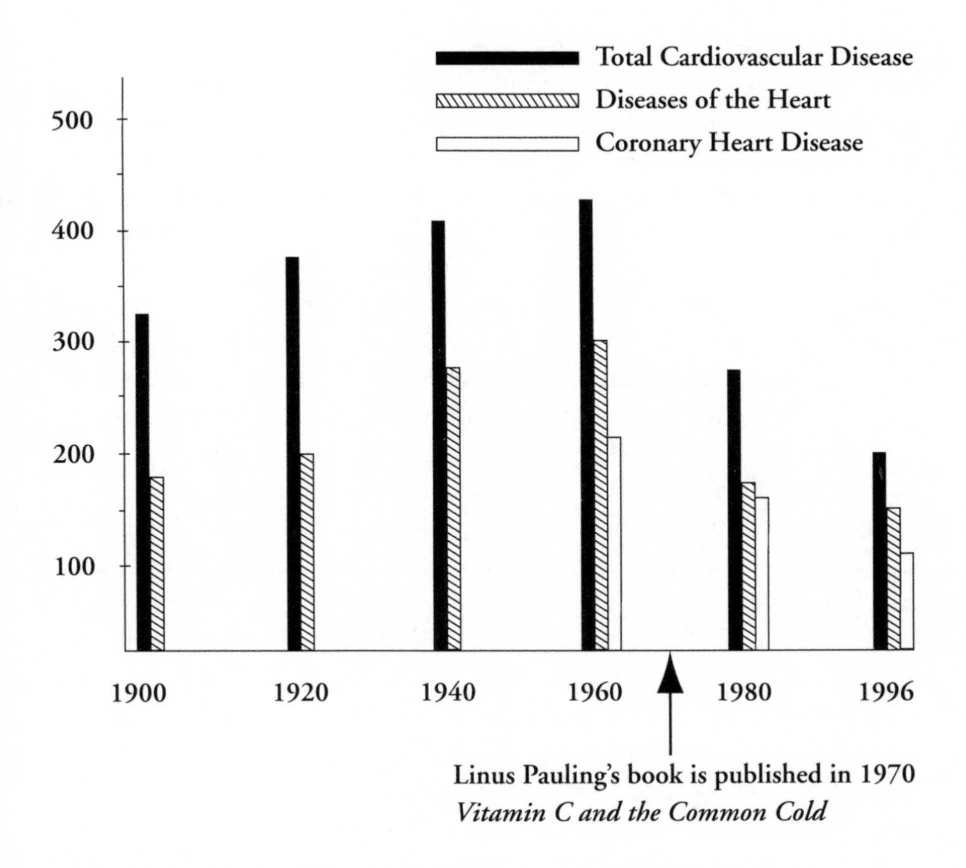

Disease per 100,000 population, standardized to the 1940 U.S. population.
SOURCE: National Heart, Lung and Blood Institute; National Institutes of Health

According to Dr. Paul Wand, a neurologist who did an in-depth analysis of published studies on vitamin C and cardiovascular disease, there is conclusive evidence that males who had taken at least five hundred milligrams of vitamin C per day had a significant decrease in heart disease, heart attacks, and stroke. The decrease shown in the previous graph happens to coincide with a 300 percent increase in vitamin C consumption following Nobel Prize winner Linus Pauling's 1970 claims about the benefits of vitamin C. This increase, as would be expected if vitamin C were a principle contributing factor, occurred only in the United States, where the vitamin C craze took hold.

However, we could go back even further and examine studies done in the 1950s to see what else is not being told. Let's look, for example, at guinea pigs, used for decades in cardiovascular research because they don't produce ascorbic acid and die a terrible death without it. When guinea pigs are deprived of vitamin C, they die of scurvy in a few weeks. When given low doses—equivalent to the recommended daily allowance—they live but develop atherosclerosis similar to that in humans. Finally, when they're fed higher doses—equivalent to the large amounts they would consume naturally—they live out their lives with no signs of atherosclerosis at all. Going back yet another decade to the 1940s, pharmaceutical companies actively promoted vitamin C because they knew how beneficial it was—until, of course, they realized the extent to which widespread use of vitamin C would actually cut into the lucrative prescription drug market.

Together with Dr. Matthias Rath, a world-famous cardiologist and pioneer of cellular medicine, Dr. Pauling proposed that the main cause of heart disease in man is his inability to manufacture vitamin C, and that this is likely the reason we are the only animal on the planet that gets heart attacks and strokes. With the small amount of vitamin C we get from our diets, we typically begin the process of atherosclerosis early in our lives, and we continue that process until we eventually develop coronary heart disease. More specifically, the most damaging LDL molecule is lipoprotein (a) [Lp(a)], a protein-covered fat-and-cholesterol package that literally sticks to our blood vessels and plugs up everything in its path. It's what seals the cracks, but it works so well that over time it becomes a sort of glue that attracts other fats and other molecules and eventually clogs up the entire blood vessel. Lp(a) is rarely found in the blood of animals that produce ascorbic acid.

But that's not all we know. What those who benefit from the hundred billion dollar heart disease industry want kept from the general public is the fact that vitamin C increases high-density lipoprotein (HDL), the other lipoprotein that gets rid of bad fats, lowers the body's production of Lp(a) and cholesterol, lowers glucose levels, and helps prevent clot formation, which can lead to heart attacks and stroke. Studies throughout the world have shown for decades that humans are the only animals that develop atherosclerosis naturally and that taking higher doses of vitamin C is a simple and economical way to actually reverse or eliminate heart disease. For example, one major five-year study published in a 1997 issue of *British Medical Journal* found that men who were deficient in vitamin C had 3.5 times more heart attacks than men who were not deficient. Another ten-year study of eleven thousand Americans found that vitamin C cut heart disease by half and prolonged life by more than six years.

This would seem like a godsend were it not for the threat it posed to a multibillion dollar economy dependent on fifty thousand dollar bypass surgeries, angioplasties, heart drugs, medical instruments, and countless other services that are totally dependent on more and more people getting heart disease. Even the least cynical among us would begin to see the rationale behind the enforceable Codex ban on vitamin C and other supplements and why some in this country would secretly be hoping that heart disease is not eradicated quite yet. Even the NIH is not immune to the charges of stifling research, with not a single major study funded to investigate the vitamin C–heart disease connection.

By classifying supplements and herbs as drugs, and by lowering potencies of vitamins to levels that are basically useless, more and more health experts are convinced that the drug industry and the medical establishment hope to accomplish two goals: (1) to gain control of the health supplement market while driving smaller companies out of business; and (2) to maintain the status quo of cardiovascular drugs and medical procedures in order to prevent the economic collapse of an industry that relies solely on people getting treated but not necessarily cured. Between 1995 and 2002, more than one hundred cardiovascular drugs were approved by the FDA, adding billions to the economy and making it that much harder to ever go back.

Is the CDC, FDA, or NIH calling for more definitive studies or demanding to know why there's such an urgency to stifle research that would answer the questions of why heart disease is an exclusively human trait and why so many individuals have been helped dramatically by vitamin C therapy? A previous history of fraud and pressure from industry suggests that we shouldn't hold our breath on this one. While the CDC does a great job in tracking epidemics and outbreaks and helping us understand diseases, it, like any other federal agency, comprises individuals who simply want to keep their government jobs and is controlled by powers that have a vested interest in maintaining the status quo.

How likely is it that any of these agencies will fight the bureaucratic, economic, and industry pressures to do the right thing? As long as jobs are on the line and fraud exists, not very. One has only to look at the numbers to see that the heart disease business is booming at a record pace, and whenever the winds of change begin to pick up and blow, they are usually snuffed out before they can gather into storms of protest. Collectively, federal agencies will ultimately be accountable for their sins of omission and what they've done to violate public trust. Yet agencies are also composed of men and women who sometimes compromise their principles, sell their souls, and, in the end, are equally guilty of crimes against their fellow man. Another agency that has not taken as much heat but also has much to answer for in this regard is the NIH.

The NIH Grant Mill and Scientific Misconduct

It's said that dishonesty is like a mushroom. It grows best where the light doesn't shine. In science, whenever the light of openness is kept from illuminating research and exposing the darkness of scientific fraud, anything can happen. Sadly, my initial taste of that darkness came first hand while a biomedical researcher at a major medical university almost twenty years ago, and my experience has taught me a lesson in human nature that I will never forget.

Having just graduated magna cum laude with a Ph.D. in physiology from Utah State University, I was more than ready to make my mark and establish myself as a biomedical researcher in the field of neuroendocrinology. I'd done everything right as a graduate student, so it was satisfying when a renowned scientist who'd just received a five-year, multimillion dollar grant from the NIH offered me a research position in his laboratory. Life, I thought, could not get any better. I had just taken my first career step, and my wife and children were as excited about our new adventure as I was. But all that enthusiasm began to change soon after I'd made the two-thousand-mile trek from Logan, Utah and found myself a few years later as one of the key witnesses in a major NIH fraud investigation.

I'd arrived at the medical center in August 1983 with the promise that I would be provided with a high-quality NIH research opportunity, excellent postgraduate training, and the potential to enhance my future career aspirations. After all, "major" NIH grant funding would assure me that the research I'd be doing would be a springboard to bigger and better things. And it didn't hurt that my superior—I'll call him Dr. A—was retiring after the project and wanted to go out in a blaze of scientific glory. So it seemed a little odd to me, and I didn't think much of it at the time, that virtually every member of the department faculty smirked and shook their heads as I was being introduced to them. What's going on? I questioned, not understanding what I had gotten myself into.

A week went by. I had cursory discussions with Dr. A about the role I was to play in the project, but was taken aback when he said, "You're not here to publish papers; you're here to work." This was supposed to be the most productive time in a young scientist's career, I thought, so why am I not going to publish anything, especially when I'm involved in a major NIH project? And since "publish or perish" was the principal game plan for any scientist who wanted to get ahead, it had to have been a joke, I assumed. It wasn't, and I quickly began to learn why.

Until I'd experienced it for myself, I couldn't possibly have imagined the kinds of people that were able to slip through the cracks and get NIH grants. Those unfamiliar with science probably assume that the profession is filled with intellectuals doing honest and worthwhile research and who would never compromise their principles. After all, such individuals had become scientists in order to seek knowledge and truth. In many cases, that's absolutely true; but in some cases, the scientists doing research are bordering on insane. I happened to have had the misfortune of discovering the dark side of NIH science and, sadly, am convinced that it's more common than most of us would want to believe.

Dr. A's laboratory at the medical center was a hodgepodge of aging, obsolete, and nonfunctioning equipment and instruments, and some that actually worked. The older equipment was typically stored in a room the faculty amusingly called the "black hole" because Dr. A conducted raids of abandoned labs and collected anything he could get his hands on and squirrel away, whether it was useful or not. The most impressive-looking equipment would be placed strategically around our lab, strictly for show, so that, in Dr. A's words, "It looked like a real working lab." Whenever we had official visitors, he unlocked his black hole, brought out even more useless equipment, hooked up wires and tubes, and arranged everything as if staging a George Lucas film. One of the other researchers who'd worked with him at another university told me that Dr. A arranged the equipment there in exactly the same way as if enacting some bizarre ritual.

Once, before an NIH inspection team came and Dr. A spread out his normally hidden equipment, he became irate that things looked too neat, barking at us that "we have to have the appearance that we do work here. No one will think we do anything unless we have a lot of equipment around the lab." If it weren't so obviously pathetic, I would have thought it hilarious. Not so hilarious was the falling glassware and equipment we had to move at the risk of injury, or that teetered dangerously on the edges of lab benches just waiting to come crashing down, all for the sake of appearances.

My assignments were to take daily blood pressure readings from stressed and aging rats and to assist in various hormone experiments conducted by other researchers on the grant. The purpose of the NIH study was to determine how physical stress during the aging process affected cardiovascular and neuroendocrine parameters. The results of the experiments were to be used as a basis for further biomedical and pharmacological studies. At one point, I was in charge of as many as four hundred laboratory rats of various ages, half of them divided into controls and the other half into experimental groups that I'd stress each day on electric grids that randomly turned on and shocked the animals, thus triggering release of stress hormones. The first sign that everything I was doing would be worthless and that the man heading up this major NIH study was perpetrating fraud was the obsessive secrecy surrounding our lab and what was going on inside.

While periodically working on other projects with other researchers—whenever Dr. A would allow such a thing—I was at least able to do some honest work, get valuable experience, and salvage what I could of my career. Back in Dr. A's lab, however, it was a different story. As NIH money poured in, it was used to fund six individual projects. The problem was that each project depended on Dr. A's ability to maintain the viability of the research animal population and the integrity of the data collection and analysis. If these were in any way compromised, NIH might as well have been giving millions of taxpayer dollars to Tony Soprano. When I think about it, Tony would probably have used the money more efficiently.

The reason all that government cash kept flowing, and the reason we will continue to have the problems with research fraud we have, is twofold: (1) There was, and still is, a close network of scientists who know each other well and who depend on one another for favors and reviews and, therefore, are hesitant to make waves, especially if the researcher seeking funds is well-known and can affect the other's career. I have attended many scientific meetings where 10 percent of the time is spent at the meeting and 90 percent is spent socializing and lobbying. One of the researchers on our NIH grant actually flew to Washington just to talk with NIH officials about funding. His trips were always fruitful. (2) If you know what you're doing, and these people are experts, it's relatively simple to hide or manipulate data in order to get the results you want. Unless someone is standing over you every time you set foot inside your lab, you can do whatever you want and get away with it.

Anyone who thinks that manipulating data is not easy would be surprised to know that it's as simple as deciding to throw out an undesirable value or eliminating an animal or two from the data pool in order to get a better statistical outcome (remember the aspartame studies). I've witnessed such activity on several occasions. For example, whenever data were collected, Dr. A eliminated any values that did not fit the normal pattern he would have predicted. If values were "supposed" to be low, he eliminated high values saying, "These are obviously not correct." If a sample had a value of zero, perhaps because of an equipment malfunction, rather than finding out what went wrong with the assay he arbitrarily assigned a low value to the data point that should have measured low or, if he thought the data point should have measured high, eliminated the sample altogether so as to keep the results in line with his expectations.

In another instance, we'd received 120 diseased rats that were to be used for critical experiments but that needed to be healthy. After getting a healthy replacement shipment a few weeks later, Dr. A, instead of destroying the diseased rats as he should have, included them as an experimental group in order to include the data with previous data he'd recorded from healthy rats. He collected data from these animals while refusing to administer antibiotics because "making them healthy," he said, "would interfere with the experiment." Even the normal shipments were subject to manipulation. When animals arrived, any that looked a little sick were selected as part of the experimental group so that they would more easily be affected by stress and would not affect the data collected from the control group. Nothing was arbitrary or random when it came to choosing how rats would be divided into groups. From the moment animals came in until the day they were sacrificed and dissected, the process could be controlled in various and deceptive ways.

During the experiments, I was told not to talk about the animals to any of the other investigators involved in the study, not to report on their health status, and not to say anything that could be of use to any of the researchers. Little wonder. The experimental procedures were a horror and would not have passed muster with a high school science teacher. Surgical instruments were rarely sterilized, and every time I attempted to perform a technique someone else had worked out and used successfully I encountered resistance because money had to be diverted for "other" things. On one occasion, Dr. A removed a piece of equipment thrown into the trash by another investigator and gave it to me to use. We lost an entire set of animals because the equipment failed during critical experiments. He used the data anyway.

At the same time we were being denied what we needed to ensure that experiments worked properly, more than fifty thousand dollars worth of equipment arrived at our lab and was never used. Instead, it was positioned next to the obsolete pieces of equipment for photo purposes and for public relations. This kind of abuse occurred annually, since NIH money that was not spent had to be returned before the end of the fiscal year. The scramble to spend every last penny before the deadline was something to behold, with researchers desperate to hold on to the public's money as if it were their own.

During the last year of my association with Dr. A, we began a study to see how stress during aging affected cholesterol levels. This, I believed, was going to be my breakout study, an extension of the Ph.D. work I had done in Utah. When I mentioned that I was going to discuss the project with one of the world's leading experts on cholesterol, who just happened to be working at our medical center four floors above us, I was told not to associate with him and that if I ever did, I would be immediately removed from the study. So much for the exchange of ideas and cooperation among scientists.

Unfortunately, nothing changed. In fact, things had gotten so bad that I was now submitting reports to the university about the fraud I was witnessing. We were well into the third year of a major NIH study and virtually everything everyone had done up to that point was at risk of being tossed into the ash heap of research history because of one man. How could this possibly happen? I kept asking myself. How could an agency responsible for dispensing so much of the public's money allow individuals to have the power to literally make up data in order to keep getting more money? Yet as I reflect on the politics of the NIH review process, the network of scientists who take care of each other, the desperation scientists feel for job security, and the ease with which one can manipulate data and get away with it, I'm convinced that fraud is more rampant than we might suspect. If it can happen at a major medical center, it can happen anywhere, especially when the only game in town is grants, publications, and then even more and bigger grants, awarded only to individuals with publications in major journals.

There is an additional factor contributing to the growing epidemic of scientific fraud, one that's usually not addressed because it's politically incorrect to do so. That is the influx of foreign and visiting scientists whose principal goal of research is not to discover but to publish papers. In some cases, these individuals are in charge of labs and research projects, whereas the head scientists are simply administrators who depend on their underlings for results. Dr. Ping Wren, a visiting scientist from mainland China with whom I'd worked for a year, told me that if he didn't publish at least one paper during the time he was with us he would kill himself because he would be too ashamed to go back home to his wife and children. Pressure like that from nations that send their scientists to this country with the expectation of publishing in peer-reviewed journals will make someone do just about anything to get the kinds of results "needed." Dr. Wren, to his credit, did not consider altering his data and, after I'd convinced Dr. Wren to search for another research lab, he found a colleague to work with in Washington State.

During my four years at the medical university, I'd witnessed an almost pathological urgency by foreign nationals to return home with results and publications. It was simply expected, no questions asked. The problems arose when research didn't go as expected and the urgency turned into a desperate attempt to salvage anything from the research that could be used to show productivity. From my experience, it seemed as if everyone—both Americans and foreigners—was on a fast track to get ahead, and everything one did was focused like a laser beam on getting positive results. In the world of scientific research, results are the prize. Without good results, there are no data; without data, there are no publications in peer-reviewed journals; without publications, the odds of funding begin to diminish until one joins the ranks of those who can't seem to land that next grant that will ensure their employment. From 1983 to 1987, I watched five of my colleagues lose their university positions because they could no longer get grants. It happens every year at universities around the country; and for those desperate enough to make sure it doesn't happen to them and their families, there's always fraud.

Since my testimony with NIH, Dr. A has retired in disgrace, nearly half the original faculty are gone, the five-year NIH study was withdrawn and proved to be a multimillion dollar waste of taxpayer money, and lives and careers have been shattered. None of this should have happened if NIH had served the public as it was obligated to do. Fortunately, *our* research did not affect the health or the lives of patients. The real fear is that what happened at our research center is not unique, and that as long as human beings are dishonest and the NIH is oblivious to fraud, the problem will only deepen. An example of one of the worst cases in the past few decades linking the NIH to fraudulent research involved the drug AZT.

NIH-Sponsored AIDS Trials: How Politics and Controversial Science Helped AZT to Market

As a growing hysteria gripped the nation over a new virus that baffled every leading expert in the world, drug companies scrambled to develop anything that would kill it. Nothing seemed to work against what virologists labeled the most complex and fastest mutating organism ever seen. Desperate for *something*, companies turned to their reject shelves and began hunting through inventories of older drugs. One of those companies, Burroughs Wellcome, hit the jackpot. Stacked alongside several hundred other chemicals was an abandoned chemotherapy drug from the 1950s known as azidothymidine (AZT). Classified as a DNA chain terminator, its mode of action was to enter cells and interfere with DNA replication. In short, its mission was to literally seek out and indiscriminately destroy cells. Against the human immunodeficiency virus (HIV), it seemed to work when nothing else would.

The main problem was that some scientists who'd worked with AZT claimed that it was too toxic at any dose to be effective. In fact, the reason it was shelved in the first place was because it was so toxic and so nonselective against the cells it attacked that it would have been useless as a cancer treatment. Earlier research had already shown that it could cause muscle loss; extreme anemia; white blood cell depression; lip, mouth, and tongue sores; bone marrow destruction; loss of speech; abnormal bleeding; cancer; and lactic acidosis, which damages liver cells. In other words, the drug was so toxic that it could kill people. The other problem was that we had an epidemic needing immediate action and a public outcry for the FDA to move quickly. That combination set the stage for what some experts say may have been an NIH-FDA mistake.

Despite warning signs, the FDA approved the drug in a record-breaking nineteen months, and Burroughs Wellcome began a massive campaign that involved not only the U.S. Public Health Service but also the CDC, which was at the forefront of an effort to convince the world that AZT was the breakthrough drug that could help stop HIV in its tracks. What the public did not fully appreciate, however, was the fact that some preliminary tests, as well as follow-up studies, showed that the benefits of AZT faded shortly after treatment began and in some cases may have accelerated the onset of AIDS or made patients sicker.

The first suspicions that something was amiss were raised on February 14, 1991. On that day, Dr. John Hamilton of the Veterans Administration (VA) presented before an Antiviral Drug Advisory Committee of the Food and Drug Administration results of an extensive AZT therapy study involving 338 HIV-positive individuals. By the time Dr. Hamilton finished his report, which raised concerns about AZT as a therapy, panic shot through both the gay community, which had been demanding a radical treatment, and Wall Street, which saw a 10 percent drop in the company's stock price the very next trading day.

To stop the bleeding, the FDA had to do something and do it fast. John Lauritsen, author of *Poison by Prescription: The AZT Story*, was in the conference room when the VA presented its findings and later when the FDA allegedly rewrote those findings in an effort to reverse the damage. Compare the two versions below:

VETERANS ADMINISTRATION AZT THERAPY STUDY 298 CONCLUSIONS, 2-14-1991	FDA SUMMARY OF AZT THERAPY STUDY 298 CONCLUSIONS, 2-14-1991
1. Early AZT treatment delayed the progression to AIDS as compared to later treatment, but no benefit for either treatment arm (early vs. late) was detected for survival or the clinical endpoints of AIDS and death.	1. Early AZT treatment was beneficial for delaying the onset of AIDS. Effects might vary among different patient groups.
2. Early AZT treatment resulted in transitory benefits in whites and neutral or harmful effects in black and Hispanic patients. Minorities treated early had a significantly higher death rate (14%) than those treated later (2%).	2. Among African-American and Hispanic patients, those who received AZT at a later stage of infection may have fared better than those who received earlier treatment.
3. Further studies are mandated in minority populations.	3. Results regarding African-American and Hispanic patients were not conclusive, and no definitive changes in practice were deemed appropriate by the committee.

The VA study was crucial in suggesting that AZT treatment had little effect on survival whether it was given early in the disease or later, and that the T4 cell count initially went up and then plummeted. In fact, the study showed that overall 6 percent of patients in the early treatment group died before progressing to AIDS, which indicated that they may have died from side effects of AZT. However, the pressure mounted and the FDA buckled. Dr. Ellen Cooper, an FDA medical investigator, believes that approval was justified, but has observed that science was only one factor in reaching the decision. In her own words, "We're all under tremendous pressure and there's no question that politics is a much greater part of AIDS drug development than approval and availability of drugs in less publicly-visible diseases."

The 1986 NIH Phase II trials that formed a basis for the FDA's final approval of AZT in 1987 were the famous double-blind, placebo-controlled trials conducted at twelve medical centers throughout the United States. When one of New York's most reputable and renowned AIDS doctors, Dr. Joseph Sonnabend, was asked his opinion on the AZT approval, he said, "I'm ashamed of my colleagues. I'm embarrassed. This is such shoddy science it's hard to believe nobody is protesting. Damned cowards. The name of the game is protect your grants, don't open your mouth. It's all about money . . . it's grounds for just following the party line and not being critical when there are obviously financial and political forces that are driving this." Dr. Sonnabend, whose patients have included many long-term survivors, has never prescribed AZT. Criticism like his was based on the following summary of events that led to FDA approval and illustrates why some researchers have since called aspects of the AZT study incredibly sloppy, poorly controlled, and of dubious validity.

1. Twelve medical centers were selected for a twenty-four-week study to test the effects of AZT treatment versus placebo. Three hundred HIV-infected patients volunteered for the "double-blind" study, which meant that the drug would be labeled but neither the patients receiving the drug nor the doctors dispensing it would know whether AZT or a placebo was being used.
2. Several weeks into the study, patients begin breaking open the capsules and tasting the contents to determine which group they were in. In some cases they actually had the contents chemically analyzed. Patients who were supposed to be receiving placebos became desperate and began getting AZT from Mexico or from other patients who agreed, for humanitarian reasons, to share some of their medication. This "unblinding" by the patients, as well as by the doctors who learned from blood tests who was in which group, should have raised questions about the study immediately.

3. Seventeen weeks into the seriously flawed study, nineteen patients in the so-called placebo group had died compared with only one patient in the AZT group. Because of this unusually high death rate, the study was immediately terminated and all patients were then offered AZT, even though many in the placebo group had been surreptitiously taking AZT. Since the death rate was so unusually high—much higher than would ever have been expected—statisticians suspected that the mortality data was inaccurate. And since the study was stopped short, it provided little proof whether AZT prolongs life.
4. Analysis of the experiments uncovered sloppy record keeping. Scientific protocols had sometimes not been followed, making statistical analysis less reliable. For example, FDA inspector Patricia Spitzig found that Patient No. 1009 being treated with AZT at the Massachusetts General Hospital Clinical Center in Boston and obviously suffering from AZT toxicity was improperly entered as a placebo patient. When he died two months later, he was counted as a death in the placebo group. No one knows how many other patients were misclassified in this way.
5. Inspection of the case report forms showed false entries designed to cover up the amount of time patients were actually in the study and to purposely misrepresent survival rates. For example, patients who joined the study after it began and were treated with AZT for only three weeks before the study was terminated had their length of time in the study artificially extended by a statistical projection technique. This made it appear as if the AZT-treated patients survived the entire experiment when in reality every one of the short-term study patients might have died within the next few weeks.
6. Records show that severe reactions were not reported on case report forms as "adverse reactions to AZT." In some cases, multiple transfusions and emergency hospitalizations were not recorded.

The FDA, say some experts who have looked at the trials, chose to ignore multiple irregularities and voted to approve AZT in March 1987. For the next several years, more than one hundred promising drugs received less attention while studies continued to show that AZT did not work. Reports from Britain and Canada concluded that HIV sometimes became even more virulent following AZT treatment. The final report on the three-year Anglo-French Concorde Trial, published in the April 9,

1994 issue of *The Lancet*, showed that of 1,749 HIV-infected patients at thirty-eight health centers in the U.K., Ireland, and France, there was no difference in the progression of AIDS between those taking AZT and placebos, that 18 percent of all patients developed full-blown AIDS or had died within three years, and that within a few months AZT was shown to be completely ineffective and left patients with fewer T4 cells than they had started with. Since the study was conducted by the British Medical Research Council, one of the most reputable medical organizations in the world, and its French counterpart, the results should have been definitive proof that AZT was not as useful as initial studies had caused researchers to believe.

But what really should have alerted FDA officials was that shortly after the main U.S. trials were terminated in 1986, the death rate accelerated in some categories of patients who were subsequently put on AZT therapy. Despite that, the patient population was expanded to include individuals who tested positive for HIV but did not have AIDS. This meant that more people would be paying the eight thousand dollars per year for a much longer period. Documents have since surfaced showing that the U.S. government and the manufacturer of AZT knew as early as 1964 that AZT sometimes triggered serious side effects. According to certain critics of AZT, some unknown number of people may have died sooner as a result of taking it. All one has to do is look at the label on a bottle of AZT from Sigma Chemical Company to see that the manufacturer is careful to disclose its extreme toxicity.

The skull and crossbones, designation for a deadly poison, sits aside an ominous description: "Toxic by inhalation, in contact with skin, and if swallowed. Target organ(s): Blood, bone marrow. If you feel unwell, seek medical advice (show label when possible). Wear suitable protective clothing." Was information about the extreme toxicity dismissed solely out of the hope of slowing the deadly epidemic? One critic, Lynn Gannett, a data manager from 1987 to 1990 on another NIH-sponsored Phase III AIDS clinical trial conducted at the Syracuse, New York clinic, broke the medical establishment's silence when she spoke out against NIH and revealed more stunning facts about AZT research.

In Gannett's own words, "AZT was never proven to be safe and effective. From the particular studies in which I was involved, it would have been impossible to prove anything. The data was such a mess! I now realize that AZT is a deadly poison. All AIDS drug trials since that time have been based on the same flawed model." Although other researchers have disagreed with her views, Gannett's outspoken condemnation about AZT was meant to warn the public about the dangers of AZT and how grossly ineffective the NIH research process was that brought it to market. She goes on to say, "The data was so inaccurate and so full of holes that I often compare it to Swiss cheese. I felt like I was trapped in the middle of an awful movie about mad scientists. If there was a rule that could be broken, they broke it!"

As part of the lone major AZT research trial, Gannett was dumbfounded at the behavior of physicians and scientists conducting the research. When she tried reporting what she'd witnessed to NIH, no one ever returned her calls or showed any concern about the accusations she claimed. As someone who'd seen the research and the results first-hand, she can't believe to this day that NIH could have allowed this to go on and that the FDA approved AZT so easily.

AIDS is, of course, a death sentence for those who go untreated, and it is understandable that researchers would want to accelerate the use of a treatment that seemed to provide benefits to some patients, even if it also carried risks. Responsible researchers and physicians believe that AZT has value and can contribute to prolonging life when used appropriately. Today, AZT is typically given in lower doses and in combination with new drugs called protease inhibitors, which have become the preferred treatment for HIV infection. Moreover, despite its potential dangers, AZT remains one of the few drugs that can cross the blood-brain barrier and attack HIV in the brain.

More so now than ever, the motives of agencies such as the FDA, NCI, CDC, and NIH are being questioned. Have their decisions been based solely on science or have economics, politics, and other factors influenced the decision-making process? Have lives been sacrificed and people made to suffer because legitimate research was quashed and treatments withheld? Have dangerous drugs and other products been brought to market simply because financial losses would be too high or the potential for financial gain too great to be ignored? It will be up to honest men and women to ensure that the public trust is not abused. In the next chapter, I'll continue the story of organized medicine and the reality that, despite what we've been told by scientists, physicians, and government agencies, not everything about medical research is what it seems.

7 ORGANIZED MEDICINE: A CENTURY OF HUMAN EXPERIMENTATION

The young couple watching wave upon wave of turquoise water lap against the pristine shoreline of Luquillo Beach never suspected that somewhere on their beautiful island of Puerto Rico more than seventy years ago, Dr. Cornelius Rhoades, funded by New York's Rockefeller Institute, had become a pioneer for what would be seven decades of human cancer experiments. As the couple make their way back to the hustle and bustle of old San Juan, they hardly notice the immense white structures and laboratories that house some of America's largest pharmaceutical companies. In some ways, the twentieth century history of Puerto Rico and its ties to the United States had set the stage not only for an influx of such corporations but for an environment of colonialism that made possible Dr. Rhoades's human experiments.

It was in 1898, under the Paris Peace Treaty and as part of the spoils of war, that Spain handed all proprietary rights to Puerto Rico over to the United States. Never mind that Puerto Rico had already been awarded its independence by Spain a year earlier, had its own currency, its own postal service, and was recognized under international law as a sovereign nation. The small independent island was, in effect, given away to its giant neighbor for nothing.

Pedro Albizu Campos, who'd witnessed the U.S. Army march through his village when he was only seven years old, never forgot that day. With the image of U.S. troops forever etched in his mind, he left his homeland and emigrated to America. After attending Harvard University, serving in a U.S. Army all-black battalion during World War I, and returning to Harvard for a law degree, Campos joined the Nationalist Party of Puerto Rico and was elected vice president. He chose as his mission to speak out against U.S. colonialism and to seek independence for his nation. Six years later, after being elected president of the Puerto Rican National Party, Campos published a shocking manuscript exposing the secret medical experiments sponsored by the Rockefeller Institute.

According to Campos, and corroborated by witnesses involved in the experiments, the Rockefeller Institute, founded in 1901 to study the science of medicine and to develop an understanding of the nature and causes of disease and methods of treatment, sponsored a cancer research project using healthy Puerto Rican citizens. During the study, unwitting human subjects were deliberately injected with cancer cells as part of a medical experiment designed to see how humans develop cancer. Dr. Rhoades himself admitted to killing at least thirteen citizens who eventually developed cancer.

Inconceivable today, Dr. Rhoades, when asked why he chose Puerto Rico to conduct the research, stated flatly, "The Puerto Ricans are the dirtiest, laziest, most degenerate and thievish race of men ever to inhabit this sphere." Despite the cancer deaths and the blatant racism, Dr. Rhoades was lauded for his research efforts and praised as a man of science. He went on to establish U.S. Army biological warfare facilities in Maryland, Utah, and Panama, and was named to the U.S. Atomic Energy Commission, where he began a series of experiments on American soldiers and civilian hospital patients.

By the 1920s, it was becoming clear that unethical medical experiments on certain groups of people, including prisoners and the mentally disabled, would not so much as raise society's eyebrows. The attitude that some individuals did not deserve to live was illustrated by no less a prominent figure than Nobel laureate Dr. Alexis Carrel of New York's Rockefeller Institute when he wrote in *Man the Unknown*, "We have already referred to the vast sums at present spent upon the maintenance of prisons and lunatic asylums in order to protect the public from anti-social and insane persons. Why do we keep all these useless and dangerous creatures alive? The ideal solution would be to eliminate all such individuals as soon as they proved dangerous."

The practice of using individuals for egregious medical experiments continued and reached the height of inhumanity with Nazi and Japanese atrocities. Most believed that following World War II human experimentation would be viewed as so abhorrent that no one would dare try it again. But as we've seen, human radiation as well as chemical, biological, and mind control experiments flourished throughout the 1940s, 1950s, and 1960s with little regard for human life. And while much of the research was military, organized medicine was enthusiastically getting into the act itself.

By 1940, physicians were eager to expand medical databases—and boost their careers in the process—even if it meant performing risky human experiments on everyone from babies to the elderly. One common medical procedure was simply to induce illness in healthy medical research subjects. In studies done from 1951 to 1952, insulin was withheld from diabetic patients for as long as two days so as to induce diabetes. Some patients became comatose. Catheters were inserted via a vein in the arm through the heart and into the hepatic artery, where blood samples were taken. In some experiments, high doses of insulin were infused to produce hypoglycemia (low blood sugar), which can lead to insulin shock, coma, and death.

Another common practice was to induce cardiovascular collapse in order to study its origins and mechanisms. Several techniques were employed. One method involved tying a tourniquet around both thighs, tilting the body, and bleeding the subject until blood pressure dropped. Another involved passing a catheter into the heart after several days of intermittent bleeding (as much as three pints of blood was removed) and measuring vessel collapse and blood pressure. At the Cleveland City Hospital in the 1950s, doctors measured changes in cerebral blood flow and vessel collapse by injecting subjects with spinal anesthesia, inserting needles into their jugular vein and brachial artery, tilting them head downward, and, after paralysis and fainting occurred, measuring the drop in blood pressure. Once the subjects came to, the studies were repeated to increase the number of data points.

Children were also used widely in medical experiments. A 1941 issue of *Archives of Pediatrics* describes how physicians transmitted Vincent's angina. Doctors would take swabs of the severe gum disease that causes ulcerations of the mouth, tongue, tonsils, and cheeks from sick children and infect healthy children in order to study how the disease is spread. In 1949, *the Lancet* reported an experiment in which eighty children as young as ten years were fed agenized flour for six months to investigate the toxic effects of agene, an ingredient used in the production of flour. A 1953 article in *Clinical Science* details an experiment in which forty-one children, aged eight to fourteen, had their abdomens deliberately blistered with cantharide to study the severity of the response to the irritant. In his own words, the author of the study describes the procedure as if he were describing an experiment with lab rats: "Blistered skin was removed with scissors, the raw area swabbed with peroxide and covered with oiled silk. Healing occurred in five to six days, leaving a small pigmented area."

That same year, premature infants at the Brooklyn Doctors Hospital were given very high doses of oxygen despite earlier studies showing that high oxygen levels caused blindness. In *Burton v. Brooklyn Doctors Hospital* (452 N.Y.S.2d875), testimony revealed that researchers continued giving infants oxygen even after observing that their eyes had swelled to dangerous levels. A few years later, the *Journal of Clinical Investigation* published a report on what today would be considered torture. In 1957, doctors at Children's Hospital in Philadelphia wanted to investigate how blood flows through children's brains. Healthy children, aged three to eleven, were selected for the experiments. The researchers, who note in their article that some of the children had to be restrained by bandaging them to a board, inserted one needle into the femoral artery of the thigh and one into the jugular vein of the neck, which brings blood down from the brain. While the children had the needles secured in their bodies, they were forced to inhale a special gas through a facemask.

Throughout the 1960s, children had become the forgotten victims of terrible human experimentation. Virtually every age group, from infants to adolescents, was used in every kind of medical research because little was known about children's physiology. New antibiotics for the management of acne, for example, were tested on children at the Laurel's Children Center in Maryland in 1962, and continued to be administered even though more than half of the children developed severe liver damage. The children typically received liver punctures to monitor their liver damage, but when liver function returned to normal they were once again given the antibiotic.

A year later, in a 1963 study published in *Pediatrics*, 113 newborn infants, aged one hour to three days, were used to measure changes in blood pressure and blood flow. The procedure was something one would expect at Dachau concentration camp, not at the University of California's Department of Pediatrics. The doctors would insert, without medication, a catheter via the umbilical artery into the infant's aorta. The infant's feet would then be immersed in ice water and the aortic pressure recorded. Another fifty infants were strapped to a circumcision board and tilted over the edge of a table so that blood would rush to their heads before blood flow and pressure were measured.

Fetuses were not exempt from medical experiments either. Countless studies had been done to test drugs and see how materials cross the placenta. In one 1967 study published in the *Journal of Clinical Investigation*, pregnant women were injected with radioactive cortisol to see if the radioactive material would cross the placenta and affect the developing fetus.

Literally every major research hospital was using children for experiments, with little or no indication of parental consent. But there was an even better source of human subjects. According to Ross Mitchell, writing in the *British Medical Journal* about early medical experiments, "Children from orphanages and foundlings were commonly used as subjects for these investigations. . . . Moreover, medicine had but recently emerged from an era in which children were little regarded, a world where foundlings were bought and sold and child labor was the rule. . . . Against such a background, the use of orphans and foundlings for experiments would hardly have seemed to require permission or justification." In some parts of the world today, children are little more than slaves or property and are often used in medical research without anyone's consent and certainly without regard to their well-being.

As recently as 2001, the U.S. Food and Drug Administration (FDA) admitted that its policy to include healthy children in human experiments may have had some unexpected consequences. On its own website, it states that the policy "has led to an increasing number of proposals for studies of safety and pharmacokenetics, including those in children who do not have the condition for which the drug is intended." With the blessings of the federal government, pharmaceutical companies can now test drugs on thousands of children who may not need the drug but who may benefit from the drug in the future. Moreover, the Department of Health and Human Services has included the term "non-medical condition" in its policy regarding the protection of children as research subjects (Policy No. 46.406) in order to broaden the criteria for recruiting healthy children in medical research studies.

Prisoners were—and still are—used as a steady source of human test subjects. In fact, some of the Nazi doctors on trial at Nuremberg cited journals that detail medical experiments done on prisoners and describe work by U.S. scientists to justify their own horrific experiments. During the early part of the twentieth century, American experiments included infecting convicted criminals with plague, inducing beri-beri, and producing pellagra, which damages the brain and causes dementia. Later, in 1944, nearly one thousand prisoners in Illinois and New Jersey were infected with malaria to test new drugs. Many became violently ill from both the disease and the experimental drugs.

Penitentiaries in Georgia, Oklahoma, Kansas, Mississippi, Ohio, and Pennsylvania became breeding grounds for human subjects who volunteered for experiments in exchange for rewards as small as a pack of cigarettes or, in some cases, a few days shaved from their sentences. Research projects included organ transplantation, medical techniques, injection of live cancer cells or blood from patients with leukemia, effects of chemicals and drugs on the human body, and exposure to radiation. The March 12, 1964 issue of *Medical News* reported that at Holmesburg Prison in Pennsylvania nine of ten prisoners were medical research subjects and that throughout the United States the number was in the thousands.

That number may be smaller today, but prisoners continue to volunteer for medical research. With not much coaxing, recruits are given a little money, a little freedom, promises of cures, and even a sense of worth in exchange for participating in human experiments. Researchers consider this an ideal population because no one really cares what happens to someone locked away for criminal offenses. A friend of mine who is a prison psychologist for the State of North Carolina told me that as many as 30–50 percent of the prison population in his district are HIV-positive and are considered dead men walking. The attitude, he said, is "if they die, it just decreases the overcrowded prison population. No big deal."

The overwhelming number of experimental drugs tested on humans today do not work and may cause harm or death. An important role of science is to determine what works and what doesn't, but all too often people being used as human subjects don't know that. They assume that the purpose of the medical research or procedure in which they're participating is to make them better; they're not interested in being used for the benefit of future generations.

Despite codes of conduct, ethical standards, and laws to ensure the presence of informed consent and understanding of an experiment, one of the biggest problems in research today, says Dr. Greg Koski, director of the Federal Office for Human Research Protections, is that the opposite is happening. "Too often, individual research participants will enter a study believing that they are being treated when in fact they need to understand that if they are participating in research, treatment may not be part of that," Koski explains. Surprisingly, based on Dr. Koski's recent surveys, as many as 90 percent of current medical research projects have a problem with informed consent and as few as 30 percent of subjects could even explain what the experiments they were involved in were about.

Thus, while we claim that human subjects are more protected against abuse than ever before, the reality is that unethical and often dangerous human medical experimentation continues to grow at an alarming rate. Dr. Rhoades may have been one of the first to use humans so callously in cancer experiments, but he certainly wasn't the last or the worst. As we'll see, the drive for fame, fortune, and career survival has only strengthened the desire by researchers to use humans rather than animals in medical research.

Guinea Pigs in the War on Cancer

As if their reputations weren't sullied enough, cancer researchers took another hit when Dr. Linus Pauling, winner of the Nobel Prize in Chemistry and the Nobel Peace Prize, said:

> The ways in which the American people have been betrayed by the cancer establishment, the medical profession, and the government are shocking. Everyone should know that the war on cancer is largely a fraud and a sham, and that the National Cancer Institute and the American Cancer Society are derelict in their duties to the people who support them.

The second half of the twentieth century saw medical progress moving at astonishing rates. Unfortunately, with that progress came record numbers of unethical experiments involving human subjects. By far, cancer researchers have committed some of the most heinous transgressions.

For example, in July 1963, after receiving funds from the U.S. Public Health Service and the American Cancer Society, researchers at the Jewish Chronic Disease Hospital sat in their offices and made their final selections regarding who would be injected with live human cancer cells. This was not some prison laboratory facility at Auschwitz; it was a New York hospital preparing to subject humans to Nazi-like experiments nearly twenty years after the defeat of Germany and three years after the election of John F. Kennedy in what was described as the dawn of a new society. During the 1965 court case against the hospital (258 N.Y.S.2d397), it must have seemed like déjà vu for Jews who'd seen it all before and sat horrified at the thought that it had happened at a Jewish hospital no less.

The testimony, at times chilling, was a reminder that such things were still very much possible. Patients with and without cancer had been chosen without their consent for a study to determine not only if foreign cancer cells would live longer in debilitated noncancer patients than in patients with cancer, but if cancer could actually be induced by injection of live cells. The subjects were not told that the intradermal injections contained live cancer cells because, as the doctors explained, they (the doctors) "did not wish to stir up any unnecessary anxieties in the patients who had phobia and ignorance against cancer." Although the doctors claimed that "oral consent" was given, hospital administrators later tried to cover up the fact that many of the subjects were physically and mentally unable to give consent or that the consent had been fraudulently obtained after the injections were already administered. To add insult to injury, the American Cancer Society elected the study's principal investigator as the vice president of their organization.

The same type of cancer injections were also given to three hundred healthy women at Memorial Sloan-Kettering Cancer Center. According to Jay Katz, author of *Experimentation with Human Beings*, gynecology patients were injected with cancer without their knowledge in order to determine how the body would respond to an invasion of live cancer cells. During testimony, several doctors admitted that they believed the injected cells might cause cancer years later but injected them anyway. The reason for performing the experiment on nonconsenting subjects was obvious. Who in their right mind would willingly consent to being injected with cancer?

Less than twenty years later, the lessons of Puerto Rico and New York had been all but forgotten, and history, as it often does, repeated itself once again. For a little more than a thousand miles to the north in Heflin, Alabama, Becky Wright had been offered a chance at a unique cancer therapy that guaranteed she'd live long enough to see her children grow old.

A housewife and mother of three, Becky Wright believed the doctors when they told her that the treatment she was about to receive at the Fred Hutchinson Cancer Research Center (FHCRC) would save her life. Though apprehensive during her flight to Seattle, Washington, Becky figured that the experts knew best. Her sister, after all, was a perfect bone marrow donor match, and the biotechnology company conducting the experimental clinical trials, Genetic Systems Corporation, was supposedly on the threshold of a cancer breakthrough.

According to the *Seattle Times*, Genetic Systems was started by David Blech, a twenty-four-year old song writer and entrepreneur who'd convinced investors that the 1980s biotech boom would make them all rich beyond their wildest dreams. The only expertise the Brooklyn native had when he began his fledgling startup was his extraordinary sales charisma and his ability to recruit physicians for what promised to be a "pioneering institution in transplanting bone marrow." And what Becky Wright traveled more than three thousand miles across the country for was to become one of organized medicine's modern day guinea pigs.

By the time Becky arrived at the research center, plans had been made to forego standard transplant treatment and instead add eight experimental proteins made by Genetic Systems to her sister's bone marrow. According to the *Seattle Times* report, the oncologists, as well as the FHCRC, had significant financial ties to Genetic Systems and had a vested interest in testing the proteins and using the cancer patients as human subjects. In fact, two of the doctors involved in the experiments owned one hundred thousand and two hundred and fifty thousand shares of company stock. After the initial treatment, researchers crossed their fingers and hoped for a miracle because they certainly hadn't had much success up until then.

There wasn't much doubt about the risks and dangers involved in the experiment known as Protocol No. 126. Even worse, during initial consultations with doctors, Becky and Pete Wright allegedly were never told that the treatment Becky would be receiving hadn't worked in other patients. Investigators for the *Seattle Times* wrote, "Transplants were being rejected at alarming rates. New cancers were appearing and old ones reappearing far more than they normally would. All were problems directly attributable to the experimental treatment." According to the newspaper's account, not a word was ever said about the dozen or so patients who had already died or about the financial links between the doctors treating Becky, the research center, and David Blech. And it wasn't until Becky

Wright eventually died, along with nineteen other human subjects, that Dr. John Pesando, a member of a committee to protect patient's rights, began publicly questioning the safety of the treatments and the ethics of the experiments.

Concerned that humans were being used as nothing more than test animals, even while better and more effective treatments were available, Dr. Pesando tried to warn federal officials about what was happening. "Many patients died at the Fred Hutchinson Cancer Research Center when the Institutional Review Board charged with protecting them was shamelessly used and abused by senior staff," he wrote. "Hutch management denied the existence of financial conflicts of interest, refused to halt the protocols, and refused to have protocols reviewed by independent outside examiners."

What disturbed Dr. Pesando most was that doctors involved in the research effectively controlled the review board set up to ensure patients' safety. Rather than living by the Hippocratic Oath they'd taken to "abstain from every voluntary act of mischief and corruption," some physicians allegedly failed to make full disclosure to reviewers and board members of their financial interest in the research.

In a letter sent by Dr. Pesando to the National Institutes of Health (NIH), Pesando reported that senior clinicians conducted clinical trials with high therapy-induced mortality rates while they were major stockholders in the company with commercialization rights to those therapies. "It soon became obvious that at least one FHCRC clinical study (Protocol No. 126) involving MAb was causing very high mortality rates in patients who otherwise stood a good chance of cure by bone marrow transplantation," he wrote. "Final numbers are unavailable, but it is safe to say that protocol 126 was directly responsible for at least two dozen patient deaths. . . . A great many people, then as now, simply will not make the personal sacrifices currently required to even attempt to correct abuses of this kind, even when human lives are at stake. Thus similar abuses are all but guaranteed to continue."

It was obvious from Dr. Pesando's letter that he believed Protocol No. 126 was continued in the hope of financial gain and despite the mounting death toll caused by the experiments, a charge that the physicians involved have denied. More than two years went by without a response from NIH. A second letter from Pesando, this one to Donna Shalala, secretary of health and human services under President Bill Clinton, went even further. "More than twenty patients were killed at the FHCRC by their NIH-sponsored physicians in pursuit of profit, yet there could hardly be less concern if laboratory rats had died instead. At the FHCRC, economic and professional self-interests were clearly best served by silence or complicity, but the NIH's silence appears to arise either because it is unable to accept the fact that leading medical researchers are capable of such behavior or because it is unwilling to face the consequences of accepting the truth."

What really shocked Pete Wright after his wife's death was learning that even in the 1980s and 1990s, doctors, in the name of science or simply for financial gain, were still effectively using humans as subjects of medical research even when they allegedly knew that alternative established therapies were available. The case of Genetic Systems is likely just the tip of the iceberg because with hundreds of millions invested in pharmaceutical research and billions riding on new biotech products, the pressure to test drugs and get them to market is simply too great.

As we've already seen, by the time human subjects are recruited for experimental treatments, the assumption is that the drug is safe. The problem is that with so many new companies being formed, and so many researchers and doctors having a financial stake in the outcome of new products, the review process—often done by individuals with financial interests as well—becomes tainted. Furthermore, each step in the review and approval process can be manipulated because it's often based on the honor system. Reviewers simply believe the data they're presented and accept what the researchers tell them.

As a researcher at a major teaching hospital, I'd seen first hand the cozy relationships between scientists doing research and their colleagues who were charged with reviewing their work. The more specialized the field, the greater the odds that reviewers and researchers know each other both professionally and socially. They see each other at meetings, they have dinner and drinks together, talk about their families, have sex, and make sure that when the shoe is on the other foot, and it's the reviewer's turn to be the reviewed or a grant is pending, the favor is gladly returned. The whole system is incestuous and often compromised by fear of payback. God help a negative reviewer whose grant renewal is up for consideration or whose own research is due for review.

When it comes to human research, the real problem lies in Phase I, when researchers test doses that have never been tested before and subjects often assume they're being given drugs that work. Ironically, the system designed to ensure patient safety ultimately depends on experiments that might kill them. Let's look at the process and see how that might happen.

Under FDA requirements, a sponsor must first submit data showing that the drug is reasonably safe for use in initial, small-scale clinical studies. Depending on whether the compound has been studied or marketed previously, the sponsor may have several options for fulfilling this requirement: (1) compiling existing nonclinical data from past *in vitro* laboratory or animal studies on the compound; (2) compiling data from previous clinical testing or marketing of the drug in the United States or another country whose population is relevant to the U.S. population; or (3) undertaking new preclinical studies designed to provide the evidence necessary to support the safety of administering the compound to humans.

During preclinical drug development, a sponsor evaluates the drug's toxic and pharmacological effects through *in vitro* and *in vivo* laboratory animal testing. Genotoxicity screening is performed, as well as investigations on drug absorption and metabolism, the toxicity of the drug's metabolites, and the speed with which the drug and its metabolites are excreted from the body. At the preclinical stage, the FDA will generally ask, at a minimum, that sponsors (1) develop a pharmacological profile of the drug, (2) determine the acute toxicity of the drug in at least two species of animals, and (3) conduct short-term toxicity studies ranging from two weeks to three months, depending on the proposed duration of use of the substance in the proposed clinical studies. The problem is that humans may respond completely differently than other species, especially mice and rats, and therefore toxicity studies are sometimes flawed.

The New Drug Application (NDA) is the vehicle through which drug sponsors formally propose that the FDA approve a new pharmaceutical for sale in the United States. To obtain this authorization, a drug manufacturer submits, in an NDA, nonclinical (animal) and clinical (human) test data and analyses, drug information, and descriptions of manufacturing procedures. An NDA must provide sufficient information, data, and analyses to permit FDA reviewers to reach several key decisions, including (1) whether the drug is safe and effective for its proposed use(s) and whether the benefits of the drug outweigh its risks; (2) whether the drug's proposed labeling is appropriate and, if not, what the drug's labeling should contain; and (3) whether the methods used in manufacturing the drug and the controls used to maintain the drug's quality are adequate to preserve the drug's identity, strength, quality, and purity.

The purpose of preclinical work—animal pharmacology/toxicology testing—is to develop adequate data to undergird a decision that it is reasonably safe to proceed with human trials of the drug. Clinical trials represent the ultimate premarket testing ground for unapproved drugs. During these trials, an investigational compound is administered to humans and is evaluated for its safety and effectiveness in treating, preventing, or diagnosing a specific disease or condition. The results of this testing constitute the single most important factor in the approval or disapproval of a new drug.

Although the goal of clinical trials is to obtain safety and effectiveness data, the overriding consideration in these studies is the safety of those in the trials. The Center for Drug Evaluation and Research (CDER) monitors the study design and conduct of clinical trials to ensure that participants will not be exposed to unnecessary risks.

The research process is complicated, time consuming, and costly, and the end result is never guaranteed. Literally hundreds and sometimes thousands of chemical compounds must be made and tested in an effort to find one that can achieve a desirable result. The FDA estimates that it takes approximately eight and one-half years to study and test a new drug before it can be approved for the general public. This estimate includes early laboratory and animal testing, as well as later clinical trials using human subjects.

There is no standard route through which drugs are developed. A pharmaceutical company may decide to develop a new drug aimed at a specific disease or medical condition. Sometimes scientists choose to pursue an interesting or promising line of research. In other cases, new findings from university, government, or other laboratories may point the way for drug companies to follow with their own research.

New drug research starts with an understanding of how the body functions, both normally and abnormally, at its most basic levels. The questions raised by this research help determine a concept of how a drug might be used to prevent, cure, or manage a disease or medical condition. This provides the researcher with a target. Although sometimes scientists find the right compound quickly, usually hundreds or thousands of compounds must be screened. In a series of test tube experiments called assays, compounds are added one at a time to enzymes, cell cultures, or cellular substances grown in a laboratory. The goal is to find which additions show some effect. This process may require testing hundreds of compounds since some may not work but may indicate ways of changing the compound's chemical structure to improve its performance.

Computers can be used to simulate a chemical compound and design chemical structures that might work against it. Enzymes attach to the correct site on a cell's membrane, which causes the disease. A computer can show scientists what the receptor site looks like and how one might tailor a compound to block an enzyme from attaching there. However, despite the fact that computers give chemists clues regarding which compounds to make, a substance must still be tested in a living subject.

Another approach involves testing compounds made naturally by microscopic organisms. Candidates include fungi, viruses, and molds, such as those that led to penicillin and other antibiotics. Scientists grow the microorganisms in a "fermentation broth," with one type of organism per broth. Sometimes, one hundred thousand or more broths are tested to establish whether any compound made by a microorganism has a desirable effect.

Once animal testing begins, drug companies make every effort to use as few animals as possible and to ensure their humane and proper care. Generally, two or more species (one rodent, one nonrodent) are tested because a drug may affect one species differently from another. Animal testing is used to measure how much of a drug is absorbed into the blood, how it is broken down chemically in the body, the toxicity of the drug and its breakdown products (metabolites), and how quickly the drug and its metabolites are excreted from the body. Increasingly, animal testing is being minimized because of animal rights issues. Instead, researchers are relying more on computer modeling, which at best is still a primitive way of predicting how human organs will respond.

Institutional review boards (IRBs) are used to ensure the rights and welfare of participants in clinical trials both before and during their trial participation. IRBs at hospitals and research institutions throughout the country make sure that participants are fully informed and have given their written consent before studies begin. IRBs are monitored by the FDA to protect and ensure the safety of participants in medical research.

An IRB must be composed of no fewer than five experts and lay people with varying backgrounds to ensure a complete and adequate review of activities commonly conducted by research institutions. In addition to possessing the professional competence needed to review specific activities, an IRB must be able to ascertain the acceptability of applications and proposals in terms of institutional commitments and regulations, applicable law, standards of professional conduct and practice, and community attitudes. Therefore, IRBs must be composed of people whose concerns are in relevant areas.

However, according to a 1998 report by Health and Human Services Inspector General June Gibbs Brown, IRBs have some serious problems. The report claims that review boards review too much, too quickly, and with too little expertise; conduct minimal continuing review of approved research; ignore conflicts (professional, financial, etc.) that threaten their independence; and provide little training for investigators and board members.

At the time they were established, IRBs were designed for a research world that no longer exists. Their intent was to monitor research conducted at a single site by a single investigator, primarily at a university or teaching hospital. Today research is done at multisites, in trials across the country or around the world, sometimes involving hundreds of researchers and thousands of research subjects. The science has become so complex that many review board members don't have the expertise to question experiments; they depend on a "paper compliance" process rooted in trust, seldom if ever visit research sites, and rarely monitor actual conduct of research and informed consent procedures.

One recent phenomenon is the growing financial interests board members have in the research products being used in experiments. As long as IRBs consist of people directly affiliated with the research institute or who own stock in the company whose product is being tested, abuses will continue while reviewers look the other way and play dumb. The charade of monitoring scientists and informed consent is often just that, with human subjects the ultimate victims in a cruel game that in some cases has proven fatal.

In a recent survey of patients that signed up for Phase I clinical cancer trials, 85 percent said they expected the treatment to work and make them better. Nothing could be further from the truth. Phase I includes the initial introduction of an investigational new drug in humans and is designed to determine the metabolic and pharmacological actions of the drug in humans, the side effects associated with increasing doses, and, if possible, to gain early evidence on effectiveness.

Phase I studies also evaluate drug metabolism, structure-activity relationships, and the mechanism of action in humans. Since they're designed to test doses that have never been tested in humans, researchers have no idea how a particular patient will react to a drug at a given dose. Cancer drugs are especially sensitive, with a small dose not working at all and a larger dose sometimes lethal. Sadly for the patient, it's pretty much guesswork at this point.

In Phase I studies, CDER can impose a clinical hold (i.e., prohibit the study from proceeding or stop a trial that has started) for reasons of safety or because of a sponsor's failure to accurately disclose the risk of study to investigators. Although CDER routinely provides advice in such cases, investigators may choose to ignore any advice regarding the design of Phase I studies in areas other than patient safety.

What most volunteers aren't told is that the odds of surviving a Phase I trial are remote because the trial is not designed to deliver a cure or treatment but simply to test a "safe" dose—which means that patients not receiving the safe dose either become sicker or die from the treatment. The reason patients are not told this is because it would discourage people from signing up for Phase I trials. God forbid we inform those on whom we experiment that what we're really doing is ascertaining how much drug will kill them.

Phase II studies include the early controlled clinical studies conducted to obtain some preliminary data on the effectiveness of the drug for a particular indication or indications in patients with the disease or condition. This phase of testing also helps determine the common short-term side effects and risks associated with the drug. Phase II studies are typically well controlled, closely monitored, and conducted in a relatively small number of patients, usually involving several hundred people.

Phase III studies are expanded controlled and uncontrolled trials. They are performed after preliminary evidence suggesting effectiveness of the drug has been obtained in Phase II, and are intended to gather the additional information about effectiveness and safety that is needed to evaluate the overall benefit–risk relationship of the drug. Phase III studies also provide an adequate basis for extrapolating the results to the general population and transmitting that information in the physician labeling. Phase III studies usually include several hundred to several thousand people.

In both Phase II and III, CDER can impose a clinical hold if a study is unsafe (as in Phase I) or if the protocol is clearly deficient in design in meeting its stated objectives (Phase II). Great care is taken to ensure that this determination is not made in isolation but rather reflects current scientific knowledge, agency experience with the design of clinical trials, and experience with the class of drugs under investigation.

Accelerated development/review (*Federal Register*, April 15, 1992) is a highly specialized mechanism for speeding the development of drugs that promise significant benefit over existing therapy for serious or life-threatening illnesses for which no therapy exists. This process incorporates several novel elements aimed at ensuring that rapid development and review is balanced by safeguards to protect both the patients and the integrity of the regulatory process.

Accelerated development/review can be used under two special circumstances: when approval is based on evidence of the product's effect on a "surrogate end point" and when the FDA determines that safe use of a product depends on restricting its distribution or use. A surrogate end point is a laboratory finding or physical sign that may not be a direct measurement of how a patient feels, functions, or survives but is still considered likely to predict therapeutic benefit for the patient. The fundamental element of this process is that the manufactures must continue testing after approval to demonstrate that the drug indeed provides therapeutic benefit to the patient. If not, the FDA can withdraw the product from the market more easily than usual.

Treatment investigational new drugs (INDs) (*Federal Register*, May 22, 1987) are used to make promising new drugs available to desperately ill patients as early in the drug development process as possible. The FDA permits an investigational drug to be used under a treatment IND if there is preliminary evidence of drug efficacy and the drug is intended to treat a serious or life-threatening disease, or if there is no comparable alternative drug or therapy available for that stage of the disease in the intended patient population. In addition, these patients are not eligible to participate in the definitive clinical trials, which must be well underway, if not almost finished.

An immediately life-threatening disease means a stage of a disease in which there is a reasonable likelihood that death will occur within months or in which premature death is likely without early treatment. For example, advanced cases of AIDS, herpes simplex encephalitis, and subarachnoid hemorrhage are all considered to be immediately life-threatening diseases. Treatment INDs are made available to patients before general marketing begins, typically during Phase III studies. Treatment INDs also allow the FDA to obtain additional data on the drug's safety and effectiveness.

Protocol No. 126 is an example of what happens when IRBs are composed of colleagues, cronies, and individuals who are either professionally intimidated by physicians or have a financial stake in the research they are overseeing. We've already seen examples of how results are manipulated and how dangerous or ineffective products have been approved and brought to market. The clinical trials Becky Wright had "volunteered" for allegedly without knowing all the facts should never have gotten past Phase I, and they should have been halted before Becky even boarded the plane in Heflin, Alabama.

The investigation by the *Seattle Times* uncovered some stunning facts and fatal flaws. Half the recipients of the experimental bone marrow transplants suffered from a disease that was at worst 5–10 percent fatal. However, researchers wanted to increase transplantation success rates by using newly developed antibodies to help boost the immune system, destroy foreign material, and fight infection. According to the published reports, participants were not told that some of them had a 60 percent chance of lifetime survival with standard therapy! A majority of the Human Subjects Review Board were Hutch employees and the Hutch had a stake in the company's success. Despite that, many of the board members initially rejected Protocol No. 126 because of inadequate preliminary animal studies, a proposal to use the healthiest rather than the sickest patients, and the informed consent form, which downplayed the risk of the experimental treatment and failed to inform patients that a second transplantation (in case the first one failed) carried a 95 percent risk of failure.

After Protocol No. 126 was revised to reduce the strength of the drugs, the committee approved the research without knowing that FHCRC had a financial interest in the research and that three of the eight antibodies used in the treatment were developed by researchers who had a financial stake in the company that produced them. Dr. Robert Nowinski, a friend of David Blech and the head of Genetic Systems, had struck a deal with FHCRC to acquire exclusive commercial rights to thirty-seven antibodies for twenty years in exchange for substantial royalties on sales. The sale gave both parties an incentive to do whatever they could to get the products through the FDA approval process and into the marketplace.

By the time Becky Wright died in 1987, one of two dozen victims of Protocol No. 126, review board members had to have suspected that the world-renowned Fred Hutchinson Cancer Research Center was conducting fatal human experimentation. Still, rather than putting an end to the research, the FHCRC was allegedly at it again, this time recruiting breast cancer patients for an experiment labeled Protocol No. 681, which involved new drugs that researchers said could protect vital organs from damage by chemotherapy. And just as they'd done with Becky Wright, doctors advised Kathryn Hamilton, a forty-eight-year-old wife and mother of three children, that the experimental treatments might save her life, this time by ridding her of cancer while protecting her from the ravages of chemotherapy.

What Hamilton's doctors didn't tell her, according to Duff Wilson and David Health of the *Seattle Times*, was that not only did it appear that the drugs were ineffective but they had been busy writing a journal article officially stating that very position. So instead of receiving the standard treatment that possibly would have kept her alive at least one or two years, if not longer, Kathryn Hamilton, bleeding internally, her internal organs failing, died forty-four days after walking through the hospital doors.

According to the *Seattle Times* investigation, the journal article was submitted six days later, and the researchers who'd spent months writing it had a financial stake in the company producing the experimental drugs. The principal authors, who were also the lead investigators in Protocol No. 681, had continued treating patients with drugs they had reportedly known could not help them. Worse still, they kept increasing Kathryn Hamilton's doses to levels higher than those received by other patients.

Families of a number of cancer patients involved in the controversial study sued the Fred Hutchinson Center alleging fraud and ethics violations in connection with the failed bone marrow transplants. The Center has denied the allegations, and at least one outside investigation found no wrongdoing. The Fred Hutchinson Cancer Research Center remains one of the nation's leading recipients of research funds from the NCI.

In science, as in most areas of life, the axiom "just follow the money" often holds true. In a recent study that looked at 789 articles published in 14 journals written by a total of 1,105 authors, Dr. Sheldon Krimsky, professor of urban and environmental policy at Tufts University, found that almost 35 percent of the papers had at least one author with financial ties to the research topic. In the last decade, royalty payments to universities and colleges—of which faculty members are often allowed a percentage—has exceeded two hundred million dollars—incentive enough to make dishonest men and women out of some struggling researchers.

However, in one of the first research studies of its kind, published in the January 8, 1998 issue of the *New England Journal of Medicine*, Dr. Henry Stelfox showed in no uncertain terms that researchers who take corporate money have totally different opinions than those who don't. The controversy centered on the use of calcium channel blockers in the management of hypertension and heart disease. After examining seventy articles on channel blockers, Stelfox found that 96 percent of authors who supported the use of channel blockers had financial ties to their manufacturers, whereas only 37 percent of authors who had no such financial ties were supporters. In fact, 100 percent of channel blocker supporters had financial ties to *some* pharmaceutical company, in comparison with just 43 percent of nonsupporters, indicating that even objective scientists alter their views to ensure that industry money keeps flowing.

In light of these revelations, one can see how even human life is cheap when fortunes or reputations or careers are at stake. We don't know how many human subjects are sacrificed each year in search of medical cures or treatments or simply to make money, but we'd like to think it's not as bad as it used to be. Becky Wright and Kathryn Hamilton are just two examples of what happens when we cross that line to the dark side of science because habits die hard. Science, especially medical science, has an established history of unethical human experimentation, and doctors, after all, are no different than anyone else. As long as we place a premium on discovery and profit, human experiments and human sacrifice will continue unabated.

Vaccinations and Genetic Mutation: Are Vaccines Changing the Human Race?

On April 12, 1955, Dr. Jonas Salk stood behind a podium at the University of Michigan and announced to the world the discovery of a vaccine that was about to save tens of thousands of people a year from the ravages of polio. His vaccine, made from polio virus that was grown in monkey kidney tissue and then killed with formaldehyde, had just been licensed by the U.S. government for distribution and would be sent around the world for injection into millions of people. Within weeks of Salk's announcement, one of the largest inoculations in the history of mankind had begun. The vaccine proved to be a miracle. Polio cases plummeted from a high of fifty-seven thousand in 1952 to about a thousand within two years, and Dr. Salk was credited with orchestrating one of the greatest medical feats of the century.

However, while people were rolling up their sleeves and eagerly allowing foreign cells into their bodies, no one at the time had any idea that those cells would become part of their DNA and cause changes in their genetic structure. Or did they?

When Bernice Eddy, a researcher at NIH, injected monkey kidney extracts—the same kind used to make polio vaccine—into hamsters, she discovered that 90 percent of the hamsters developed large, malignant tumors! Upon further examination by Drs. Maurice Hilleman and Ben Sweet of Merck Laboratories, Salk's vaccine was found to be contaminated with simian virus 40 (SV40). By then, the vaccination program was so much a part of the health care system that nothing could have slowed it down. Even tests by the U.S. Public Health Service in 1961, showing that as much as a third of the stockpiled vaccine was contaminated, did little to stop the inoculations. Rather than going public, the agency quietly requested that manufacturers eliminate SV40 from newly produced vaccines while insisting that the virus caused no ill or permanent side effects.

Another polio vaccine, developed by Dr. Albert Sabin and tested in Russia and Eastern Europe at about the same time, also contained SV40. This oral version was given to millions of Russians and Europeans and contained a "live" but weakened strain of polio virus. Both vaccines were used in a three-year human experiment during the 1960s to test the effects of different vaccine dosages.

Concerned about reports of possible contaminated vaccine given to children in 1976, Dr. Joseph Fraumeni of the NCI looked at a group of 1,073, mostly black, newborns inoculated at Cleveland Metropolitan General Hospital with doses of polio vaccine as much as one hundred times the normal dose given to adults. At the time, no one suspected that their babies were being injected with cancer-causing viruses. Almost fifteen years later, when a link between vaccines and cancer was suspected, Dr. Fraumeni tried to find the children that had been inoculated to see if they'd come down with cancer. Since most of the children—now teenagers—could not be located, the study was deemed inconclusive and the cancer link was never proven. Anyone having a stake in the inoculation program was satisfied with that, and the program continued despite a stunning discovery about vaccines in 1971.

According to a September 22, 1971 article in *World Medicine*, foreign biological substances entering the bloodstream of a human being could become part of the human genetic structure by shedding DNA, which is then taken up by human cells. This process, known as "transcession," according to scientists, happens all the time in the human body and may trigger malignant cell transformations (cancer). That discovery may have been startling to many but in fact squared with earlier studies dating back to the height of polio vaccinations throughout the 1960s.

In 1961, for example, an article appeared in the December 15 issue of *Science* describing how common viruses, including those in vaccines, "acted as catalysts in producing cancer when given to mice in combination with other organic carcinogens in amounts too small to induce tumors themselves." Pesticides, herbicides, chemicals, and drugs approved by the Environmental Protection Agency (EPA) and the FDA because they were thought safe at low doses may, in fact, trigger disease when exposed to latent viruses. Later, a 1965 study reported an increase in multiple sclerosis following rabies vaccines. And in an October 22, 1967 *British Medical Journal* article, German scientists found that multiple sclerosis seemed to be triggered by vaccinations against smallpox, typhoid, tetanus, polio, tuberculosis, and diphtheria. A decade later, an article appearing in the April 4, 1997 issue of *Science* claimed that most cases of polio in the United States since 1972 were caused directly by the polio vaccine. More recently, researchers have admitted that 100 percent of polio since 1980 has been caused by vaccines, which they also say may be responsible for the major increases in leukemia.

The theory behind all this disease formation is that many vaccines given to humans are contaminated with *RNA viruses*, which contain enzymes called *reverse transcriptases*. Reverse transcriptase allows an infecting RNA virus to form strands of DNA that then integrate into the host DNA and lie dormant for long periods, perhaps years, before triggering disease. As the theory goes, vaccinations actually infect humans with RNA viruses that might remain latent, only to cause disease later in life. What we have, then, is a worldwide population infected with known dormant viruses that experts believe have been spreading and are responsible for many of the diseases in the world today.

The definitive evidence came with the development of polymerase chain reaction (PCR), which amplified previously undetectable bits of DNA and opened up an entirely new molecular world to skeptics. In 1988, Dr. Michele Carbone, a researcher at the NIH, discovered that many of the hamsters he'd accidentally infected with tiny amounts of SV40 had developed a rare form of cancer known as mesothelioma. When he injected another group of hamsters with SV40 to see if the virus was causing the cancer, he found that every single animal developed mesothelioma. Most disturbing was the fact that whereas only a few humans had ever come down with mesothelioma before 1950, by 1988 there were several thousand cases a year. Also, although asbestos had been blamed for much of that increase, with lawsuits piling up against companies who'd produced or used it, it was hard to explain how more than 20 percent of victims had never been exposed to asbestos at all. Some epidemiologists now believe that SV40-infected vaccines may be wreaking havoc decades later.

Thanks to new molecular technology and diagnostic tools, we know that SV40 is found in a host of cancers, including bone and brain tumors. Recent studies by the NIH have identified it in people with kidney disease and in 60 percent of patients with collapsing renal failure, a disease that until 1980 was quite rare. Italian scientists have even found SV40 in semen and suspect that it is being passed from person to person, generation to generation, hiding in DNA for years until it awakens and begins to multiply again. When it does, it transforms DNA, produces new proteins, triggers disease, and continues to spread like a silent killer through an unsuspecting global population. Evidence that SV40 is being transmitted without vaccinations is mounting based on a significant increase in SV40-infected tumors in people never inoculated with contaminated vaccines.

As recently as 2000, researchers continued to find SV40 in tumors where it was previously not found. Pituitary, thyroid, and lymphatic tumors can now be added to the growing list of potential targets. In 2001, scientists at a conference at the University of Chicago presented paper after paper reporting SV40 in human tissues and tumors. After years of stonewalling and denials, the NCI finally acknowledged the link in an official July 2001 statement: "SV40's interaction with tumor suppressor proteins indicates possible mechanisms that could contribute to the development of cancer." Coming from an agency that did all it could to downplay the link, the statement is one of the strongest ever made that the world's population is being contaminated with a cancer-causing virus.

Modern Vaccines and Human Experiments

Though they have saved countless lives, vaccines have been at the center of heated debate and controversy for much of this century. Many vaccine experiments have been conducted in secret and are only now being disclosed to the shock of parents whose children were used as guinea pigs. In the Australian state of Victoria, it was recently discovered that hundreds of children, some of them babies as young as three months, were human subjects from 1945 to 1970 in experiments to test vaccines against herpes, whooping cough, and influenza. Even though researchers knew the vaccines produced severe reactions, abscesses, and vomiting, and that some had actually failed safety tests done on animals, they gave healthy children full adult doses anyway. Asked about the experiments, Dr. David Vaux, who was a spokesman for the Walter and Eliza Hall Institute years after the events, said, "Doctors should be seen as heroes saving lives rather than as using children as guinea pigs." He went on to say that "If similar studies were carried out today, the experimental protocols would have to be approved by a human ethics committee, and informed consent would have to be obtained."

In England, evidence has just been uncovered about a secret whooping cough vaccine experiment conducted from 1948 to 1956 involving thousands of babies without their parents' consent. And in a 1960 experiment at the Fountain Hospital in London and the Queen Mary Hospital in Surrey, babies and young children with Down's syndrome were used to test an experimental measles vaccine. Some of the children, aged one to eleven years, had marked reactions; one died seven days later.

But it wasn't as if researchers weren't aware of the dangers they were putting children through. As early as the 1920s, the fact that vaccines were triggering outbreaks of disease was well know but kept hidden from the public. In 1922, for instance, smallpox vaccinations in England caused an epidemic of encephalitis, resulting in paralysis and death. Apparently the vaccines had interfered with nerve and nerve membrane development, causing inflammation of the brain and serious neural disorders. It wasn't until twenty years later that the public heard anything about it. Yet, even though researchers knew it all along, they thought it best to keep silent for fear that people would risk smallpox rather than get vaccinated.

Despite evidence that contaminated vaccines have intentionally been used, that some vaccines have actually triggered disease, and that new experimental vaccines may actually suppress immunity, vaccine research on humans continues nonstop. The question medical experts are asking is whether effects are only now beginning to show up and whether vaccines in the long-term will do more harm than good. Some are going so far as calling the experiments a disguised attempt at genocide and ethnic cleansing, since it seems that many human vaccine trials are conducted in Third World countries one researcher called "human offshore laboratories."

Following the polio and experimental hepatitis C vaccine programs, scientists were at it again. In the name of science and disease prevention, the U.S. Agency for International Development and the World Health Organization (WHO) began an experiment in the mid-1980s to test a high-potency measles vaccine (Edmonston-Zagreb or "EZ") on babies as young as four months. The experimental sites were a smorgasbord of poverty-stricken, mostly black nations with out-of-control populations and rampant disease: Haiti, Senegal, Guinea Bissau, Mexico, Cameroon, Gambia, Bangladesh, Rwanda, Sudan, South Africa, and Zaire, to name a few.

In the United States, the U.S. Centers for Disease Control and Prevention (CDC), in a joint effort with Kaiser Pharmaceuticals of Southern California, injected more than 1,500 six-month-old black and Hispanic babies from 1990 to 1991 in Los Angeles even though it was clearly known that children younger than one year may not have developed enough myelination around their nerves and that vaccines may seriously impair neural development. There were questions whether parents, thinking their children were being protected from disease, had received adequate disclosure that the vaccine was experimental or of all the foreseeable risks.

The CDC halted the human experiments after African countries reported a significant number of deaths. In some countries, researchers were injecting infants with vaccine as high as five hundred times the normal dose in order to break down naturally occurring maternal antibodies and replace them with vaccine-induced antibodies. However, rather than providing immunity, the vaccine did the opposite, causing suppression of children's immune systems for up to three years and making them more susceptible to diseases than if they received no vaccine. Curiously, African girls were given twice the dose of boys. Just before trials were halted, WHO had placed an order for enough vaccine to inoculate a quarter billion Third World children.

Dr. Alan Cantwell, who'd worked for Kaiser during the measles vaccine experiment, never heard about it until he read the report in the *New York Times* five years later. He wonders how many other such experiments have been conducted or are now being conducted without the public's knowledge. Other experts are questioning the epidemic rise in diseases since 1980, which they believe has been triggered in part by experimental and approved vaccines. According to the CDC, for example, the number of asthma cases has exploded from 6.7 million in 1980 to more than eighteen million today. Also on the rise in staggering numbers are immune and neurological disorders, hyperactivity, learning disabilities, systemic lupus erythematosus, multiple sclerosis, and certain types of cancers, all associated with immunosuppression and neurological damage possibly caused by tainted vaccines.

In vaccine efficacy and safety a myth? Are vaccines causing more harm than good? Are other factors more responsible for the decline in disease? Consider the following facts compiled by researchers around the world.

- The FDA's Vaccine Adverse Effects Reporting System (VAERS) gets some twelve thousand vaccine incident reports annually, of which more than 1% are vaccine-related deaths. Since less than 10 percent of cases are ever reported, the actual numbers of incidents may be as high as one hundred and fifty thousand and the number of deaths may exceed fifteen thousand. According to the National Vaccine Information Center, almost 98 percent of vaccine-related deaths and disabilities go unreported in New York. Many experts are now admitting that some vaccines may actually be causing more deaths than the disease would have without the vaccine.
- In countries where vaccination programs have decreased, death rates from sudden infant death syndrome (SIDS) and pertussis have also decreased. For instance, after Japan raised the vaccination age from two months (the age when vaccinations in the United States begin) to two years, the incidence of SIDS plummeted. In England, pertussis death rates dropped significantly following a 50 percent reduction in

vaccination rates. Countering the suggestion that vaccines have been the main reason for a drop in childhood diseases, a report by the British Association for the Advancement of Science stated that 90 percent of the decrease in diseases occurred before widespread vaccination programs, much of it due to improvements in hygiene and sanitation. Even in the United States, infectious disease rates declined by 80 percent prior to mandatory vaccinations, giving credence to critics who say that factors such as diet, hygiene, and sanitary conditions may be as important in prevention as vaccines.

- Doctors do not consider culture or genetic makeup when inoculating children, but this is one example where racial and/or genetic profiling may save lives, because not all races react to vaccines in the same way. Mortality rates in Australian aborigine infants increased by 50 percent following immunization programs. Some of the children went into shock shortly after getting their shots. A 1994 study published in the *New England Journal of Medicine* reported that Romanian children contracted polio at eight times the rate when they had received antibiotics within a month of receiving polio vaccine. Ten or more antibiotic treatments increased their risk of getting polio to 182 times that of those not receiving any polio vaccine at all. The public never hears about these cases, and the FDA, along with vaccine researchers who indiscriminately inject human subjects with experimental vaccines, never takes genetic differences into account, though it may be one of the greatest risk factors.
- Recent evidence has pointed to vaccines as a possible cause of violent crime. Statistics show that individuals who commit the most crimes and who keep returning to prison are those who have also had significantly higher incidences (compared with the rest of the population) of conditions associated with adverse reactions, especially encephalitis and disruption of neural myelination. Even the U.S. Public Health Service admits that vaccinations can cause encephalitis and long-term residual effects, citing studies that show a lowering of IQs in the 1960s when the vaccinated children of the 1950s became adults and started committing more crimes. More and more scientists are now beginning to see a relationship between neural effects of vaccines, especially in populations genetically susceptible, and the rise in violent crime.

Even so-called safe, FDA-approved vaccines are no guarantee against disease since, as we've seen, the FDA approval process is sometimes flawed and by no means foolproof. As recently as October 1999, RotaShield, a vaccine against a virus that causes severe diarrhea in children, was taken off the market after more than a million infants were inoculated and many developed life-threatening bowel obstructions. Eight months earlier, data presented at an FDA advisory meeting showed that the rotavirus vaccine not only increased the risk of severe bowel obstruction but that some infants had died after receiving the vaccine.

Most people assume that when their children receive vaccines against diseases like polio, measles, or hepatitis, they're being given a clean drug free of infected cells. Nothing could be further from the truth. First of all, vaccine production must begin with animal parts. Monkey kidney cells, chicken embryos, or human cell lines, some of them derived from cancer cells because they're immortal, are the incubators in which vaccines are grown. The HeLa cell line, for example, comes from Henrietta Lacks, a black woman who died from cervical cancer in 1951 but whose donated cancer cells continue to grow outside her body. Kept viable with a diet of various animal and human extracts, the HeLa cells became contaminated and, in turn, contaminated other cell lines used in viral research. Even vaccine experts like Jonas Salk didn't think it mattered because they believed that cancer cells would naturally be rejected by the body.

Rather than receiving a sterile vaccine, children actually receive living but weakened microorganisms. Moreover, sterilization would destroy the noninfectious proteins that make vaccines work; therefore, along with the virus, a vaccination contains other ingredients that could be the source of numerous health problems. Some of the other ingredients in a vaccine include heavy metals such as aluminum and mercury (used as adjuvants, which are chemicals that enhance immune response); preservatives such as alcohol, thimerosal, and formaldehyde (a carcinogen used in embalming fluid); antibiotics such as neomycin and streptomycin (which may cause allergic reactions); growth medium, including chicken embryo, horse or bovine serum, human fetal cells, monkey kidney cells, or cancer cells; and other contaminants such as genetic material, tissue and fluids from diseased animals, and fecal material.

Together with the live component of vaccines, these chemicals and contaminants can interfere with production of myelin, the insulation that surrounds nerves. Myelin production begins before birth and continues until adulthood. Scientists as early as 1953 were beginning to observe that epidemic diseases infecting children and for which children were getting vaccinated were increasingly attacking the nervous system. In 1978, after more than twenty years of inoculations, British scientists admitted that demyelination diseases were increasing at alarming rates because vaccines were abnormally sensitizing the nervous system.

Behavioral scientists are now convinced that many of today's mental health and behavioral disorders are linked to a delay or breakdown in the myelination process. Ignored by the medical establishment are studies showing that immature myelination in infants and children aged four days to three years causes delayed brain and neural development. In his book *The Nervous System*, Peter Nathan says that the large frontal lobes are not fully developed until a person is twenty years old. In addition, Leslie Hart, in her book *Human Brain and Human Learning*, says that prefrontal portions of the cerebrum have a profound effect on human behavior. If true, then vaccines, which attack unmyelinated and therefore unprotected nerve fibers, can seriously disrupt learning, memory, concentration, and behavior.

In my opinion, vaccine ingredients may be only part of the problem. Why, for example, are so many children not adversely affected by vaccines while others develop serious health and behavioral problems, sometimes for the rest of their lives? This is especially true when minority children or children living in inner cities are involved. Perhaps it's because many clinical trials and vaccine experiments are performed on minorities (in the Bronx and in Harlem, hospitalization rates for asthma are more than twenty times higher than for those living in affluent areas of New York). And since vaccinations may actually lower resistance by making immune cells less able to respond to infections other than those targeted by the vaccine, I believe that a breakdown in immunity is caused by a combination of vaccine, environmental factors, diet, genetics, and stress, which profoundly depresses immunity and can transform an otherwise harmless vaccine into a disease machine.

In light of recent findings on vaccines and immunosuppression, one would think that doctors injecting foreign cells and dangerous chemicals into an infant or young child would consider the consequences of their actions. Unfortunately, vaccinations have become such an integral part of life that nothing short of death will slow down current experimental vaccine programs.

The Willowbrook Scandal

Some had likened it to a death camp; others called it a place of unspeakable cruelty, human suffering, shame, and degradation. For the victims whose nightmares still remind them of their treatment and the medical experiments performed on them at a New York State institute for the mentally ill named Willowbrook State School, the term "crimes against humanity" isn't quite strong enough. Dr. William Bronston, a California physician who'd spent three years on the Willowbrook staff, said, "This is an evil place. It has an evil history." Part of that history included a series of human experiments in which doctors, without proper consent from parents, deliberately infected children with hepatitis.

Little has been mentioned about the hepatitis study done on mentally retarded children from 1963 to 1966. Behind the steel doors and barred windows of Willowbrook, doctors performed unconscionable acts on healthy, nonconsenting subjects in the name of science.

Located on New York's Staten Island, Willowbrook, like many other mental institutions, was grappling with an influx of patients and the problems of overcrowded conditions. At one point, more than six thousand New Yorkers—many of them children—were locked away in filthy, roach-infested rooms where they often wallowed in their own excrement. Disease was rampant. Hepatitis, as common among the patients as the flu in winter, had been receiving a great deal of attention because of strides made in hepatitis vaccine research. Therefore, it wasn't surprising that the one area of the institution not experiencing the overcrowding and the squalor was a separate facility designated especially for hepatitis research studies.

After its doors were officially closed to new patients, the parents of retarded children, desperate for their mentally ill youngsters to be admitted *somewhere*, were given an offer they found hard to refuse. In exchange for allowing their children to be part of a special hepatitis study, the children would be admitted to Willowbrook, no questions asked. With few other options available to them, parents eagerly agreed. Papers were signed, departing hugs exchanged, and the children whisked away to isolation wards where parents figured they would be treated far better than they would be in other parts of the institution. They *were*, after all, assigned to a "special" study.

Once the parents had driven off and were out of sight, the doctors took over. Like other medical studies before this one, the object was to learn about the natural history of disease—in this case hepatitis—and to test the effects of such treatments as gamma globulin. The only way to do that was to deliberately expose the subjects to disease. However, consent forms signed by parents had improperly stated that children were to receive a vaccine against hepatitis, not that they would first be infected with the disease.

Healthy children with no signs of hepatitis were taken to sterile rooms and prepared for the experiments. Initially, the procedure was quite crude. Stools were taken from hepatitis-infected individuals and used to make extracts that were then fed to healthy individuals. Since many of Willowbrook's patients had low IQs and were not toilet trained, obtaining feces was not a problem. As the study progressed, the procedure was refined. More purified extracts were produced and subsequently injected into subjects to give them hepatitis. When asked to justify the experiment, researchers claimed that the children would have gotten hepatitis at Willowbrook within six months anyway, so it was better to have them participate in a controlled study where at least they would be monitored and given treatment.

When the experiments became public, there was nationwide condemnation. Once again it seemed as if nothing was out of bounds as long as it advanced medical progress. The experiments were over by 1966, but the drive to develop a hepatitis vaccine was too strong to allow vocal critics of human experimentation to win. For the next decade, researchers quietly went about their business, biding their time until an epidemic of hepatitis hit the gay community and sent doctors back into a frenzy. As though Willowbrook had never happened, another human vaccine experiment was started. This time, say an increasing number of experts, the result may have been the source of the virus that triggered an AIDS epidemic across two continents.

Unfortunately, the lessons learned from Willowbrook—eventually closed down due to the weight of its atrocities—was short lived. There were too many career-oriented researchers with too much grant money and too few human subjects available. The two bright spots on the medical research horizon, though, were the number of people with mental disabilities who could be used as nonconsenting subjects and the eagerness by parents to allow their children to participate in studies in exchange for money.

Dr. Adil Shamoo, professor of biological chemistry at the University of Maryland Medical School and a member of the state attorney general's working group to draft legislation to protect the rights of decisionally incapacitated people, said, "Abuses are no accident. Researchers who recruit subjects in secret and under less-than-honest pretenses know exactly what they're doing. Few people knowingly expose themselves to these types of experiments." Dr. Shamoo claims that human experiments are being conducted all across America on people who don't have the ability to make informed decisions and that "the attitude of current psychiatric researchers is no different from those who conducted the Tuskegee Study."

As recently as 1992, for example, experiments on children took place at two other New York hospitals. One hundred males, mostly black and Hispanic, were recruited by the New York State Psychiatric Institute (NYSPI), affiliated with Columbia University, and the Mount Sinai School of Medicine, to study the effects of the subsequently FDA-banned drug fenfluramine (fen-fen) on brain activity. The children, all between the ages of six and ten, and all of them younger brothers of delinquents, were given ten milligram per kilogram body weight doses of fenfluramine to test the theory that violent or aggressive behavior is linked to low levels of certain brain chemicals like serotonin. In exchange for their children's participation, parents each received one hundred twenty-five dollars, which included a twenty-five dollar gift certificate to Toys 'R' Us.

While participants in the NYSPI study did not report significant side effects, an adult study conducted elsewhere, in which the dose was one-tenth the amount used in the NYSPI study reported causing, in some instances, headache, muscle pain, gastrointestinal problems, mental cloudiness, anxiety attacks, and other symptoms so severe that subjects could not continue normal functions, like going to work. Researchers in the adult study concluded that a dose no larger than 0.6 mg/kg was enough.

An investigation by the Office for Protection from Research Risks ultimately exonerated NYSPI, finding no wrongdoing on its part. Nevertheless, critics expressed concern about the way participants in the study were selected.

According to Vera Sharav, president of Citizens for Responsible Care in Psychiatry and Research, "These racist and morally offensive studies put minority children at risk of harm in order to prove they are generally predisposed to be violent in the future." "When it comes to protection from zealous researchers," says Sharav, "animals have more rights than people." If we stop and think about it, we've probably heard more from People for the Ethical Treatment of Animals (PETA) than from activists protesting against human medical research. The NYSPI denied all allegations that it acted in an unethical or racist manner.

Railing against the overmedication of children and disgusted by the growing trend of medical experimentation on children, Dr. Peter Breggin, director of the International Center for the Study of Psychiatry and Psychology, in comments that were not directed at the NYSPI study, condemns the practice. "These experiments are worthless and they constitute high-tech child abuse," he says. "They are an extreme example of what happens when we look for biological solutions to the psychological, social, and educational problems that children are experiencing in America today." Of particular concern is that many of these children are used in early-stage research designed to supply information about negative reactions, not to provide any therapeutic benefit.

Studies like those described are not really new. Experimentation on the mentally ill has been going on for many decades because it's been so simple to get around informed consent. In many cases, doctors have literally had their way in doing whatever they wanted to do. At one mental institution, psychotic patients were given lumbar punctures followed by tuberculin injections into their spines for several days to ensure infection. In the *Journal of Pathology*, the doctors wrote, "Ten hours after the injection the temperature begins to rise, pulse quickens, and a little later the patient usually vomits. The disturbance reaches its maximum about twenty four hours after the infection. By this time the temperature is anything from 101 to 104 degrees F, the patient is pale, listless, photobiotic [profoundly disturbed by light] and reluctant to eat, while the neck is often stiff." The doctors reported that inflammation occurring to the brain and spinal chord often lasted for days, but changes in the cerebrospinal fluid sometimes persisted for months.

In another study at the Walter E. Fernald State School in Waverly, Massachusetts, in 1956, researchers studying calcium metabolism gave mentally retarded children radioactive calcium both orally and by intravenous infusion. Radioactive materials were also injected into mentally deficient infants aged one to nine months, after which needles were inserted through the skull into the brain, through the neck, and into the spine to collect and analyze cerebrospinal fluid.

To measure blood flow in patients with dementia, doctors at St. Elizabeth Hospital in Bethesda, Maryland inserted long needles into patients' jugular veins and femoral arteries in the groin. A drug was then injected to cause a sudden and dangerous drop in blood sugar. At the New England Center Hospital in Boston, Massachusetts, schizophrenics were used to study brain changes during experimentally induced alkalosis and acidosis. After needles were inserted into jugular veins and femoral arteries, patients were injected first with ammonium carbonate to produce toxic alkalosis and raise blood pH, and then with ammonium chloride to produce toxic acidosis and lower blood pH.

In recent times, many unethical psychiatrists have simply been lured into performing human experiments by commercial contracts that pay them far more than they would otherwise receive in their regular practices. Such was the case of Dr. Faruk Abuzzahab, a respected psychiatrist and former president of the Minnesota Psychiatric Society. The incentives offered by Abbott Laboratories to test its experimental antipsychotic drug, Sertindole, were enough for Dr. Abuzzahab to recruit, in his words, "disturbed and vulnerable" patients he knew did not meet eligibility requirements. He also knew from their medical histories that if he took the subjects off their own medication for the purposes of the study, their condition might deteriorate drastically.

According to investigative correspondent Robert Whitaker, Dr. Abuzzahab ignored experimental protocols and falsified records. Suicidal patients like Susan Endersbe, a forty-one-year-old with schizophrenia, were not excluded from the study, as they should have been. Several weeks after being taken off the medication that was controlling her mental illness so as to wash it out of her system and test Sertindole, Endersbe's symptoms got progressively worse. On June 11, while Abuzzahab was claiming that his patient was fine and experiencing absolutely no side effects, Susan Endersbe leaped to her death in the Mississippi River. Sertindole, the drug that was supposed to cure Endersbe's psychosis, was eventually withdrawn by the FDA for safety reasons.

An extensive *Boston Globe* investigation uncovered a troubling pattern concerning the safety of psychiatric human experiments. Based on FDA records, among 12,176 patients taking four experimental antipsychotic drugs, including Sertindole, there were eighty-eight deaths and thirty-eight suicides, a rate five times higher than is normal for any clinical trial. One of the main reasons for this kind of horrendous track record is the urgency for FDA approval of antipsychotic drugs, which are undergoing remarkable growth in the overall drug market, with sales surpassing two billion dollars in 2002. The faster patients are recruited, the sooner trials can begin and the faster a drug can get to market and produce revenue (as much as two million dollars per day).

According to Whitaker, "Physicians who do commercial drug research full-time regularly report generating more than one million dollars in annual revenue and profits exceeding three hundred thousand dollars." Monetary incentives, reports Whitaker, are so great that physicians leave no stone unturned in recruiting as many patients as possible. In many cases, physicians are given bonuses as though they were sales people hunting down potential targets. A 1997 criminal case in Georgia illustrates how the lure of pharmaceutical riches can create monsters out of doctors.

From 1989 to 1996, Drs. Richard Borison and Bruce Diamond tested schizophrenia drugs for various drug companies such as Eli Lilly and Novartis. During that time, they recruited heavily and received more than ten million dollars from clinical trials. Their office staff was paid bonuses and offered cars to coax schizophrenic patients into becoming human subjects. Untrained laboratory workers drew blood samples and adjusted doses of the experimental drugs even though they were not qualified to do so. At the trial, the workers testified that Dr. Diamond paid little attention to his patients' care. In the court transcript, Diamond is reported to have said, "We don't care about how the patients are doing. We just want to know how many people you have enrolled in the past week or couple of weeks." For their part, Diamond and Borison were sentenced to five to fifteen years in prison, respectively, and ordered to pay millions of dollars in fines.

Psychiatry is unique in that in order to test how effective an antipsychotic drug is, researchers need actively psychotic human subjects. If a patient is taking an existing drug that is already controlling her mental illness, she must be taken off of it to induce psychosis. Since healthy patients in psychiatric studies are useless, they must be forced into relapse and sometimes into dangerously emotional conditions. In the case of stabilized schizophrenics, their delusions and hallucinations must return to them. Only after mental illness is induced is the experimental drug given. The researchers then roll the dice and hope the patient survives, since studies have shown that this technique of "washing out" a drug can make the psychosis even worse than if the patient had never been treated.

Such was the case of Abigayle McIntyre, a young girl with schizophrenia who was recruited to test Janssen Pharmaceutical's new antipsychotic drug, resperidone. After Abigayle entered Camarillo State Hospital in California, her medications were withdrawn. Almost immediately, she became violent and noticeably psychotic. As part of the experimental procedure, she was then given Haldol, which triggered terrible headaches and raised her blood pressure to dangerous levels. Two weeks later, Abigayle committed suicide by swallowing a large number of aspirin tablets. According to transcripts of an investigation by the California Department of Health, her mother, a psychiatrist, said:

> I thought research was the best treatment in the country. Today it is not. It is the most dangerous. Abigayle wouldn't have gotten sick like that if she hadn't been in research. She ended up outside my control and outside the control of a good doctor who would have done something about her illness. She was actively neglected in research.

Though researchers insist that withholding treatment or inducing mental illness is a legitimate experimental protocol, the fact is that it's not done in other, nonpsychiatric medical experiments. Dr. Leonard Glantz, a professor of health law at Boston University School of Public Health said, "You don't make psychotic people psychotic. You don't put diabetic people into shock. You don't give people with heart disease heart attacks. It's not an appropriate use of human beings." I'm sure we would be outraged if antibiotics were withheld from someone with a severe infection in order to see if another medicine would work. Yet that's exactly what we continue to do to this day with the mentally ill. In published articles, doctors have reported administering drugs designed to actually intensify symptoms such as hallucinations but offered no therapeutic benefits; other articles describe studies in which doctors withheld medication for the sole purpose of measuring how quickly schizophrenic patients would become schizophrenic again.

Psychiatric patients are especially at risk because in many cases they're incapable of understanding what they are consenting to. The worldwide increase in the use of mentally retarded or incapacitated subjects is staggering, with much of Europe subscribing to guidelines that permit this kind of research even if it doesn't benefit the subject directly. In the

United States, the FDA admitted that more than half of the projects they monitor do not have proper informed consent and that many of the researchers hold back information so that patient consent is not really "informed."

The problem in most cases of washout and "closely monitored" forced relapse is not very serious. The only realistic way to test a new drug and isolate its effect is to have a withdrawal of existing medication. The ethical dilemma lies with doctors who recruit nonconsenting or unqualified people as if they were lab rats and leave them to suffer, become violently ill, or commit suicide simply to increase the trial numbers and get paid by pharmaceutical companies to recruit as many warm bodies as possible. With only fourteen federal officials in charge of monitoring thousands of such psychiatric research studies and more than five thousand review boards each year, it's little wonder that the system is overwhelmed, often depending on reviewers unqualified to review, and allowing patients to die in experiments that should never have been allowed to continue.

In one of the more illuminating examples of just how incompetent and lax review board members can be in discharging their duties, Dr. Dorothy Rosenthal, the head of a University of California–Los Angeles review board overseeing a schizophrenia study, admitted that she was not familiar with the 1964 Declaration of Helsinki or the 1947 Nuremberg Code, two of the cornerstones of medical ethics and human rights. According to a 1998 article in the *Boston Globe*, Rosenthal, when asked about the Declaration of Helsinki, said that she had a "vague recollection of that as a title but didn't have a clue as to what was contained in it." A June 1998 federal report found that some review boards examined as many as two thousand research protocols per year, with some receiving an average of two hundred reports a month. One board member claimed that at smaller IRBs a single staff member may be responsible for all IRB activity. He added that when he attends a hearing, he reviews the continuing review summaries to see if a patient has died. If no patient has died, he doesn't bother to raise any questions.

Throughout this century, countless human experiments have been performed covertly and overtly, with and without consent, on babies, children, and adults, in many cases ending in death or disability. The rationale is always the same: Without human experimentation, medicine would never have progressed as far as it has; and because of the sacrifices made by human subjects future generations are the beneficiaries. No one doubts that. However, as we cross new thresholds into fields such as molecular genetics, cloning, and gene therapy, experiments are becoming more complex and much more dangerous.

Today more than ever, people need to be aware and informed about what they're volunteering for, because in this century human experimentation will involve experiments on the most basic molecule of life itself, DNA. Unlike testing a new drug, there's not much margin of error when it comes to manipulating or altering one's genes. With past experience on its side, perhaps society will be more cautious this time around; for when it comes to medicine in the twenty-first century, the consequences of failure may be more frightening than anyone realizes.

8 ETHNIC WEAPONS: THE NEW GENETIC WARFARE

Hidden by the blackness of the night sky and its own technology, a stealth bomber approaches its coordinates, opens its underbelly, and releases a bomb that finds its mark 20,000 feet below on the outskirts of a large city. Other than the handful of residents unlucky enough to be sleeping near the target area, the majority of city dwellers suspect nothing. In a few weeks, more than a quarter of the population will either be dead or incapacitated. The carnage begins at a moment's notice: one individual is reading the local newspaper and drinking a cup of coffee, another standing next to him collapses and begins to convulse. Soon, the city is divided into the healthy and dying; the only difference between them is race and a slight variation in their genetic code. Halfway across the country, a special agent is busy setting the fuse on another biological weapon timed to disperse a genetically engineered microorganism that will sweep through the population but incapacitate only those with the targeted genes.

Once the stuff of science fiction, this nightmarish scenario in which certain ethnic groups are eliminated while others are left unaffected is not far from becoming reality. We've known for some time that genetic differences exist within populations and between ethnic groups, and that a very small number of genes can cause a large number of functional differences. Selectively targeting those specific differences is the rationale behind ethnic weapons or what some have called "race bombs." The United States began serious research into these weapons as early as the 1970s, and it's suspected that several other nations have also developed them or are actively in pursuit.

Thus far, one of the major drawbacks in using biological agents has been their indiscriminate nature. Detonate a bioweapon, for instance, and risk infecting your own people as well as those targeted. Infect the water supply and soil of a target region and forget about sending in troops any time soon. But design a plague bomb or anthrax strain that infects people carrying a certain set of genes and you now have a weapon that would make everything else obsolete.

Dr. Carl Larson, in a 1970 *Military Review* journal article entitled "Ethnic Weapons," stated that "human populations can be characterized by frequencies of distinct genes. Sometimes, gene frequencies agree fairly well between widely dispersed populations, but more often there are great differences." Head of the Department of Genetics at Sweden's Institute of Genetics at the time, Larson went on to say, "Surrounded with clouds of secrecy, a systematic search for new incapacitating agents is going on in many laboratories. . . . Psychochemicals would make it possible to paralyze temporarily entire population centers without damage to homes and other structures. . . . It is quite possible to use incapacitating agents over the entire range of offensive operations, from covert activities to mass destruction." Keep in mind that Dr. Larson's predictions were made more than three decades ago, before the explosion of research in biotechnology and molecular genetics. If it was conceivable then, it's more conceivable now.

Although Dr. Larson's 1970 article was the first public acknowledgment that scientists were considering the possibility of ethnic weapons, the idea was not novel. In their book *Gene Wars,* Charles Piller and Keith Yamamoto wrote of Larson's review, "In the military's private circles it was old news." They claim that twenty years before Larson's article, the Naval Supply Depot at Mechanicsburg, Pennsylvania, was the site of experiments in which blacks were tested for their susceptibility to infectious agents. According to Piller and Yamamoto, "The site was chosen because within this system there are employed large numbers of laborers, including many Negroes, whose incapacitation would seriously affect the operation of the supply system." The problem was that blacks were more susceptible to a strain of valley fever than were whites, and by infecting them with a substitute microorganism instead of the actual virus, the government was trying to field-test an attack on various ethnic groups.

In a 1996 article appearing in the reputable London journal *Foreign Report,* Bo Rybeck, former head of Sweden's defense research, was quoted as saying, "Genetic weapons might be around the corner." His contention was that it wouldn't be much longer before diseases like influenza were designed to attack only blacks, for example, or toxins developed to kill only Serbs. Rybeck wasn't alone in thinking the unthinkable and in speculating that we may be much closer than anyone had realized to having the capability to engineer microbes virulent to some people but not others. William Cohen, secretary of defense during the Clinton administration, agreed with Rybeck, saying that there'd be no protection for genetically engineered biological agents and that "these are the weapons of the future and the future is coming closer and closer." A June 1997 issue of *Jane's Defense Weekly* quoted Cohen as saying, "I'd seen reports about certain types of pathogens that would be ethnic specific so that they could just eliminate certain ethnic groups and races."

However, Cohen's prediction came five years after *Defense News* reported that "genetic engineering may enable us to recognize DNA from different people and attach different things that will kill only that group of people. You will be able to determine the difference between blacks and whites and Orientals and Jews and Swedes and Finns and develop an agent that will kill only a particular group." A report by the British Medical Association titled "Biotechnology, Weapons and Humanity" added fuel to the fire by stating, "There are distinguishing DNA sequences between groups, and if these can be targeted in a way that is known to produce a harmful outcome, a genetic weapon is possible."

Researchers deploying today's modern biotechnology can engineer viruses that would spread diseases like wildfire through populations or produce designer toxins that only attack individuals with gene sequences specific to their ethnicity. Designer germs could be used "as a deterrent against attacks by extremists," according to Craig Venter, cofounder of a genetic research company called Celera Genomics. Some of these weapons might be so subtle in their mode of action, perhaps taking years to reach full effect, that a population may not even realize it had been targeted until it was too late. For example, Dr. Vivienne Nathanson, a bioethicist with the British Medical Association, claims that a delayed ethnic weapon might be designed to affect only future generations and that "it will unfortunately be possible to design biological weapons of this type when more information on genome research is available."

Molecular biologists believe we're not very far from taking the basic building blocks of life and making viruses from scratch. Laboratories around the world are busy manipulating various microbes to determine which genes they can live without in order to locate exactly where other genes can be inserted as replacements and thereby produce bugs to order. Military scientists are trying desperately to stay ahead of regimes that are improving bioweapons by making them more virulent or more resistant to antibiotics or vaccines. Other avenues of research are the development of new microorganisms that target vital crops or attack cell membrane receptors, and the development of chemicals that interfere with enzymes specific to certain populations. Since enzymes (which can respond differently depending on genetic inheritances) catalyze biochemical reactions critical to life processes, interference with those enzymes would be an effective weapon.

One of the more significant life activities enabled by enzyme reactions is muscle relaxation. For instance, if a certain enzyme were blocked, the victim would become paralyzed, even to the point of death by asphyxiation. The discovery of organophosphates by Germany in the 1930s, first used as insecticides and then to exterminate Jews, worked this way. Another potential target would be individuals lacking certain enzymes. For example, Southeast Asian populations historically have lactose intolerance due to virtual absence of the enzyme lactase. A weapon that takes advantage of the genetic variance could incapacitate entire regions or populations, leaving the invaded country vulnerable and the invading armies relatively unscathed. In addition, genetic weapons could be used as vectors, i.e. to break down resistance and immune defenses so much that natural microorganisms are allowed in to finish the job.

However, ethnic weapons may be developed for a simpler yet more dastardly role: population control and ethnic cleansing. Jonathan Moreno, a former official in the Clinton administration who helped investigate U.S. government involvement in human radiation experiments and is now the director for biomedical ethics at the University of Virginia, told an audience at a recent meeting of the American Association for the Advancement of Science that South Africa's old apartheid regime was heavily involved in DNA research. He was later quoted in a *Popular Mechanics* article, claiming, "The South African Defense Force conducted research for the possible development of biological agents that could be used against the black population. They were particularly interested in seeking ways to sterilize women of color."

Most experts agree that we can forget about attacks by nuclear missiles costing millions to produce and deploy. These require a degree of sophistication most in the world don't possess and would be met with instant and devastating retaliation. Instead, what could be simpler than an insidious attack on DNA? So are we really on the brink of being able to wage this kind of war on groups of people and eliminate them simply because they carry certain genes or lack the gene for a single protein? If we are, then one can only imagine such weapons in the hands of terrorists or states that espouse genocide. The same experts who believe research is ongoing predict that the question is not if but when, and that it is only a matter of time before the term "ethnic cleansing" will be seen in an entirely new light.

Arabs, Jews, and Ethnic Weapons

For some time, suspicions have swirled around classified Israeli military research. Although it's no secret that Israel has developed nuclear weapons and probably has hundreds of warheads, the fact that it still has not ratified the Chemical Weapons Convention or signed the Biological Weapons Treaty has many convinced that it has developed an arsenal of extremely powerful weapons of mass destruction. As a defense, Israel points to enemies and terrorist groups surrounding it on all sides, wanting nothing short of eliminating the State of Israel from the face of the earth. In an interview with the Egyptian newspaper *Al-Ahram Weekly,* Major General Abdel-Rahman El-Hawwari said that Israel ranks fourth among the world's nations in the production of chemical and biological weapons and is very advanced in the field. Notwithstanding the obvious threat to Israelis, the idea sends shudders through those who view any use of these weapons for any reason as an attack on humanity.

In response to what it perceives as another potential holocaust, the Israeli government is thought to have discovered specific differences in Arab genes and may be close to developing an ethnic weapon that targets Arab populations, especially Iraqis. It should come as no surprise that Israeli scientists would be so involved in genetic research because there are many unique and varied ethnic variations in the Jewish population and because of the genetic disorders unique to Israelis. The National Laboratory for the Genetics of Israeli Populations in Tel-Aviv, initiated in 1994 by the Israeli Academy of Sciences and Humanities, was established as a national repository for human cell lines and has been actively involved in the Human Genome Project (HGP). However, other facilities operating throughout the country are thought to be involved in chemical and biological weapons programs designed to stay one step ahead of Israel's enemies.

One such facility, the Institute for Biological Research, is located in the suburb of Nes Tziyona twelve miles south of Tel-Aviv. According to a statement in the October 4, 1998 *London Times*, a former senior biologist for Israeli intelligence claimed, "There is hardly a single known or unknown form of chemical or biological weapon which is not manufactured at the institute." Whether this was disinformation for defensive purposes or bluster for Arab consumption is debatable, but a month later, when the *London Times* reported that Israel may have pinpointed a specific Arab gene and was close to developing an ethnic bomb specific to the Iraqi people, Saddam Hussein backed away from his threats of an all-out war against the Jews.

Based on more recent reports from South African and Israeli sources, the two countries had worked closely on nuclear and genetic research. With biotechnology advancing at breakneck speed, the Israelis, who have become experts in the field, are believed to have uncovered clear genetic differences between Ashkenazi or East European Jews, Jews of Arab origin, and Arabs. For Israelis, however, one of the obstacles in developing ethnic weapons had been the fact that since many Jews have Arab origins, unless the weapon targets a gene so specific that it would affect an incredibly slight variation in gene structure, Israelis would be victims as well. Were those slight variations discovered, it would only be a matter of time before a chemical or biological agent was developed that targeted the gene. Asked to comment on the possibility that such ethnic weapons exist, most Israeli officials, rather than denying their existence, either dismiss the question or respond by saying that the State of Israel would not hesitate to use anything at its disposal if it was seriously threatened.

In the final analysis, the Middle East may be the breeding ground for genetic warfare. More Jews than ever are leaving Russia and other countries around the world, including the United States, and emigrating to Israel, among them top scientists who are bringing with them the expertise to help develop the next generation of bioweapons. If the Arab states believe they will someday drive Israelis into the Mediterranean Sea, they are seriously mistaken. Right now Israel has the power to literally wipe entire nations from the human race. Soon they will be able to do so with even more terrifying weapons, weapons that simply sterilize populations out of existence or unleash diseases that will eliminate the ethnic hatred that threatens their existence. Since there's obviously no turning back from the science that brought all this about in the first place, Arabs and Jews have no other option than to set aside their blind prejudices and live together. If not, the wars of the past will pale in comparison.

The Human Genome Project: Molecules of Life and Death

Mapping the three billion or so bases and thirty-five thousand genes in human DNA was one of the great accomplishments of twentieth century science. It is also one of the main prerequisites in developing ethnic weapons, since without knowledge of genetic differences between races the whole issue becomes moot. Of course, the other prerequisite is developing the biotechnology to alter and/or recombine genes in order to develop new and unique microorganisms. That biotechnology is now firmly in place and is being advanced and refined as I write this chapter. But to see exactly how the Human Genome Project (HGP), headquartered at Cold Spring Harbor, New York, former home of the Eugenics Research Office, could be used for developing and producing ethnic weapons, let's first look at what the HGP is and what its principal goals are.

In 1953, while much of the world was still recovering from World War II and preoccupied with ending the war in Korea, an incredible discovery was announced. The molecule of life—deoxyribonucleic acid (DNA)—had taken the scientific community by storm and was giving mankind a glimpse at how genes inside of cells work to create all life on earth. From that moment on, nothing would be the same, and the speed at which science progressed seemed almost miraculous. By the 1970s, molecular biologists were using recombinant DNA techniques to splice, remove, and rearrange genes in cells of different organisms. Thus, the science of genetic engineering was born, allowing scientists to literally create new life forms and drive a revolution with unlimited and frightening possibilities.

Begun officially in 1990 as an international, fifteen-year effort to locate and identify the entire set of human genes, HGP has expanded to include applications for improving agricultural products, finding new energy sources, countering environmental pollution, and customizing medicines by "distinguishing variations among populations." In fact, the U.S. government's own 2001 primer on genomics includes the following statement:

> Although more than 99 percent of human DNA sequences are the same across the population, variations in DNA sequence can have a major impact on how humans respond to disease; to such environmental insults as bacteria, viruses, toxins, and chemicals; and to drugs and other therapies. Methods are being developed to detect different types of variation, particularly the most common type called single nucleotide polymorphisms (SNPs), which occur about once every one hundred to three hundred bases. Scientists believe SNP maps will help them identify the multiple genes associated with such complex diseases such as cancer, diabetes, vascular disease, and some forms of mental illness.

In other words, a principle goal of HGP is to identify sequence variations among populations for the purpose of developing effective drugs. "Within the next decade," states the U.S. Department of Energy (DOE), "researchers will begin to correlate DNA variants with individual responses to treatments, identify particular subgroups of patients, and develop drugs customized for those populations." Since more than one hundred thousand people die each year from adverse reactions that are beneficial to others, focusing on genetic differences that cause these deaths is a high research priority. The problem with what is termed "pharmacogenomics" is that if human variations are identified for the purpose of modifying treatments specific to those individuals, then those same molecular fingerprints can be used to develop chemical or biological agents to cause harm. At this writing, more than twenty countries have established genome research projects, including some that would certainly use the research for malevolent purposes.

The question scientists have been grappling with is whether there are enough genes to actually indicate ethnic differences. One of the surprises coming out of HGP is that humans have one-third fewer genes than the ninety thousand or so once thought. However, it's now believed that the incredible diversity of the human race lies not in the number of genes but in how genes are used to build different products through a mechanism called "alternative splicing" (a process that yields different protein products from the same gene), and the thousands of chemical modifications made to proteins once they are formed. It is *these* variations that may ultimately prove to be the component on which ethnic weapons are based.

So far, scientists have identified about 1.4 million locations among the three million chemical nucleotide bases where single-base DNA differences occur in humans. The daily avalanche of genetic information, according to the DOE, principal sponsor of the HGP, is currently leading scientists in five main directions:

Transcriptomics involves large-scale analysis of messenger RNAs (molecules that are transcribed from active genes) to determine when, where, and under what conditions genes are expressed.

Proteomics or the study of protein expression and function can bring researchers closer than gene expression studies to what's actually happening in the cell.

Structural genomics initiatives are being launched worldwide to generate the three-dimensional structures of one or more proteins from each protein family, thus offering clues to their function and providing biological targets for drug design.

Knockout studies are one experimental model for understanding the function of DNA sequences and the proteins they encode. Researchers inactivate genes in living organisms and monitor any changes that could reveal the function of specific genes.

Comparative genomics analyzes DNA sequence patterns of humans and well-studied model organisms side by side in order to identify human genes and interpret their function.

One of the breakthroughs of the genetic revolution has been gene therapy, a process in which faulty genes are replaced with good ones. Although still in its experimental stages, the principles of gene therapy may be adopted to exploit genetic weaknesses and target victims by attacking specific bits of DNA found only in certain individuals. That kind of information is being put together under a parallel program called The Human Genetic Diversity Project, designed to identify and map the significant differences among populations. According to the British Medical Association's Report on Biotechnology, Weapons and Humanity, "The research which develops specific therapeutic agents is scientifically (but not ethically) indistinguishable from research to develop a lethal or disabling agent targeted at specific clusters of genes or alleles." So as we congratulate ourselves on one of mankind's greatest achievements, we're also reminded that our discoveries, while monumental in helping scientists save the world, may just as easily help in destroying it.

HIV: Natural Pandemic or Inadvertent Ethnic Weapon?

Kary Mullis, winner of the 1993 Nobel Prize in Chemistry for inventing polymerase chain reaction (PCR), a technique used to detect the human immunodeficiency virus (HIV), must not have been too popular with AIDS researchers after making the following statement: "Where is the research that says HIV is the cause of AIDS? There are 10,000 people in the world now who specialize in HIV. None has any interest in the possibility HIV doesn't cause AIDS because if it doesn't, their expertise is useless. I can't find a single virologist who will give me references that show HIV as the probable cause of AIDS. If you ask a virologist for that information, you don't get an answer, you get fury." That view, despite the risk it brings to one's reputation, is increasingly shared by many of the world's leading scientists.

The human immunodeficiency virus belongs to a class of viruses known as retroviruses, which do not have DNA, only RNA. As early as the 1960s, scientists knew that certain types of leukemias were caused by retroviruses and even attempted to develop vaccines against them. The science of virology had taken giant steps from 1960 to 1969, when Dr. Howard Temin of the University of Wisconsin finally unraveled the mechanisms of viral reproduction and described how a retrovirus binds to its host's DNA and uses it to make copies of itself.

That same year, in June 1969, government scientists testifying before a senate appropriations committee made the now famous and often quoted statement, "Within five to ten years, it will be possible to produce a synthetic biological agent that does not naturally exist and for which no natural immunity could be acquired." Perhaps they'd known more than they were willing to admit at the time because biowarfare laboratories were at least five years ahead of other laboratories in the techniques of gene manipulation and, less than a decade after that testimony, a series of events began to occur that literally changed the world.

In the year that statement was made, a serious outbreak of AIDS-like epidemics broke out at U.S. primate centers working on experimental techniques that transferred viruses among various primate species. A few years later, in 1974, chimpanzees began to die from leukemia and pneumocystis pneumonia (a common illness associated with AIDS) when they were weaned on milk infected with bovine C-type virus, and from a type of "cat AIDS" created from HIV-like cat retroviruses. Much of this work was being done under the auspices of the U.S. government's Special Virus Cancer Program (SVCP), whose mission was to grow large amounts of cancer-causing viruses, adapt them to human cells, and ship them to laboratories around the world. Among the famous scientists at the SVCP were Dr. Robert Gallo, who later discovered HIV, and Dr. Peter Duesberg, a Nobel Prize nominee at the University of California at Berkeley, who today claims that HIV does not cause AIDS. In her book *The Coming Plague*, Laurie Garrett says that monkeys at primate research centers in 1976 and 1978 died of severe T-cell immune system depression, lymphomas, and AIDS-like opportunistic infections.

By 1975, the virus section of Fort Detrick, formerly the U.S. Army's biological warfare research unit, had been converted to the Frederick Cancer Research and Development Center under the direction of Dr. Robert Gallo, who by then had already isolated a virus similar to HIV named HTLV-1. It was later discovered that the molecular structure of HIV appears very much like a combination of HTLV-1 and visna, a fatal sheep virus that destroys T cells—exactly what HIV does. In fact, when Professor Jakob Segal, professor of biology at Humboldt University in Berlin, analyzed HIV's genetic structure, he discovered that it is more similar to visna than to any other retrovirus and that, when combined with HTLV-1, it enters T4 lymphocytes and causes immune deficiency.

From 1977 to 1978, homosexual men in New York and San Francisco were actively recruited for a hepatitis B vaccine study and injected with a 1:10 dilution of hepatitis infective serum developed in chimpanzees, the only animal susceptible to human hepatitis B virus. By 1979, an outbreak of unknown origin was making news in Manhattan. The victims developed symptoms of a new illness that the U.S. Centers for Disease Control and Prevention (CDC) would later call AIDS. According to some reports, by 1984 nearly 70 percent of the homosexual populations in New York and San Francisco were HIV-infected in comparison with only 10 percent of populations in certain cities that were not involved in the vaccine study. Manhattan gays had the highest incidence of HIV anywhere in the world, including Africa, where AIDS did not appear until 1982. That same year, Dr. Gallo's laboratory at the National Cancer Institute (NCI) was credited with discovering the AIDS virus.

Since the discovery of AIDS, scientists have been debating its origins with no conclusive answers but are in agreement that no virus known to man has been as complex or mutated as quickly as HIV. Was it a natural virus all along, hidden by the ravages of other diseases, or had it jumped species only recently? Were human experiments done with accidentally tainted hepatitis vaccine? Was HIV inadvertently spread by polio vaccinations in the 1950s or by smallpox inoculations in the late 1970s and early 1980s? Is HIV a synthetic virus produced by genetically engineering a T-cell attacker and a T-cell destroyer to make the perfect killing machine? Or did something happen to transform a naturally existing but harmless virus into a silent killer? Some leading experts, including Nobel laureates, do not believe that events surrounding the appearance of HIV are coincidental.

On April 30, 2000, the *Washington Post* issued a somber warning: "The U.S. government formally designated AIDS a security threat that could topple foreign governments, touch off ethnic wars, and undo decades of work in building democracies abroad." A few months later, the CDC ominously announced that the rate of people dying of AIDS or being diagnosed with HIV in the United States was no longer falling but rather beginning to reverse. The most shocking revelation came on February 25, 2002, when the CDC reported that the number of Americans living with HIV would hit one million by the end of the year and that as many as half of them would not even know it. Still, while those numbers seemed incredibly high, in the rest of the world the count was an astounding forty million; HIV was quickly becoming a pandemic that threatened the entire global population.

Even more ominous was a claim by Dr. Max Essex of Harvard University, an expert who created a type of feline AIDS, that different strains of HIV affected ethnic groups differently. For instance, when Dr. Essex tested HIV strains from Thailand, he discovered that the Asian strain infected women's genital cells whereas the strain that infected gay white men in the United States did not. Since then, molecular biologists have identified at least eight substrains of HIV that infect people around the world differently and make certain ethnic groups more susceptible than others.

The most surprising study was done in 1997 by Stephen O'Brian and Michael Dean of the Genomic Diversity Laboratory at the NCI. Their study showed that one of ten white people have AIDS-resistant genes, whereas Africans have none. Their article in *Scientific American*, entitled "In Search of AIDS-Resistant Genes," raised a few politically correct eyebrows but was the first admission that HIV may have applications as an ethnic weapon.

Adding to the controversy, Peter Duesberg claimed that HIV alone cannot be the source of the AIDS epidemic because pure HIV injected into chimpanzees or accidentally infecting healthy humans does not cause AIDS. Duesberg insisted in the *Proceedings of the National Academy of Science* that "no such virus or microbe would require almost a decade in some cases to cause primary disease nor could it cause the diverse collection of AIDS diseases. Neither would its host range be as selective as that of AIDS, nor could it survive if it were as inefficiently transmitted as AIDS."

How could a virus spread so quickly, seemingly out of nowhere, and infect so many people almost simultaneously in three separate areas of the world? To trace how we may have gotten to this point and to determine whether the pandemic is natural or somehow man-made as a result of the introduction of other agents, we need to first examine the theories and history of AIDS and then go back in time to a Siberian labor camp during the height of World War II. For after spending decades trying to unravel the mystery and origins of HIV, some experts believe that in the midst of that frozen tundra lies the secret to how it all began.

Monkeys, Chimps, and SIV

One of the first theories proposed for the origin of AIDS was that in the late 1970s a monkey virus similar to HIV, called simian immunodeficiency virus (SIV), jumped species, perhaps by a bite, into the human population. Following this single event, the virus spread from one infected individual to millions of people through sexual promiscuity, blood transfusions, and dirty needles. Sometime during the epidemic, the virus migrated out of Africa and into Haiti and the United States. To accept this theory, one needs to make several assumptions, all of which have yet to be proven.

First, SIV and HIV are dissimilar enough that when HIV is injected into the green monkey it has no effect at all. Second, experts believe that it is statistically impossible for HIV to have spread from a single individual who may have been bitten to so many individuals in such a short time. To counter this problem, other theories have been offered suggesting that the transmission might have occurred a hundred or more years earlier but that the disease may not have been recognized. One scenario claimed that HIV was found in the stored blood of a sailor from Manchester, England who had traveled to central Africa, thus proving that HIV had been around for at least forty years. However, when more sophisticated blood tests were performed (earlier tests could not distinguish between HIV and lymphadenopathy associated virus, or LAV), the tests proved not only that the seaman hadn't died from AIDS as previously believed but that many blood samples initially thought to contain HIV were "false positive." In fact, the oldest known case of HIV identified by rigorous genetic testing so far is from 1976, and no stored African tissue from the early 1970s has tested positive for HIV.

Although HIV and SIV were thought to be closely related, another monkey wrench was thrown into the theory, so to speak. A second AIDS-related virus was discovered in Senegal and named HIV-2 to distinguish it from the original HIV-1. Upon further analysis, scientists discovered that HIV-1 mutated much faster than HIV-2 or SIV and that they were only similar in 50 percent of their gene sequences. According to Harvard AIDS specialist and coauthor of *The Origins of the AIDS Virus* Max Essex, "This 50 percent similarity in the genetic sequence is not close enough to make it (HIV-1) a descendent of SIV." Today scientists concur that most SIVs are so different in genetic makeup that no amount of evolution could have transformed them into HIV. Rather, if HIV mutated from a closely related virus, it probably did so from a harmless HIV-1 progenitor that was somehow converted to a virulent strain.

The most recent theory involves chimpanzees rather than green monkeys. An international team of investigators led by Dr. Beatrice Hahn of the University of Alabama at Birmingham has purportedly identified SIV-cpz in chimpanzees. The following factors have convinced some experts that this may be the evidence the world has been looking for. First of all, the samples taken from chimpanzees strongly resemble the different subgroups of HIV-1; and second, the natural habitats of these chimpanzees coincide with the pattern of the HIV-1 epidemic in that part of Africa. Based on molecular analysis, Dr. Hahn concluded that chimpanzees are the natural reservoir of HIV-1 and have been the source of cross-species transmission to the human population.

If anything illustrates how a theory such as the monkey or chimpanzee hypothesis can develop on the basis of false assumptions and misleading information, it is the story of a University of Arkansas professor who found traces of HIV in tissue samples from a dead chimpanzee and led experts to believe that a definitive origin of AIDS had been uncovered. In her book *The Monkey Wars*, published by Oxford University Press, author Deborah Blum recounts how the dead chimpanzee was actually born in the United States twenty-three years earlier, lived its entire life inside a U.S. military research facility, and was used for experiments on immunosuppressive disease development. Apparently when storage space at the government laboratory was running out, the military sent out requests for anyone who wanted a dead chimpanzee. Jumping at the chance of obtaining anything she could, the professor quickly replied. When she tested the chimp for HIV, she had no idea that what she'd actually discovered was an HIV-infected chimpanzee born and raised in the very laboratory from which it was shipped. So despite the fact that of all naturally occurring SIVs, SIVcpz most closely resembles HIV-1, the chimpanzee theory needs additional scrutiny before it can be given serious weight.

Indeed, there are a number of competing, as yet unproven, sometimes conflicting, theories concerning the origin and spread of AIDS.

Polio Vaccine Theory

Anyone who'd ever seen a crippled child hobbling on crutches or lying helplessly in an iron lung must have hailed the development of the polio vaccine as one of the greatest humanitarian contributions of the century. It all began in 1952 when Jonas Salk developed a vaccine using a mixture of three types of poliovirus grown in monkey kidney cells. When massive testing began in 1954 throughout the United States and Canada, no one suspected that Salk's vaccine might have been contaminated with Simian Virus 40 (SV40), a cancer-causing monkey virus. But according to Dr. Edward Shorter, commissioned by the National Institutes of Health (NIH) to write a one-hundred-year history of NIH's medical accomplishments, live monkey viruses were discovered in polio vaccines shortly after the testing began. Max Essex also reported that the monkey species used to produce live polio vaccine in the United States was a reservoir of SIV. The problem was that testing was not sensitive enough to pick up SIV contaminants in the range of one hundred viruses per dose.

This would not have been so bad if the viruses only caused cancer in birds as was previously thought but, to the horror of investigators, the polyoma virus caused cancer in virtually every animal tested. Nonetheless, in 1955, the U.S. government granted permission to distribute the vaccine to every American child, and for the next decade the tainted vaccine was administered to tens of millions of individuals. More significantly, many children younger than one month were given the vaccine before their immune systems were fully developed, and in some cases were given much higher doses to ensure that the vaccine worked properly.

Within the scientific community alarm over SV40 increased because it was frightening to consider how many individuals were infected and the problems that would ensure. Dr. Ben Sweet, author of *The Vacuolating Virus: SV40* and one of the scientists who developed and field-tested the vaccines, was especially concerned that researchers "had no idea what this virus would do." In an interview with *Chronic Ill Net*, he said, "First, we knew that SV40 had oncogenic properties which was bad news. Second, we found out that it hybridized with certain DNA viruses—like adeno virus—such that the adeno virus would then have SV40 genes attached to it. We couldn't clean up the adeno virus vaccine seed stocks grown in monkey kidney cells only."

Possibly tens of millions of people were exposed to the virus regardless of whether it was live or killed polio vaccine. Dr. Sweet later recognized that papers written about formalin-killed vaccines being free of SV40 had been incorrect and that updated information never was published because by then the mass inoculation had already begun. The other concern is that SIV could conceivably have been present in the original vaccines. Even more frightening, according to microbiologist Dr. Howard Urnovitz, who presented a paper at the Eighth Annual Conference on AIDS in Houston, Texas, is that HIV-1 may have originated from contaminated polio vaccines through recombination of human genes. Urnovitz says, "It is very likely that HIV-1 may have been a result, and that it may in fact be a monkey–human hybrid."

As we learn more about SV40 and SIV and develop more advanced techniques in molecular biology, we could find that anyone who ever received contaminated polio vaccine is susceptible to a host of diseases. Barbara Loe Fisher, president of the National Vaccine Information Web, warns consumers that some vaccines are still grown on African green monkey tissues and companies cannot be certain whether cross-species transfers are going on. "With two hundred vaccines in the research pipeline," she says, "more than one hundred in clinical trials, and scores on the brink of being licensed, vaccine research had better get back to basic science before another AIDS epidemic is created in a vaccine lab."

In the past few years, contrary evidence has surfaced refuting the theory that polio vaccine was contaminated or had any role in the spread of AIDS, and scientists on both sides are still searching records and research data in hopes of settling the controversy once and for all.

WHO Smallpox Vaccine Theory

One of the worst scourges of Third World nations, global smallpox was essentially eradicated by the early 1980s. The United States saw its last case of smallpox in 1949. However, some experts looking into the massive smallpox vaccine eradication program sponsored by the World Health Organization (WHO) in the late 1970s now suspect that the vaccines may have been contaminated. Alan Chase, author of *Magic Shots*, found that WHO used "two hundred thousand people in forty countries, most of them non-doctors trained by seven hundred doctors and health professionals from over seventy participating countries, spent $300 million, and used forty million bifurcated vaccinating needles to administer 2.4 billion doses of smallpox vaccine."

All of the vaccines used in the program had been supplied by the United States. According to C. Piller and K. R. Yamamoto, the vaccinia virus used to make smallpox vaccine can be easily manipulated by genetic engineering. In their book *Gene Wars: Military Control Over the New Genetic Technologies*, they wrote, "Researchers have been able to splice genes coding for the surface coats of other viruses, such as influenza, hepatitis, and rabies into vaccinia virus DNA. The result: a broad spectrum vaccine with a coat of many colors."

Five years later, the *London Times* published a front-page story about WHO's smallpox program and its alleged involvement in the AIDS epidemic. In that 1987 article, Dr. Robert Gallo, the man credited with discovering HIV, is quoted as saying, "The link between the WHO program and the epidemic is an interesting and important hypothesis. I cannot say that it actually happened, but I have been saying for some years that the use of live vaccines such as that for smallpox can activate a dormant infection such as HIV (the AIDS virus)." It was reported that in some areas of

Africa, India, Nepal, and Pakistan, as many as 60 percent of those inoculated with smallpox vaccines developed AIDS within five years.

One fact of which many people are unaware is that health agencies have been busy testing older and no longer sexually active people in Africa and have not found HIV infections in these individuals. If, indeed, AIDS has been around for many years (some say hundreds of years), then some of these older individuals would certainly have been infected at some point in their lives. Instead, there were three simultaneous outbreaks: New York, Africa, and Haiti. According to some accounts, the infections in Haiti were the result of some fourteen thousand returning Haitian UN workers on assignment in central Africa, who coincidentally all received WHO smallpox vaccinations in the early 1970s. Since AIDS seemed to have emerged so suddenly in the late 1970s and early 1980s, one possible explanation for the explosion of the disease in Africa theoretically could be the mass inoculation program.

That said, WHO has dismissed the theory and said there is no causal link between AIDS and the eradication of smallpox. Also, no clinical proof of the theory has been presented.

The Mycoplasma Connection

When Dr. Garth Nicolson first reported mycoplasma infections in returning Gulf War veterans, a collective shudder must have shot through the community of scientists who knew that it wouldn't be long before the secret about mycoplasma got out—because it wasn't just Gulf War illness they were concerned about. Investigators looking into the origin of AIDS had been pointing to past mycoplasma research, which began in the 1950s as part of the U.S. biological weapons program. However, mycoplasma research took a giant step forward in 1962 when Dr. Len Hayflick set up a mycoplasma laboratory at Stanford University, in setting the stage for the NCI's Special Virus Cancer Program.

In 1969, the Pentagon asked for funds to develop an immunosuppressive agent, and in 1971, less than two years later, an article appeared in *Federation Proceedings* entitled "Viral Infections in Man Associated with Acquired Immunological Deficiency States." Was it coincidental that a paper about an AIDS-like virus came out at a time when mycoplasma research was focusing on immune function? Perhaps. However, considering the history and uses of viruses and mycoplasma by the United States military, some experts are not so sure. During that same time period (1972), WHO proposed in their annual bulletin that "an attempt be made to ascertain whether viruses can in fact exert selective effects on immune function, e.g., by affecting T-cell functions as opposed to B-cell function. The possibility should also be looked into that the immune response to the virus may be impaired if the infecting virus damages more or less selectively the cells responding to the viral antigens." The description sounds very much like the characteristics of HIV.

Of the two hundred or so species of mycoplasma, the smallest and simplest of self-replicating bacteria, only five are pathogenic, and a few of those have been genetically engineered for increased virulence. One of those pathogenic mycoplasmas was issued U.S. Patent No. 5,242,820 on September 7, 1993. Joint holders of the patent are the U.S. military and Dr. Shyh-Ching Lo, senior research scientist at the Armed Forces Institute of Pathology. According to Dr. Maurice Hilleman, chief virologist with the pharmaceutical giant Merck, mycoplasma is probably carried by most people throughout the world. Other experts claim that the highly infectious agent may be responsible for cancer, multiple sclerosis, chronic fatigue syndrome, type 1 diabetes, Parkinson's disease, rheumatoid arthritis, Alzheimer's disease, and AIDS.

The most sinister characteristic of mycoplasma is that it enters individual cells where it can lie dormant for as long as thirty years until it's triggered by a physical event such as vaccination, stress, or disease. If it invades nerve tissue, the individual develops neurological disorders; if digestive tissue is attacked, diseases such as Crohn's colitis may result. Mycoplasmas have derogatorily been nicknamed the "crabgrass of cell cultures" because, like crabgrass, they're persistent, infectious, and difficult to get rid of. According to a statement made by Dr. Charles Engel of Walter Reed Army Medical Center at a February 7, 2000 NIH meeting, "I am now of the view that the probable cause of chronic fatigue syndrome and fibromyalgia is mycoplasma."

It's believed that the mycoplasma disease agent developed by the U.S. military was engineered from the *Brucella* bacterium. Some experts claim that in an attempt to create a "weaponized" form of the bacteria, scientists combined it with visna virus and reduced it to a crystalline form for storage and future deployment as an aerosol or through a vector such as an insect. Dr. Donald MacArthur, speaking for the Pentagon before a 1969 congressional committee, testified that mycoplasma at a certain strength develops into an immune suppressor that bypasses the body's natural defenses and kills the infected individual. A less virulent form simply causes chronic fatigue and wasting but cannot be detected because it only crystallizes at a pH of 8.1 whereas the body's pH is 7.4.

As early as 1949, evidence existed that *Brucella* might be linked to debilitating diseases like multiple sclerosis. Drs. Kyger and Haden in their article "Brucellosis and Multiple Sclerosis" claimed that "multiple sclerosis might be a central nervous system manifestation of chronic brucellosis." Of the 113 MS patients they tested, 95 percent were infected with *Brucella*, suggesting that many diseases may very well be triggered by the rampant spread of *Brucella* and/or mycoplasma. Individuals who want to be tested and receive treatment may contact the Institute for Molecular Medicine, 15162 Triton Lane, Huntington Beach, CA 92649, (714) 903-2900 (www.immed.org).

In the case of AIDS, recent evidence has shown that a significant number of HIV-infected patients are also infected with mycoplasma. At the Sixth International Conference on AIDS, Dr. Luc Montagnier, codiscoverer of the AIDS virus, hypothesized that mycoplasma is a major cofactor in the development of AIDS and reported that he'd found mycoplasma in about 30 percent of AIDS patients. One *in vitro* study found that the probability of showing HIV-1 expression in a cell group infected with both HIV-1 and mycoplasma is forty times greater that when the cell group is infected with HIV-1 alone. This remarkable experiment demonstrates the significance of mycoplasma as an infectious disease agent and the rationale for using tetracyclines such as doxycycline as a treatment.

So far, every mycoplasma cultivated and identified has been a parasite of humans, animals, or plants. Dr. Garth Nicolson's discovery of a mycoplasma strain in returning Gulf War veterans containing part of the HIV protein coat was not that much of a surprise to those familiar with mycoplasma or HIV research. In the 1990 issue of *Current Topics in Microbiology and Immunology*, for example, Dr. P. O. Brown stated that retroviruses such as HIV have been widely used as vectors for genetic engineering and are likely to be the first vectors for introducing foreign genes into cellular chromosomes. Moreover, the *Journal of Virology* and the international journal *Cell* have published detailed articles describing new gene therapy techniques in which parts of the HIV-1 are altered and packaged as a system for delivery of components into human cells.

Could a relatively benign mycoplasma have been genetically modified into a highly invasive and pathogenic microorganism? Certainly the technology was there. And based on the government's track record of research into other weapons systems, why wouldn't they make use of an ideal vector that would transport dangerous microorganisms into cells? Dr. Garth Nicolson's lab continues to investigate the role of mycoplasma in Gulf War veterans, but other researchers are also finding suspicious links between a new strain of mycoplasma, *Mycoplasma fermentans incognitus*, and a rash of epidemic diseases. Hopefully, an answer to how mycoplasma could have spread so quickly and infected so many will come sooner rather than later.

Special Virus Cancer Program

Soon after declaring a "war on cancer" in 1971, President Richard Nixon quadrupled the NCI's budget and converted Fort Detrick, the U.S. Army's biological weapons research facility, into the world's foremost cancer research laboratory. Part of the NCI's budget was used to continue the Special Virus Cancer Program (SVCP) that began in 1964 and whose mission was to identify existing cancer-causing viruses and to develop new viruses in order to study how viral infections trigger tumor growth. However, despite the fact that Nixon renounced germ warfare research except for defensive and medical purposes, the army's bioweapons program remained intact and continued under a special and secret section of the NCI.

Like a beehive of scientific activity, men and women in white lab coats were busy inoculating thousands of primates—monkeys, chimpanzees, and marmosets—imported from western Africa and Asia. Following experiments, the animals either died from cancer or immune diseases, were transferred to other laboratories around the world, or, according to some reports, were rehabilitated and released back into the wild, even after they were infected with viruses.

By 1972, the research facility had produced enough cancer-causing viruses to fill a sixty-thousand-liter tank. Cooperation between the NCI and private pharmaceutical companies, though already a decade old, had intensified, resulting in the production of large quantities of immunosuppressive monkey viruses and other animal viruses to be grown in human cell lines. Before long, scientists learned to transfer cancer viruses from one species to another. In his book *Queer Blood*, Dr. Alan Cantwell wrote, "Chicken viruses were put into lamb kidney cells; baboon viruses were spliced into human cancer cells; the combinations were endless. In due process, deadly man-made viruses were developed, and new forms of cancer, immunodeficiency, and opportunistic infections were produced when these viruses were forced or adapted into laboratory animals and into human tissue cell cultures." Four years later, Dr. Seymor Kalter, an NCI scientist, managed to blend viral genes from mice and baboons to create a new virus that caused cancer in monkeys and chimpanzees. Once this "species barrier" was broken, an entire new world had suddenly opened up for virologists.

So chilling was this new development that Lawrence Loeb and Kenneth Tartof of the Institute for Cancer Research in Philadelphia, Pennsylvania, called for an outright ban on the work. They are quoted in *Science* as saying, "The production of malignant tumors in a variety of primate species suggests the possibility of creating viruses that are oncogenic for humans. Therefore, we urge that all experiments involving co-cultivation of known oncogenic viruses with primate viruses be immediately halted until the safety of such experiments are extensively evaluated."

The SVCP's 1971 report, *The Special Virus Cancer Program: Progress Report No. 8*, which is almost impossible to find, states that several contracts were awarded indicating that some of the research was used to develop AIDS-like viruses.

The flowchart included in Progress Report No. 8 is a road map to what the NCI scientists were doing and what their ultimate goals were. When former Congressman James Trafficant saw the report for the first time and immediately called for a General Accounting Office investigation into the SVCP, his regional director said, "The information is very shocking and very revealing. Our government is supposed to be sensitive to the people's concerns and the flowchart was very revealing to me." Even today, not many congressional leaders have ever seen Progress Report No. 8.

Dr. Boyd Graves, who presented the flowchart to the congressman, claims that Phase IV hints at a cure for AIDS and that on page 24 of the report itself the drug *n*-demethyl rifampicin was shown to stop the virus in its tracks. Page 24, paragraph 2 of Progress Report No. 8 states:

> Intensive investigations have now revealed polymerase activity in cells of patients with acute lymphoblastic leukemia; more preliminary evidence has shown the enzyme is in cells of sarcomas, Burkitt's lymphoma, and breast cancer. Since the RNA-dependent DNA polymerase is apparently always present in the RNA tumor viruses of animals, its discovery in the human tumor cells offers good supportive evidence that viruses are associated with cancers in men. The RNA-dependent DNA polymerase of human leukemia cells is inhibited by a drug, *n*-demethyl rifampicin, which also inhibits the enzyme activity found in the type C RNA tumor viruses of animals. Studies are underway to explore the action of this drug and the various modifications of it. These investigations could provide new approaches to the treatment of malignancies in man.

Included in Progress Report No. 8, page 61, is a flowchart of experiments, illustrating the government's plan to produce viruses that suppress immunity and induce and maintain cancer. For reasons unexplained, the flowchart was not included in the report until 1999.

To suppress the immune system and ensure that primates would develop cancer, researchers needed to prime the animals with drugs, chemicals, radiation, or immunosuppressive agents. In many cases, the thymus, which produces T cells, was removed to weaken resistance. Viruses were then injected to trigger tumor growth.

After thousands of such experiments, the SVCP was shut down in 1977, but not before it had developed groundbreaking techniques in molecular biology. Anyone who believes that we did not yet have the ability to manipulate viral genetic material when AIDS was first discovered never heard of the SVCP as the real birthplace of genetic engineering and the possible source of a virus that, as a Pentagon official told Congress in 1969, does not naturally exist and for which no natural immunity could be acquired.

AIDS and the Hepatitis Vaccine Experiments

Dr. Wolf Szmuness, born in Poland in 1919, was attending medical school in Lublin, Poland when Nazi forces invaded his country in the summer of 1939. Following the partition of Poland by Germany and the Soviet Union, Szmuness was taken prisoner and, because he was living in eastern Poland at the time, was sent to a Soviet labor camp in Siberia. Luckier than most of his family members who'd died at the hands of the Nazis, Szmuness survived exile in Siberia, was released in 1946, and finished medical school in Tomsk in central Russia. It was there, at a time when his Soviet wife had nearly died of hepatitis, that Szmuness vowed to devote the rest of his life to hepatitis research.

In 1959, Szmuness returned to Poland as an epidemiologist working for several health departments. Ten years later, as fate would have it, he was allowed to take his wife and daughter to a scientific meeting in Italy. At that time of political unrest, it was impossible to leave Russia and take family members abroad for fear of defections, so when the opportunity presented itself, Szmuness promptly defected and emigrated to the United States. Penniless and with no more than a few suitcases to his name, Szmuness made some contacts and landed a job as a laboratory technician at the New York City Blood Center. What is so remarkable about this story is how quickly Szmuness was made head of the epidemiology department at the blood center and a professor of epidemiology at the Columbia University School of Public Health. At the same time, his vow to devote his life to the study of hepatitis had made him a worldwide authority on the disease and the recipient of millions of grant dollars for hepatitis research.

The highlight of Szmuness's career was no doubt the introduction in 1978 of a hepatitis B vaccine, which was the subject of an experiment in the New York City gay community. A controversial theory has been put forward that the vaccine inadvertently contributed to the spread of AIDS in America. Most experts later discounted the hepatitis vaccine theory because HIV, according to reports, had ultimately been discovered in blood samples taken prior to the 1970s vaccine program. Moreover, the CDC has also indicated that there was no evidence of vaccine contamination. Nevertheless, the theory remains interesting for what it tells us about what could conceivably happen.

The experiment began in November 1978 following a newspaper advertising campaign to recruit only sexually promiscuous gay men under the age of forty. One of the more prominent ads read, "Last Chance for Gay Men to Join the Hepatitis B Vaccine Program. Enrollment closes in June, after which the vaccine may not be available for several years." More than ten thousand men signed up, but only the most promiscuous were selected. Heterosexuals were excluded from the study altogether.

New York public health officials knew they had a serious problem when they saw the astonishingly high rates of venereal disease and hepatitis B in the gay community. Some reports had the rate as high as 50 percent, a health time bomb just waiting to explode and infect even the heterosexual population. The sexual revolution of the 1970s included large numbers of gays coming out and becoming more sexually active with multiple partners. It was one of the reasons, writes Dr. Alan Cantwell in *Queer Blood*, that gays were a more despised minority than even blacks and Jews.

A few months after 1,083 homosexuals were injected with the experimental hepatitis vaccine, physicians began noticing that men from Greenwich Village were showing up with purple skin lesions all over their bodies. Kaposi's sarcoma, a disease not seen in young men in the United States before AIDS, was suddenly targeting only young, gay, promiscuous white males. In addition, victims were also infected with *Mycoplasma penetrans* and a new strain of herpes virus closely related to a cancer-causing herpes virus isolated from primates. The hepatitis B theory held that the mycoplasma and the new herpes virus were somehow accidentally introduced along with HIV into the gay community.

By 1980, a year before AIDS was acknowledged, 20 percent of the subjects were HIV-positive, which at the time would have been the highest incidence of HIV in the world, including central Africa where AIDS supposedly originated. Of course, few volunteers would have appreciated that the vaccine they were administered had been developed in chimpanzees and made from pooled blood serum of hepatitis-infected homosexuals. According to Dr. Alan Cantwell, molecular biologists later discovered, with the help of advanced techniques, that the particular HIV strain infecting homosexuals had a remarkable affinity for rectal tissue cells as opposed to the HIV strain in Africa, which had an affinity for vaginal and cervical cells in women and cells of the penis foreskin in men.

March 1980 marked the start of experimental vaccination programs in Los Angeles, San Francisco, St. Louis, Denver, and Chicago. In San Francisco alone, more than seven thousand gay men were recruited. At about the same time, the first AIDS case was observed in that city; and by 1981, the CDC was looking into twenty-six AIDS cases, all of them in previously healthy homosexuals living in New York, Los Angeles, and San

Francisco. A year later, 30 percent of the volunteers were HIV-positive, an astoundingly high rate that exceeded anything even seen in Africa. And ten years after the study was terminated, a very large percentage of the young men participating in the vaccine study were either infected or had died of AIDS. The mainstream theory of AIDS transmission would, of course, hold that this high incidence of AIDS infection was attributable to the fact the vaccine experiment volunteers were selected precisely because they were sexually promiscuous. Most researchers now believe the hepatitis vaccine experiment and the onset of the AIDS epidemic were simply coincidental.

A related alternative theory, proposed by Mathilde Krim of the American Foundation for AIDS Research, speculates that homosexuals may have received contaminated gamma globulin, a blood product used to temporarily boost immunity against a disease like hepatitis. The gamma globulin supposedly was produced from contaminated blood obtained from prisoners in Africa and the Caribbean during the 1970s. The problem with this theory is that, once again, there are no samples of HIV-positive blood prior to 1978 in either Africa or the Caribbean. This would tend to disprove Krim's theory.

What biologists were doing in the mid-1970s was no longer earth shattering, but it *was* becoming dangerous. According to Horowitz, "Specific enzymes and other biochemical processes needed to induce immune system collapse were identified. In 1971, Fujioka and Gallo designed experiments in which tumor specific cell tRNA was added directly to normal human white blood cells. To achieve this, simian monkey virus (SV40) and mouse parotid tumor (polyoma) virus were routinely employed to deliver foreign cancer-causing tRNA into these normal human white blood cells. . . . Gallo and other researchers commonly modified monkey viruses enabling these viruses to induce AIDS-like immunosuppression, cancer, and wasting and death in primates and lower animals."

The argument for a "vaccine-induced" origin of AIDS becomes more plausible when we also look at three contaminants possibly found in the experimental vaccines: SV40, SIVagm from the African green monkey, and simian foamy retroviruses. Researchers studying simian foamy retroviruses say that they can easily cross species barriers and mutate.

So the proponents of the SIV theory, in which SIV mutated naturally into HIV, may have gotten it only partially correct. It is at least possible that SIV may have gotten some help from agents introduced into experimental vaccines, inadvertently contributing to the most devastating mutation and cross-species infection in the history of the human race.

This brings us back to the original question, "Could AIDS inadvertently have been an ethnic or genetic weapon?" It theoretically could have been, albeit unknowingly. But for every theory proposed so far, there are opposing points of view that lead us in so many new directions that it may be a long time before we know for sure when and where AIDS really began. We also know that research into the feasibility of ethnic weapons is ongoing and progressing even faster since the completion of the Human Genome Project. What lies ahead may be unimaginable; for what science is discovering each day is bringing us closer to unraveling the molecular secrets of life itself. A decade from now, we may all wish that we hadn't been so anxious to unravel those secrets so quickly.

9 WHAT THE FUTURE HOLDS: HUMAN EXPERIMENTATION IN THE TWENTY-FIRST CENTURY

In Aldous Huxley's 1932 novel, *Brave New World*, individuals of every caste are given a drug called soma to ensure that no one ever feels pain or is unhappy. From the moment an egg is fertilized, it's conditioned to become part of Utopia in the year 632 A.F. (After Ford). Collectively, society is governed by science and technology, populations are dependent on artificial birth, babies are mass produced in reproductive hatcheries and bred in conditioning centers, and family life is a thing of the past. It's a world gone mad, ruled by those with little use for feelings and emotions and whose vision of life is as sterile as the hatching tubes from which they'd come.

Likewise, who could forget the scene in H. G. Wells's *The Island of Doctor Moreau* when shipwrecked gentleman Edward Prendick stumbles across the monstrous man-animal creations of a doctor who later tells him, "The study of nature makes a man at last as remorseless as nature." In his relentless and sadistic drive to reshape and transform life itself, Doctor Moreau symbolizes the ultimate goals of science and medicine: to improve upon nature and to use knowledge for the betterment of mankind, no matter the cost or the consequences. It was a theme repeated in Michael Crichton's *Jurassic Park*, where genetic engineering goes terribly awry and creates a population of out-of-control dinosaurs.

The bizarre and sometimes ghastly worlds that Huxley, Wells, and Crichton invented now seem like windows into the future, where advances in science and technology promise to take us to places no one had thought we would ever go. Gene therapy, cloning, stem-cell technology, tissue harvesting, DNA vaccine technology, nanotechnology, cryogenics, and cross-species transplantation are only the tip of the iceberg in a sea of technologies that scientists will make routine within a decade. Because of the remarkable advances we've made over the last twenty years, we're on the verge of being able to direct and shape our own evolution.

A glimpse into our twenty-first century world was recently offered by the U.S.-based BioTransplant Incorporated and the Australian firm Stem Cell Sciences when they announced a process they had jointly developed to produce a "human–pig" hybrid embryo. The technique, called nuclear transfer, involves taking a pig egg cell, removing the nucleus, inserting human DNA in its place, and then growing the cells in medium to allow them to divide and produce harvestable human stem cells. By replacing the original pig nucleus with a human nucleus, the cell, according to Peter Mountford of Stem Cell Sciences, reprograms itself to begin dividing as if it were back in the very first stages of development. He added that the company's research has proven that human and animal cells could be successfully fused for therapeutic cloning.

The fact that they sought a government patent for their technique (Patent No. W099/21415) indicates that the methodology had been well established and that research on fusing human and animal cells was well underway. However, most of those shocked by the announcement were not aware that another company, Advanced Cell Technology of Worcester, Massachusetts had already fused human and cow cells to create a "human–cow" embryo for the purpose of obtaining organ and tissue transplants. Most people today are also not aware that, although U.S. government law forbids federal funding for such work, private companies and privately funded scientists can do whatever they please and are no doubt doing so in hundreds of laboratories around the world.

Another disturbing reality, cross-species transplantation, hit the front page in 2000 when French scientists implanted jellyfish genes into a rabbit embryo to see if mammals could take on characteristics of other species. The result was a white rabbit that glowed slightly green under a blue light, much like the bioluminescence exhibited by jellyfish. When examined under a microscope, the rabbit's cells also glowed like jellyfish cells. Halfway around the globe, Amrad, an Australian biotechnology company, has acquired a patent for creating cross-species embryos containing cells from humans, mice, sheep, pigs, cows, and goats.

Huxley's brave new world and Wells' volcanic island are no longer fiction. The speed at which science is moving has taken many of us, even in the scientific community, by surprise. Experiencing the most astonishing growth is the field of molecular genetics, which has brought us in a few short years from the discovery of DNA to the potential cloning of a human being. Things once thought impossible are now trivial. And unless we consider the consequences of future unrestrained research, our drive for scientific discovery may very well lead us to our own Jurassic Park.

Human Chop Shops: The Booming Business of Tissue Harvesting

The sun had barely risen over a small Midwestern city when a technician outside the women's health clinic slung two plastic bags filled with yesterday's discarded fetal body parts above her head and into an open dumpster. That's because the sink disposal, where leftover fetuses are usually ground up and flushed into the city's sewer system, was not working properly. The green bags thumped against the metal wall, falling into a heap with the rest of the week's garbage.

Inside the clinic, another technician was busy thumbing through a computer-generated list of order forms from researchers around the world. Today's requests included kidneys, brains, a spinal cord, lungs, one leg with hip attached, eyes, a thymus gland, and two livers. The orders are carefully matched with patients scheduled to come in that day and have abortions. Throughout the rest of the morning and afternoon, fetuses, some as old as thirty weeks, will be extracted (sometimes killed outside the womb if necessary), dissected in assembly line fashion, packaged in dry ice, and shipped by UPS, FedEx, and Airborne to labs, pharmaceutical companies, universities, and clinics for use in medical experiments.

Although federal law prohibits the outright sale of human tissue or body parts, President Clinton's 1993 executive order overturning the ban on taxpayer funding of research on aborted fetuses opened the floodgates for a booming new growth industry: the harvesting and sale of baby body parts. Less than a year after the executive order was signed, the U.S. National Institutes of Health (NIH), which operates its own twenty-four-hour collection service at sponsored abortion clinics, issued matter-of-fact guidelines and information about its harvesting services. The following is taken from the March 11, 1994 NIH guide, *Availability of Human Fetal Tissue*:

> Human embryonic and fetal tissues are available from the Central Laboratory for Human Embryology at the University of Washington. The laboratory, which is supported by the National Institutes of Health, can supply tissue from normal or abnormal embryos and fetuses of desired gestational ages between 40 days and term. Specimens are obtained within minutes of passage and tissues are aseptically identified, staged and immediately processed according to the requirements of individual investigators. Presently, processing methods include immediate fixation, snap fixation, snap freezing in liquid nitrogen, and placement in balanced salt solutions or media designated and/or supplied by investigators. Specimens are shipped by overnight express, arriving the day following procurement. The laboratory can also supply serial sections of human embryos that have been preserved in methyl Carnoy's fixative, embedded in paraffin and sectioned at 5 microns.

To avoid the appearance of impropriety or to hide evidence that a sale is being made, abortion clinics skirt the law by renting on-site lab space to harvesting companies that basically serve as middlemen between doctors performing abortions and researchers needing body parts. The abortion clinic is paid a site fee to allow harvesters to set up a "chop shop" where they collect and dissect "donated" aborted babies. Harvesters, in turn, donate the body parts to researchers. So, rather than selling tissue directly, the harvesting companies are paid for "retrieval services" and "shipping fees" and not for body parts.

Picked apart and sold in pieces, a single baby could fetch as much as fourteen thousand dollars. One such company, Opening Lines, Inc. of West Franklin, Illinois processes more than fifteen hundred fetuses a day and openly advertises "the highest quality, most affordable, freshest tissue prepared to your specifications and delivered in the quantities you need, when you need it." Their prices, according to corporate president Dr. Miles Jones, are determined by market forces and by how much buyers are willing to pay for human tissue. The company brochure, which encourages abortionists to "turn your patient's decision into something wonderful," offers a detailed fee schedule that includes the following rates:

Unprocessed Specimen (> 8 weeks)		$70
Unprocessed Specimen (< 8 weeks)'		$50/
Livers (< 8 weeks)	*30% discount if significantly fragmented*	$150
Livers (> 8 weeks)	*30 % discount if significantly fragmented*	$125
Spleens (< 8 weeks)		$75
Spleens (> 8 weeks)		$50
Pancreas (< 8 weeks)		$100
Pancreas (> 8 weeks)		$75
Thymus (< 8 weeks)		$100
Thymus (> 8 weeks)		$75
Intestines and Mesentery		$50
Mesentery (< 8 weeks)		$125
Mesentery (> 8 weeks)		$100
Kidney—with / without adrenal (< 8 weeks)		$125
Kidney—with / without adrenal (< 8 weeks)		$100
Limbs (at least 2)		$125
Brain (< 8 weeks)	*30 % discount if significantly fragmented*	$999
Brain (> 8 weeks)	*30 % discount if significantly fragmented*	$150
Pituitary Gland (> 8 weeks)		$300
Bone Marrow (< 8 weeks)		$350
Bone Marrow (> 8 weeks)		$250
Ears (< 8 weeks)		$75
Ears (> 8 weeks)		$50
Eyes (< 8 weeks)	*40% discount for single eye*	$75
Eyes (> 8 weeks)	*40% discount for single eye*	$50
Skin (> 12 weeks)		$100
Lungs and Heart Block		$150
Intact Embryonic Cadaver (< 8 weeks)		$400
Intact Embryonic Cadaver		$600

(> 8 weeks)	
Intact Calvarium	$125
Intact Trunk (with / without limbs)	$500
Gonads	$550
Cord Blood (Snap Frozen LN_2)	$125
Spinal Column	$150
Spinal Cord	$325

Prices in effect through December 31, 1999

One reality kept hidden from the public is that babies often need to be manipulated into the proper position and slowly butchered alive during the harvesting process to ensure that the valuable goods are not damaged. A 1990 article in *Archives of Neurology* describes abortion techniques that take three to four times longer than normal in order to preserve tissue and obtain the best samples possible. The longer the procedure, the longer the baby is subjected to the torture.

Here's a partial transcript of testimony from a July 1997 civil court case brought by University of Nebraska contract harvester Dr. Leroy Carhart, challenging Nebraska's prohibition on certain abortion techniques.

Carhart: My normal course would be to dismember the extremity and then go back and try to take the fetus out either foot or skull first, whatever end I can get to first.

Attorney: How do you go about dismembering that extremity?

Carhart: Just traction and rotation, grasping the portion that you can get a hold of, which would be usually somewhere up the shaft of the exposed portion of the fetus, pulling down on it through the os, using the internal os as your counter-traction and rotating to dismember the shoulder or the hip or whatever it would be. Sometimes you will get one leg and you can't get the other leg out.

Attorney: In that situation, when you pull on the arm and remove it, is the fetus still alive?

Carhart: Yes.

Attorney: Do you consider an arm, for example, to be a substantial portion of the fetus?

Carhart: In the way I read it, I think if I lost an arm that would be a substantial loss to me. I think I would have to interpret it that way.

Attorney: And then what happens next after you remove the arm? You then try to remove the rest of the fetus?

Carhart: Then I would go back and attempt to either bring the feet down or bring the skull down, or even sometimes you bring the other arm down and remove that also and then get the feet down.

Attorney: At what point is the fetus . . . does the fetus die during that process?

Carhart: I don't really know. I know that the fetus is alive during the process most of the time because I can see fetal heartbeat on the ultrasound.

Attorney: At what point in the process does fetal demise occur between initial removal . . . removal of the feet or legs and the crushing of the skull, or I'm sorry, the decompressing of the skull?

Carhart: Well, you know, again, this is where I'm not sure what fetal demise is. I mean, I honestly have to share a concern, your Honor. You can remove the cranial contents and the fetus will still have a heartbeat for several seconds or several minutes, so is the fetus alive? I would have to say probably, although I don't think it has any brain function, so it's brain dead at that point.

Attorney: So the brain death might occur when you begin suctioning out of the cranium?

Carhart: I think brain death would occur because the suctioning to remove contents is only two or three seconds, so somewhere in that period of time, obviously not when you penetrate

> the skull, because people get shot in the head and they don't die immediately from that, if they are going to die at all, so that probably is not sufficient to kill the fetus, but I think removing the brain contents eventually will.

The gruesome secret of fetal dismemberment and harvesting was revealed publicly in 1999 when *20/20* televised an investigative report and nationally syndicated columnist Mona Charen described a typical day at one firm trafficking in body parts. Interviewing a technician—sometimes referred to as a fetal tissue procurement technician—Charen describes how the young woman collected fetuses from late-term abortions and then dissected them in order to obtain needed parts. According to the technician, almost all of the specimens were "perfect" and many were at least seven months old.

However, nothing could have prepared the medical technician for what she was about to experience one day when a set of seven-month-old twin fetuses were brought to her in a metal bucket. Looking down at the pink babies, she must have recoiled in horror when she saw both of them moving, gasping for breath. She had to have been even more horrified when the doctor suddenly appeared and, according to Charen, said, "I got you some good specimens—twins" before pouring a bottle of water into the bucket to drown what until then were two living human beings.

Disgusted by the process, the technician said there were many such live births. The doctors would simply break their tiny necks or kill the fetuses by beating them with metal tongs. In some cases, the technician revealed, they would begin a dissection by cutting open the chest, assuming the baby was already dead, only to find that the heart was still beating. She added that the manner in which abortions were performed had also been altered, that is, done more deliberately and slowly to ensure that the baby remained intact, even if that meant a live birth or prolonged pain for the baby before it died. When the pace of abortions quickened, babies taken alive would sometimes have their parts removed before they were dead.

Training for the position of fetal harvester is amazingly simple. One technician explained it this way: "The training consisted of on the job, when I was there, of them bringing back a huge plate—a placenta, blood clot—and showing me how to sift through all the stuff that was in there in order to find limbs, liver, pancreas, kidneys—what to look for, what the identification markers were in all that mess."

At a March 9, 2000 congressional hearing before the Subcommittee on Health and Environment, procurement technician Lawrence Dean Alberty, Jr. left congressmen stunned as he described a routine day at the facility where body parts were harvested as if they were crops:

> Upon taking the job as a fetal tissue procurement tech, I was under the impression that what I was going to do would make life better for Parkinson's patients, Alzheimer's, and cancer patients. Never was I led to believe that the tissue would be anything but helpful for those in need. What changed my mind was watching later-term abortions, seeing their eyes looking at me as I cut through their skull to extract their brain for Parkinson's and Alzheimer's patients, cutting open their chest cavities, only to see a beating heart moving ever so slowly until it stopped, all the while drawing blood from their heart, or watching fetuses in a metal pan covered with blood, moving and breathing, only to find myself in a place with no doors, no exits, thinking all the time, 'My God, what have I done to see this?' Night after night in my sleep, the fetuses were there. Hearts were beating, the screams of their mothers as the babies were pulled out of their bodies. These dreams turned into nightmares of the ends of the world. Apocalyptic nightmares would wake me up in a cold sweat. I felt sick every day, never wanting to leave the comfort of my home.

Later during the hearing, Alberty was questioned by one of the congressmen about why he eventually called the FBI. His answer left many in the chamber speechless. "The reason why I called the FBI was that one day I saw two twin fetuses twenty-four-plus gestational weeks born alive and brought back to me in a pan. When the person removed the drape and showed me what it was, it very much disturbed me to the point where I did not know what to do. In my eyes, seeing two twin fetuses moving and kicking and breathing in a pan really upset me. I'm not a doctor. I've never, ever claimed to be a doctor, and I couldn't tell you if these twins had any genetic problems. All I saw was they were untouched, meaning there were no clamp marks on them, they weren't bleeding, they were two twins cuddling each other in front of me. And I walked out the door."

Perhaps the most gruesome testimony came from Brenda Pratt Shafer, a registered nurse assigned by her agency to a harvesting site. Very pro-choice at the time, nurse Shafer assumed her assignment would be routine and that a valuable service was being performed by using discarded fetal tissue to advance medical science. According to the *Ashville Tribune*, which carried her story, what nurse Shafer witnessed changed her forever.

> I stood at the doctor's side and watched him perform a partial-birth abortion on a woman who was six month pregnant. The baby's heartbeat was clearly visible on the ultrasound screen. The doctor delivered the baby's body and arms, everything but his little head. The baby's body was moving. His little fingers were clasping together. He was kicking his feet. The doctor took a pair of scissors and inserted them into the back of the baby's head, and the baby's arm jerked out in a flinch, a startle reaction, like a baby does when he thinks that he might fall. Then the doctor opened the scissors up. Then he stuck the high-powered suction tube into the hole and sucked the baby's brains out. Now the baby was completely limp. I never went back to the clinic. But I am still haunted by the face of that little boy. It was the most perfect, angelic face I have ever seen.

Some fear that eventually, if not already, women will choose to get pregnant and have selective abortions simply to earn money for fetal tissue, in the process meeting the demands of researchers who need a steady supply of fresh body parts for medical experiments. It's predicted that the market, which had grown at an annual rate of 14 percent, is now worth more than a billion dollars a year, excluding whatever enormous profits will come from related patents and company products. As the number of research programs at universities, biotech companies, and pharmaceutical corporations increase at a record pace (NIH alone awards more than twenty million dollars a year for fetal tissue research), human chop shops will continue to flourish; and babies only weeks from birth will be sacrificed so that researchers can learn from the tissues and organs they harvest how to improve the lives of future generations lucky enough to have survived.

Gene Therapy

When Jesse Gelsinger, a happy-go-lucky eighteen-year old with a serious genetic disorder known as ornithine transcarbamylase deficiency (OTCD), signed the consent form volunteering for a new gene therapy experiment, his doctors failed to inform him that the same therapy he was about to receive had already killed monkeys and that two previous patients developed serious side effects. Jesse, whose disease was being well controlled by diet and a daily regiment of drugs, flew from his home in Arizona to the University of Pennsylvania hoping that someone had finally developed a treatment that would cure him of a disease that often leads to coma and death in children due to their inability to break down ammonia. Because his son nearly died from a coma induced by liver failure a year earlier, Jesse's father thought that perhaps this was a way to ultimately save the boy's life.

On September 13, the doctors gave a thumbs-up and prepared for the experiment. Jesse was apprehensive but excited to be part of a possible cure that would help other children. He was also fairly healthy despite his disease. After a final rundown of the protocol, doctors infused a massive amount of genetically engineered adenovirus directly into his liver's blood supply. Within twenty-four hours, Jesse's immune system began reacting violently to the foreign virus. Shortly after, he developed jaundice, his blood started to clot, and one by one his kidneys, lungs, and then his other organs failed. Four days after walking into the hospital, Jesse was dead.

What is so tragic about Jesse Gelsinger's death is that he was not the typical, desperately ill candidate who needed the experimental therapy as a last resort. With diet and pills, he was being treated quite successfully and could have led a relatively normal life. The most egregious action by gene therapy researchers, at the University of Pennsylvania as well as other institutions, has been their alleged failures to report adverse reactions associated with gene therapy trials.

Put simply, gene therapy is a technique that delivers a corrected version of a person's DNA in order to restore normal cell function. Instead of injecting a drug (pharmacological therapy), a doctor injects healthy copies of missing or altered genes via a carrier or vector, the most common being viruses because they're so good at getting into a cell's DNA. The basic steps involved in gene therapy are as follows: First, remove cells from the patient's blood or bone marrow and grow them in the lab. Second, expose the cells to viruses that have been disabled and then splice them with replacement genes. During incubation, the virus becomes incorporated into the cell's DNA. Third, inject the new cells back into the patient, which then continue to reproduce and take over from the original genetically deficient cells.

Let's take a simple example. If the genetic disorder affects the lungs, as in cystic fibrosis, then the organism to use would be a virus like influenza, which attacks lung tissue. Through recombinant DNA techniques, a good human gene is inserted into the virus to replace the gene causing cystic fibrosis and the virus is then injected into the patient. Once the virus infects the lungs, it delivers the good gene, which then replicates and hopefully corrects the defect. A variety of different approaches are currently being studied and will be ready for human trials within a few years.

Since more than one hundred thousand patients die each year from adverse drug reactions and side effects, researchers are trying to develop "designer drugs" that will match each person's DNA and, thus, minimize risk. Likewise, scientists believe that because gene therapy involves injection of a person's own unique cells, it holds great promise as a treatment. As a result of genetic research, we've identified about four thousand five hundred human diseases and disorders that are gene based and that would be candidates for gene therapy. And though gene therapy has so far been plagued by setbacks, experts believe that within twenty years it will become a routine part of medicine.

According to an October 2000 FDA consumer report, gene therapy researchers, who had launched more than four hundred clinical trials worldwide since 1990, were not reporting unexpected adverse events associated with human gene therapy experiments. Even worse, claims the report, scientists were asking that problems not be made public, especially the unreported deaths attributed to genetic treatments. LeRoy Walters, former chairman of NIH's Recombinant DNA Advisory Committee, said, "The clearest evidence of the system to protect research subjects not working is that only 35 to 37 of 970 serious adverse events from a common type of gene therapy trial were reported to the NIH. That is fewer than 5 percent of the serious adverse events."

Since the first gene therapy experiment in 1990, little has worked at all. "We haven't even taken one baby step beyond that first clinical experiment," laments Abbey Meyers, president of the National Organization of Rare Disorders. "It has hardly gotten anywhere." For all practical purposes, gene therapy has received a black eye, although new experiments and dozens of clinical trials have resumed at several institutions. Researchers, anxious to test their ideas on human beings, are applying for grants and forging ahead with human trials. Will things be different this time? Dr. Arthur Caplan, University of Pennsylvania's director of bioethics, doubts it. "Reporting of adverse events is a joke," he says. "It hasn't worked for years. Gene therapy is getting close scrutiny at the moment, but human subjects research has had serious problems of non-compliance for ages."

As recently as April 2000, there had been accusations that deaths in gene therapy experiments were being hidden. According to the *New York Times,* researchers at St. Elizabeth's Medical Center in Boston allegedly had failed to report the death of a patient from the treatment and never reported the fact that gene therapy may have caused cancer in another. Admitting that the experiments had problems, Jack Cumming, president of Vascular Genetics, was quick to ensure that his company would correct the protocol and monitor its remaining human trials more closely.

Although gene therapy sounds like a simple enough approach, the risks are compounded every time viruses are introduced into the body. For instance, a virus can infect more than one type of cell; thus, when vectors carry genes into the body they might alter more than just the intended cell or get incorporated into DNA in such a way as to mutate or cause cancer growth. They may also trigger an adverse immune response or be transmitted to another individual.

While medical research has come a long way, as long as researchers are pressured to deliver results and hide failures, they will continue to put unwitting human subjects at risk. Already there are more than one hundred approved protocols for human experiments that may have significant problems. The number of gene therapy clinical trials is greater than six hundred; the majority of these were carried out in the United States. No doubt progress will be made. Along the way, though, there will be adverse reactions and even deaths. Hopefully, there's been enough of an outcry over previous experiments that review boards will do their jobs and scientists will think twice before playing God.

DNA Vaccines

The goal of standard vaccines is to stimulate the body's production of antibodies or T cells by introducing an antigen without causing the disease. Until now, most vaccines have been produced by either use of the entire dead organism (typhoid, for example), or use of a live but weakened or attenuated form of the pathogen (tuberculosis, polio, measles, mumps, rubella, chickenpox, smallpox, and yellow fever). The problems researchers face are that standard vaccines made from dead pathogens are not effective against many microbes that find their way into cells, and that attenuated vaccines can actually cause disease or mutate over time so as to become even more virulent than the weakened strain.

The answer scientists have come up with is a genetic or DNA vaccine in which subjects are injected not with the antigen but with a plasmid (a circular piece of bacterial DNA containing a human gene) that codes for the antigen. Steps in the process include the following:

1. The plasmid is taken up by the cell.
2. The inserted gene is read and then translated into a protein (the antigen).
3. The antigen is broken down into smaller proteins called *peptides*.
4. The peptides leave the cell and stimulate the immune system.

One of the most recent developments in the field is an HIV-1 DNA vaccine. Earlier trials in 1995 involved the infusion of HIV genes into a patient who was already HIV positive. A year later, genes coding for HIV proteins were injected into healthy individuals for the first time. In 2001, enrollment began for Protocol No. 01-1-0079 calling for the recruitment of healthy men and women aged eighteen to sixty years who were not infected and had a low risk of becoming so. Since this was a Phase I clinical trial, the experimental DNA vaccine was being tested for safety and for whether or not it caused an immune response to HIV proteins. The vaccine is produced from the DNA of two HIV proteins and, when injected into a human being, instructs the body to make a small amount of HIV, which does not produce AIDS. Theoretically, the body recognizes the proteins and triggers an immune response against them.

Dr. Gary Nabel, director of the National Institute of Allergy and Infectious Diseases Vaccine Research Center, and his fellow researchers began developing the vaccine in September 2000. The key elements are two pieces of HIV's DNA blueprint: Gag, which is HIV's core protein, and Pol, which includes three enzymes crucial for HIV replication. Both are modified in a way that makes them nonfunctional yet capable of evoking an immune response.

Previous DNA-based vaccines have caused adverse reactions, so that researchers are being more cautious this time around. However, the question scientists have not been able to answer is whether the vaccine, if it doesn't work, will cause a weakening of the immune system and even greater susceptibility to disease later on. All eyes are on the researchers and the clinical trials to make sure that mistakes are not made and people are protected from unethical experimentation.

Stem Cell Research

In 1998, after more than two decades of trial, error, and futility, researchers at the University of Wisconsin–Madison isolated a remarkable type of self-renewing, unspecialized cell that has the potential to develop into more than two hundred different cell types composing almost every tissue in the body. These first "stem cells" were taken from human embryos or aborted fetuses and then grown in culture for as long as two years. The promise of the research was that the cells would be transplanted in patients to repair or replace damaged tissue such as heart, pancreas, liver, and brain.

Recently, scientists have discovered stem cells in adult tissue but found that important differences between adult and embryonic stem cells may prove to be a stumbling block in stem cell research. For one thing, embryonic stem cells, unlike their adult counterparts, can proliferate indefinitely. Embryo or fetal tissue is the only known source of what are known as "pluripotent stem cells," which have the ability to develop into the three human germ layers (ectoderm, mesoderm, and endoderm). During the past year, at least six laboratories have been successful in extracting pluripotent stem cells from human embryos and fetuses, and an increasing number of companies have been successful in establishing rare adult stem cell lines.

What exactly is stem cell research all about and where is it going? The principle is fairly simple. Let's say we're trying to cure type I (juvenile) diabetes, which is caused by the destruction of insulin-producing beta cells in the pancreas. Scientists would extract stem cells from a human embryo, direct them to differentiate and become specialized cell populations that can regenerate into pancreatic tissue, transplant them into the diseased organ, and allow them to divide and generate replacement cells that would then produce insulin and cure diabetes.

Similar transplants could be done for Parkinson's and Alzheimer's diseases, multiple sclerosis, kidney or liver failure, heart disease, spinal cord injury, and a host of other diseases and disorders. Another avenue of research is to seek ways to use stem cells as a vehicle for delivering genes to specific tissues in the body or for delivering chemotherapy drugs to specific cancer cells.

On June 13, 2002, delegates from Saudi Arabia announced at the annual biotechnology conference in Toronto that, with its newly built BioCity in Jeddah, their goal is to become the biotechnology capital of the Middle East. Dr. Sultan Bahabri was quoted as saying, "We believe biotechnology could some day be the new oil of Saudi Arabia. We are very well positioned to lead in biotechnology in our region of the world." When asked about his country's laws against the use of embryos for stem cell research, Dr. Bahabri said he is hopeful that eventually his government will allow fertility clinics to donate embryos and that the day could come when Saudi Arabia allows cloning techniques.

The objection many people have with any kind of embryonic stem cell research is that embryos must necessarily be destroyed in the process of extracting cells. There's also concern that because some diseases are embryonic in origin, human development will be monitored and, if the embryonic stem cells show any abnormality at all, the fetus will be aborted as a preventive measure. Finally, in the worst-case scenario, embryos might be grown for the sole purpose of providing the rest of us with a steady supply of stem cells for therapy and for use as spare parts in regenerating diseased tissues and organs.

This last scenario addresses the high probability that an individual's immune system would be attacked by foreign stem cells much like a noncompatible organ is rejected following transplantation. Creation of a "therapeutic clone" would allow a patient to grow a human spare and then harvest his or her own cells, thus preventing the problem of tissue rejection. Scientists around the world have already pledged their commitment to stem cell research and therapeutic cloning. How far they go depends on how far societies are willing to allow human research to continue so that the rest of us can benefit.

Cloning

Dolly, the much-celebrated sheep cloned at Roslin Institute in Scotland a few years ago, opened our eyes to the possibility that just about anything man can think about doing can be done. However, the breakthrough feat also aroused our worst fears: that the Frankensteins of the scientific world are alive and well and conducting research that will forever change the world, if not for the better then certainly for the worse. An outcry to stem the tide of cloning research was like trying to stop a runaway train. The moment an article about Dolly appeared in the 1997 issue of *Nature*, the genie was unleashed from the bottle for good, and an irreversible momentum was set in motion. To think that we could turn back the clock would be like saying that if we want to take back the discovery of nuclear power all we have to do is wish for it to be so.

What started out as plant and animal breeding is now approaching an exact science. Since Dolly, researchers have literally sprinted back to their labs to replicate earlier cloning experiments and outdo the Scottish scientists. Cloned mice were next, followed by cows, pigs, goats, cats, and even monkeys, which, according to a researcher at Advanced Cell Technology, was a "gallery of horrors." By 1998, researchers from South Korea had grown a cloned human embryo and then destroyed it at the four-cell stage. As a method to save endangered species, cloning got a boost in 2000 when Advanced Cell Technology used a surrogate cow mother to implant the embryo of a guar, a rare Indian ox species. In early 2002, Chinese scientists at Shanghai Medical University cloned human embryos using human DNA and rabbit eggs. Not to be outdone, a research team at a competing medical college in Xiangya created dozens of human embryos for the purpose of tissue transplantation.

A few years ago, the British Broadcasting Corporation (BBC) reported one of the most controversial cloning experiments to date. The parents of Molly Nash, a young girl suffering with Fanconi's anemia, a debilitating and life-threatening bone marrow disease, decided to use cloning to save their daughter's life. With the help of genetic scientists, Molly's parents began the process of selecting an embryo for in vitro fertilization to produce a healthy donor baby. After fourteen rejected embryos and four in vitro fertilization cycles, Adam was born and donated stem cells from his umbilical cord. In this case, Adam was simply a living donor, but after that success there's fear that future embryos will be sacrificed if cells are needed from other tissues or organs.

However, lest we think that the United States, with all the restrictions and regulations in place, has put itself above human cloning, think again. Private companies have made the decision to go full speed ahead regardless of public opinion. Experiments had already begun at Geron Corporation at Menlo Park, California even as the Chinese were creating cloned embryos. In May 1999, Geron acquired Roslin Biomed, the laboratory that produced Dolly, in order to combine the two companies' efforts in therapeutic cloning research. On July 13, 2001, Advanced Cell Technology announced that they had created cloned human embryos as a way to stockpile embryonic stem cells for sale.

To understand how these scientists are doing what they're doing, let's look at what cloning really is and why it's done. The reason to clone, say researchers, is to develop ways to alter embryos genetically (add or repair genes, for example) and reproduce them reliably so that they can be used for therapy and to produce whole organs for transplantation. There are three different types of cloning:

1. **Embryo cloning:** To duplicate an organism, a cell is induced to split soon after fertilization in order to produce a twin with identical DNA. Twinning or "blastomere separation" occurs naturally, and this technique has been commonly used on various animals.
2. **Reproductive Cloning:** Also known as adult DNA cloning or somatic cell nuclear transfer, this was the technique used to produce Dolly. DNA from an embryo is first removed and replaced with DNA from an adult. The embryo is then implanted into the uterus of another animal where it develops and produces a clone. Individuals cloned in this way have the same genes as the adult but, because of environmental effects, would not be exact copies.
3. **Therapeutic Cloning:** This type of cloning involves the growth of an embryo, which produces stem cells during early development. The stems cells are removed, with destruction of the embryo in the process, and transplanted back into the person who supplied the original DNA for the embryo. With this cloning method, the supply of cells would be unlimited, making waiting for tissue and organ transplants and tissue rejection a thing of the past.

This last type of cloning is what twenty-first century health care hopes will be the rule and not the exception. It's basically a form of stem cell therapy that would create perfectly matched cells for transplantation into patients with genetic material identical to that of the implanted cells. Here's how it's done:

Patient

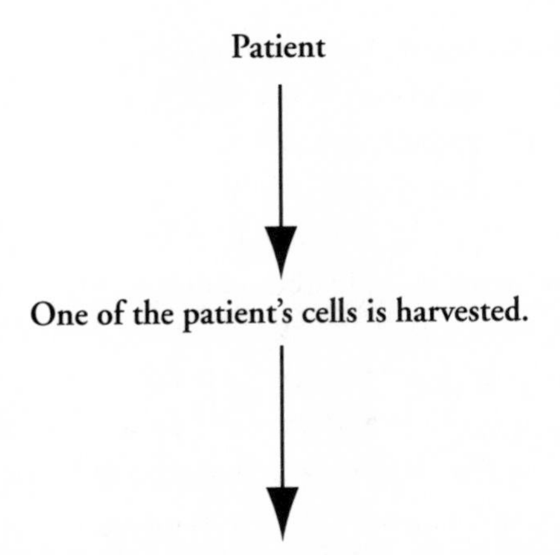

One of the patient's cells is harvested.

The nucleus is removed from the cell and transplanted into an egg that has had its nucleus removed.

The nucleus-egg cell hybrid is artificially stimulated and develops for several days.

Embryonic stem cells are removed from the dividing blastocyst.

Stem cells are directed to create cell types (brain, blood, skin, etc.).

Differentiated cells are transplanted back into the patient and replace defective cells.

However, animals like Dolly are the exception, not the rule. Very few reproductive cloning attempts have been successful; in many cases, clones have either died in utero or been born with severe birth defects. In fact, regardless of the species, less than 5 percent of cloned embryos survive to term. Despite that, Dr. James Grifo of New York University has developed a technique to produce a hybrid egg to help infertile women have children. The technique involves removing genes from a donor woman's egg and transplanting them into another woman's egg. The procedure, some reproductive experts fear, could produce abnormal fetuses with physical and genetic defects.

Most recently, Dr. Severino Antinori, whose claim to fame was using in vitro fertilization to impregnate a sixty-two-year-old woman, announced at a health care ethics conference in the United Arab Emirates on April 3, 2002 that he had implanted a cloned embryo into a woman and that she was eight weeks pregnant. He refused to disclose his patient's name or where she was living. However, a year earlier, Antinori, together with Dr. Pavos Zavos, announced the start of experiments at a secret location that would allow infertile couples to clone themselves as a form of self-reproduction. In the name of equality, Dr. Calum MacKellar of the University of Edinburgh said that male homosexual couples could one day use cloning techniques to produce a child using their own DNA and a surrogate woman to carry the offspring.

Since the United States' denouncement of human cloning and a ban on federal money, there's been a noticeable shift in cloning research away from Western nations to China, Southeast Asia, India, and parts of the Middle East. With no cloning laws to speak of, researchers in these countries can pretty much clone at will. To keep from being shut out, U.S. scientists and U.S. companies are setting up shop in various parts of the world, circumventing the Ban on Human Cloning Act (H.R. 1260) that imposes criminal penalties on anyone who attempts to clone a human being with the intent of implantation. Researchers like Dr. Zavos believe that the same thing will happen with cloning as happened when the United States banned in vitro fertilization when that technology first came out. Within a few years the ban was lifted and researchers spent the next decade catching up with their counterparts in other developed nations.

As of June 2002, proposed bans on human cloning appeared dead in the U.S. Senate for at least a year, despite public outcry against the procedure. An alternative bill, which would place a two-year moratorium on human cloning, will no doubt languish in committee as long as legislators are preoccupied with terrorism and the economy. The debates and inaction play right into the hands of cloning proponents and offer scientists a window of opportunity to continue research and develop new ways to clone human beings. So as arguments rage from Washington to communities around the nation, we'll simply move closer and closer to the day when cloning is reality and no amount of legislation will ever reverse it.

Nanomedicine and Life Extension

Imagine the day when needles are obsolete, having been replaced by implantable microchips so small as to be nearly invisible to the naked eye. Imagine next a computer that could fit inside a human cell and direct the life process, or a biorobot, armed with knowledge of a person's DNA, that patrols the span of the human body on a search-and-destroy mission for foreign invaders. Going back to movie analogies, this time to the sci-fi film *Fantastic Voyage*, those of us who saw it remember how the microscopic submarine set sail through a sea of capillaries, erythrocytes, and white blood cells to find its target. It was the stuff of science fiction back in the fun days of the 1970s. Today nanomedicine, a branch of bioengineering called *nanotechnology* or *molecular manufacturing*, promises to create molecule-sized medical devices within two decades.

The term *nano* refers to nanometer or 10^{-9} meters. "It is truly a magical unit of length," says Eugene Wong, the National Science Foundation's assistant director for engineering. "It is the point where the smallest man-made things meet nature." If you have trouble envisioning something that small, picture yourself peering through an electron microscope at a membrane on the outer edges of a human cell. As incredibly thin as that membrane is, you'd be able to insert into that space a device ten nanometers in diameter. Another way to look at it is that a single strand of DNA is twice the size of a nanometer. To carry molecules or devices that small, "nanobots" will be produced that are no larger than the period at the end of this sentence.

Considered the manufacturing technology of the twenty-first century, this breakthrough science will allow scientists to build molecular tools, machines, medical robots, microsurgical instruments, and drug delivery systems to target cancer cells, repair tissue, sample blood, monitor physiological changes, bring medicines to specific organs, and take over nonfunctioning organelles. The NIH has already awarded researchers millions of dollars for nanomedicine. One of the largest grants has been awarded to Dr. James Baker, director of the Center for Biologic Nanotechnology at the University of Michigan, to enable him to develop ways to insert DNA and therapeutic proteins into tumor cells. The beauty of an artificial delivery system, according to Baker, is that the body will not respond to it as it would to an antigen and, thus, will not send its immune system into attack mode.

The body is constantly building, repairing, and maintaining. It uses oxygen and food to construct molecules, cells, and tissues such as bone, muscle, and nerve. Every second of every day, our body uses molecular machines like enzymes to keep us alive and healthy. Nanotechnology will allow us to build artificial molecular machines to replace broken or inefficient ones. Much like robots assemble cars in Detroit, nanobots will be used to build microdevices for medical implants and even to slow down or reverse aging by fixing the physiological mechanisms that trigger or speed up the aging process.

A seemingly impossible feat of science, these molecular devices would be assembled atom by atom in a sort of mechanical reproductive act until trillions of atoms are arranged in whatever form is required. So far, IBM, Bell Labs, NASA, and a host of biotech companies are investing millions of dollars on what they're certain will be the medical breakthrough of the century. The federal government's current spending on nanotechnology is about three hundred million dollars per year.

Realistically, one of the first uses for nanotechnology will be diagnostics. A nanorobot weaving its way through the cardiovascular system will be able to sample tiny amounts of blood and test for hundreds of different factors. More complex nanorobots will have the ability to respond if something is wrong. For diabetics, the robots would monitor blood sugar; for heart patients, levels of enzymes that might signal an imminent heart attack.

The most likely application of nanomedicine, writes Dr. Robert Freitas, a research scientist for Zyvex and author of a three-volume text on the subject, "will consist of an injection of perhaps a few cubic centimeters of micron-sized nanorobots suspended in fluid." His description of the technology sounds like science fiction but is close to becoming reality. "The typical therapeutic dose may include up to 1 to 10 trillion individual devices. The nanorobots are going to be doing exactly what the doctor tells them to do, and nothing more, barring malfunctions. They will have multiple-redundant systems, like the five consensus computers on board the space shuttle. The only physical change you will see in the patient is that he or she will very rapidly become well again."

C Sixty Inc., a Toronto-based nanotechnology company, is currently developing carbon molecules for cancer, AIDS, and other diseases. Each of the microscopic spheres, which look like soccer balls with hexagonal patterns, will contain drugs or radioactive atoms that target specific cells. "Think of a smart bomb," says Dr. Uri Sagman in a *Newsweek* article. "Conventional chemotherapy is like carpet bombing. You drop it from 60,000 feet and hope for the best. This goes precisely to the target."

Scientists at Houston's Rice University are working on an alternate process. Rather than carbon, their drug delivery system, called a "nanoshell," is made of silica and gold, which gives it unique properties. The sphere is larger than Sagman's, hollow, and surrounded by a polymer that contains the drug. When the whole system is injected and then heated, the nanoshell melts the polymer, which then releases the drug at the exact site of the tumor or infection.

One of the great dilemmas we face in medicine today is drug resistance. Bacteria and viruses are unequaled in their ability to undergo endless mutations in their quest to circumvent treatments against them. Enter a nanorobot programmed to sense an invader before it causes any damage and to eliminate it. No need for antibiotics, vaccines, or invasive treatments when our biological sentinel can work twenty-four hours a day just looking for trouble. If the nanorobots pick up invaders like bacteria or viruses, they would become "immune machines," surrounding, attacking, and puncturing the invaders to death.

Scientists are optimistic that within a decade or two nanocomputers will be coded with their owners' DNA. Anything detected in the body that is not part of that DNA code (cancer, bacteria, virus, etc.) will be recognized as foreign and destroyed. While they circulate, hunt, and destroy, these little immune machines will escape being engulfed by white blood cells because they will carry on their surfaces molecules that the body recognizes. They may even be programmed to have a set life span, neutralizing foreign invaders for a few hours or days before falling apart and being eliminated as waste. These marvelous machines, if theory is put into practice, will literally give new meaning to the term "medical miracle."

Cryonics

For a one-time payment of twenty-eight thousand dollars, Bill, who was suffering from incurable pancreatic cancer, had just signed up for a whole-body suspension with a company that advertises cryonics as "the only hope for the elderly or terminally ill, or for those who die suddenly." After mulling over a number of options, Bill made his decision. As soon as possible after his death, a rapid-response team would remove his body, cool it to a temperature where physical decay stops, and hopefully in the near future, when medical science allows, reanimate the tissue and cure the disease. Theoretically, at a low enough temperature, all molecules in the body are immobilized to the point that they no longer move and react with each other. In this state, individuals can remain unchanged indefinitely.

Cryonics, a movement that began in the 1960s, has come a long way since Doubleday first published several successful editions of Robert Ettinger's *The Prospect of Immortality*, a book describing suspended animation as a way to delay death until a treatment or cure for disease comes along. But the problem with cryonics has always been with the suspension process, which lowers body temperature so much that it halts metabolic decay but does far too much damage along the way. Therefore, in reality, there's no such thing as suspended animation because we don't yet have the technology to revive an individual once he or she has been frozen. The promise cryonics makes is that soon we will have that technology; and for the price of a new car all one has to do is look at the alternative and be willing to take the chance.

Since 2000, new advances in cryobiology have enabled cryoprotectant solutions to penetrate cells four times faster than older methods and cause "vitrification," the changing of liquid to a gaslike solid without formation of the damaging ice crystals that destroy cell structure. A company called 21st Century Medicine had recently won a court battle with the FDA to continue their research and development in this area. They have come up with a technology that literally inhibits ice formation and eliminates the need for liquid nitrogen as a method of long-term storage.

Besides freezing diseased bodies until they can be cured, cryonics boasts of real-world applications such as preserving transplant tissue and sperm. In preliminary tests, sperm died after being frozen and thawed in normal one molar glycerol solution, but survived in 21st Century's new VX cryoprotectant. Tests are now being done to see if similar success can be achieved with human corneas, which represent a two billion dollar per year market. However, the most exciting prospect, according to cryobiologists, is the success they have had in vitrifying brain tissue. In rabbits, for instance, scientists claim that they have achieved complete vitrification and subsequent rewarming with virtually no structural brain damage. The only concern so far is possible toxicity from the cryoprotectant. As of this writing, cryonics still has a long way to go before it can be considered a reality.

Molecular Medicine by the Year 2020

Before the middle of this century, nothing that science fiction writers could have dreamed up will compare with what science and medicine will have actually accomplished. At the U.S. government's Los Alamos National Laboratory in New Mexico, the use of supercomputers and advanced data analysis is speeding up the development of vaccines and drug treatments that resist evolutionary changes in pathogens. Together with a consortium of academic institutions and commercial drug and biotech companies, Los Alamos researchers have developed computer models that will unravel molecular sequences in order to fashion designer drugs for combating disease. The real biological warfare in the twenty-first century will take place on hospital battlefields and inside doctors' offices where deadly strains of microbes will become as innocuous as a mild cold.

Some of the computer modeling tools were initially developed to analyze rapidly mutating viruses, such as HIV, and then later expanded to include the new field of "molecular diversity." A highly promising area of genetics, the goal of molecular diversity is to control and direct evolution by creating an environment in which molecules evolve through artificial selection rather than random natural selection. According to *Dateline Los Alamos*, by generating billions of diverse molecules of DNA, RNA, proteins, and other organic molecules at random to see which do best at fitting into a receptor on, say, a viral coat, top candidates are identified and reproduced with mutations to accelerate a laboratory version of evolution. This artificial selection process lets scientists develop and test an enormous number of variants in a matter of hours and generate a tremendous number of molecular diversity data.

Los Alamos has been selected by the WHO's global program on AIDS to characterize the molecular structure of HIV strains in Uganda,

Rwanda, Thailand, and Brazil, where AIDS vaccine trials are currently underway. Also underway at Los Alamos is a project to generate a database for human papilloma virus, the leading sexually transmitted disease in the world. With the Center for Human Genome Studies, established to help decode the DNA of all twenty-three pairs of human chromosomes, Los Alamos National Laboratory is at the cutting edge of research that should take us farther than we ever thought we'd go in such a short time.

Today we can look at a cell's genetic blueprints—its DNA—and tell not only what color eyes and hair a child will have but what medical disorders he or she might develop after birth. Forecasting medical problems, especially genetic disorders, is considered a godsend for some but a nightmarish scenario for many. Critics say that while knowing more about our genes will surely save lives, it will invariably lead to a society in which we will choose who lives and who dies based on DNA.

As an increasing array of genetic technologies become commonplace, so will gene testing. Visiting a medical lab to examine one's DNA for a disease or disorder is already a reality for a variety of diseases, but by 2020 it will be routine to identify everyone for risk of getting a disease before any symptoms appear. There are tests already available for dozens of such diseases, with more gene tests coming to market as more disease-causing genes are discovered. Completion of the Human Genome Project should create an explosion in our ability to test for virtually every genetic disorder known to man.

The dark side of all this is that individuals who might otherwise be born and develop into great thinkers, world leaders, scientists, and entrepreneurs could end up being eliminated because a prenatal gene test has turned up a disorder. Even "predictive" gene tests, which assign a probability of getting a disorder based on family history, would be used to terminate a pregnancy that may or may not result in the disorder. Assessing risk and then beating the odds of passing on defective genes will become the name of the game. Eugenics, which horrified so many in the past, is really not that far from becoming standard practice for many couples looking for their perfect baby.

By 2020, molecular medicine will bring together pharmacology and genomics to produce tailor-made drugs adapted to each person's DNA. Knowing an individual's unique genetic makeup will be the key to creating personalized drugs that are safer and work more effectively. This revolutionary field, called *pharmacogenomics*, is already looking toward human trials and promises many benefits. The U.S. government's Office of Biological and Environmental Research, Human Genome Program has issued a list of anticipated advances we'll be seeing in the not-too-distant future:

- **More powerful medicines:** Companies will be able to create new drugs targeted to specific diseases based on proteins, enzymes, and RNA molecules. Accurate drug production will not only maximize therapeutic effects but will also eliminate damage to nearby healthy cells.
- **Better, safer drugs the first time:** Rather than using trial and error to match patients with the right drugs, doctors will simply analyze a patient's genetic profile and prescribe the best treatment from the beginning. Guesswork is eliminated, recovery is accelerated, and adverse reactions are eliminated.
- **More accurate methods of determining proper doses:** Dosages based on weight and age will be replaced with dosages based on a person's genetics—how well the body processes the medicine and the time required to metabolize it. This will maximize the therapy's value and decrease the likelihood of overdose.
- **Advanced screening for disease:** Knowing one's genetic code will allow a person to make adequate lifestyle and environmental changes at an early age to avoid or reduce the severity of genetic disease. Likewise, advance knowledge of susceptibility to a particular disease will allow monitoring and treatment at the appropriate stage to maximize successful therapy.
- **Better vaccines:** Vaccines made of genetic materials (DNA or RNA) would offer all of the benefits of existing vaccines without all the risks. They would activate the immune system but would not cause infections. Such vaccines would also be inexpensive to produce, stable, easy to store, and capable of being genetically engineered to carry several strains of a pathogen at once.
- **Improvements in drug discovery and approval:** Pharmaceutical companies would be able to make discoveries more easily using genome targets. Previously failed drug candidates might be revived as they are matched to a particular population that would respond to a designer drug. Clinical trials would be less expensive and more effective

because they would only target people capable of responding to a certain drug.

- **Decrease in cost of health care:** Decreases in the number of adverse drug reactions (recall there are one hundred thousand deaths per year), the number of failed drug trials, the time it takes to get a drug approved, the length of time patients are on medications, the number of medications patients have to take to find effective therapy, and an increase in the range of possible drug targets all add up to a significant drop in the cost of health care.

The next generation will be the true recipients of genomics and a host of scientific advances and discoveries that are growing at a seemingly exponential pace. It will also have to deal with the controversies and unexpected consequences that come with the territory. Nothing we've seen previously will compare with what our children will see in their lifetimes. It will be up to them to keep watch and to make sure that ethics and human rights play a key role in twenty-first century molecular medicine.

HAARP: Climate Control or Mind Control?

Thousands of visitors to Washington, D.C. each year pass right by 4555 Overlook Avenue, S.W., a massive facility that employs several thousand scientists, military personnel, and ancillary office staff. What most passers-by don't know is that the U.S. Naval Research Laboratory is also the headquarters of a secret project named HAARP (High Frequency Active Auroral Research Program), which began more than a decade ago to study the properties and behavior of the ionosphere (upper atmosphere) and the effects of high-power, high-frequency transmissions on communications and surveillance. Although the military will not officially admit that HAARP has other purposes and has conducted human experimentation, experts in the know claim the following additional applications:

- Alteration of global weather patterns through ionosphere changes and disturbances by high-frequency radio waves. HAARP basically zaps the upper atmosphere between forty and six hundred miles above the earth's surface, where it's relatively unstable. Once the ionosphere is disturbed, the atmosphere below it also becomes disturbed.

- Induction of harmful or debilitating biological effects on targeted populations with radio waves that bounce back to earth.
- Use of different frequencies for psychological disablement and mind control. These can be targeted at both military personnel and civilian populations. U.S. Air Force documents allude to research for disrupting mental processes through pulsed radio frequency radiation over large geographic areas.

Captain Paul Tyler, author of a chapter in David Dean's book *Low Intensity Conflict and Modern Technology*, wrote, "The potential applications of artificial electromagnetic fields are wide-ranging and can be used in many military or quasi-military situations. Some of the potential uses include dealing with terrorist groups, crowd control, and antipersonnel techniques in tactical warfare. In all cases, the electromagnetic systems would be used to produce mild to severe physiological disruption or perceptual distortion or disorientation. In addition, the ability of individuals to function could be degraded to such a point that they would be combat ineffective. Another advantage of electromagnetic systems is that they provide coverage over large areas with a single system."

Dr. Richard Williams, a physical chemist and consultant to Sarnoff Laboratory at Princeton University, added in a May 1998 article that "the U.S. military is developing high-powered microwave weapons for use against human beings." His concern was that "such weapons do not simply attack a person's body, they reach all the way into a person's mind. They are meant to disorient or upset mental stability." He concluded with a stunning revelation that may explain why individuals living near HAARP facilities have such high rates of memory loss, leukemia, birth defects, cancer, and brain disorders: "While a government study on the bioeffects of HAARP radiation concluded that chronic exposure may not necessarily be harmful, other government documents warn that such radiation is powerful enough to explode highway flares in passing vehicles a quarter mile away and disrupt cardiac pacemakers in jet passengers flying overhead. The Pentagon has already decided that HAARP's radio interference is too intense to allow it to be located near any military facilities."

Originally located in remote areas of Alaska, HAARP may now have as many as forty sites, including Montauk Air Force Station in Montauk, Long Island near Brookhaven National Laboratory. Revealing the government's real intent in a publication more than thirty years ago, Zbigniew

Brzezinski, former national security advisor to President Jimmy Carter, quoted geophysicists as saying, "Artificially excited electronic strokes could lead to a pattern of oscillations that produce relatively high power levels over certain regions of the Earth. In this way, one could develop a system that would seriously impair the brain performance of very large populations in selected regions over an extended time. No matter how deeply disturbing the thought of using the environment to manipulate behavior for national advantages, to some the technology permitting such use will very probably develop within the next few decades."

Those next few decades are now upon us. But the history of HAARP actually begins in the 1940s with "Project Rainbow" and the "Philadelphia Experiment," forerunners of today's experiments with electromagnetic fields and stealth technology. By the 1950s, the two had become the "Phoenix Project," an attempt to develop advanced technologies for both weather manipulation and mind control. Given the U.S. government's interest in mind control at the time, it's not surprising that the military would try to produce a weapons system that would neutralize an enemy without having to fire a single shot.

However, the real predecessor of HAARP was the "Montauk Project," which took over from Phoenix and was set up at the reopened Montauk Air Force Base in 1971. Some of the tests measured how high-frequency pulse rates and amplitudes affected biological functions in humans. Interestingly, when the federal government turned over the base to New York State to use as a public park in 1984, it retained the right to reclaim the land at any time for national security reasons. An even more interesting stipulation in the deed is that the U.S. government retains all rights to all property located beneath the surface of the land. Since Montauk is actually situated on top of an undersea mountain with enough bedrock to establish an underground facility, since much of the park is off-limits to the public, and since individuals have actually monitored high-frequency transmissions emanating from the area, it's safe to say that HAARP is alive and well.

More recently, the area has been equipped with new high-capacity power lines, radar towers, particle beam units, and other equipment that observers say has been used to send transmissions to civilian populations in the surrounding communities. Newspapers in both Montauk and East Hampton have been reporting highly irregular activities at the air force base, including heightened security, strange radio signals, and electronic interference. In addition, according to some investigators, there were particle beam operations in the area on July 17, 1996, the day TWA Flight 800 crashed off the shore of Westhampton, Long Island.

Most people have never heard of HAARP, which some experts claim is on the verge of breakthroughs in modern warfare and population control technology. Since the Vietnam War, the United States has been determined to develop fighting capabilities that would minimize military casualties while rendering the enemy—even hidden in bunkers—helpless. The prediction is that by the next decade such weapons will be available thanks to HAARP research. When it happens, much of the population of Long Island, much less the rest of the country, will have no idea that they were probably a part of the research all along.

The twenty-first century will surely be one of unexpected breakthroughs, medical miracles, extended life spans, and better living. But as science breaks new ground and enters uncharted territories, the dangers that await us may be more terrifying than we can imagine. We've entered the age of genetic engineering, biological warfare, a renewed call for eugenics and population control, and medical advances so radical that no one knows how they will impact the human race. For our children's sake, we need to think through the consequences of our actions and consider how our arrogance will affect future generations. We also need to realize that we're merely at the water's edge when it comes to scientific advances; and while there's an entire ocean of new discoveries waiting to be uncovered, we're only now beginning to feel the salt spray in the air. All of us deserve a guarantee that life will be better, not worse. Let us hope that the history of previous centuries has taught us to be wise stewards of the knowledge we gain.

APPENDIX I

THE NUREMBERG CODE: DIRECTIVES FOR HUMAN EXPERIMENTATION

1. The voluntary consent of the human subject is absolutely essential. This means that the person involved should have legal capacity to give consent; should be so situated as to be able to exercise free power of choice, without the intervention of any element of force, fraud, deceit, duress, over-reaching, or other ulterior form of constraint or coercion; and should have sufficient knowledge and comprehension of the elements of the subject matter involved as to enable him to make an understanding and enlightened decision. This latter element requires that before the acceptance of an affirmative decision by the experimental subject there should be made known to him the nature, duration, and purpose of the experiment; the method and means by which it is to be conducted; all inconveniences and hazards reasonable to be expected; and the effects upon his health or person which may possibly come from his participation in the experiment. The duty and responsibility for ascertaining the quality of the consent rests upon each individual who initiates, directs or engages in the experiment. It is a personal duty and responsibility, which may not be delegated to another with impunity.
2. The experiment should be such as to yield fruitful results for the good of society, unprocurable by other methods or means of study, and not random and unnecessary in nature.
3. The experiment should be so designed and based on the results of animal experimentation and a knowledge of the natural history of

the disease or other problem under study that the anticipated results will justify the performance of the experiment.

4. The experiment should be so conducted as to avoid all unnecessary physical and mental suffering and injury.
5. No experiment should be conducted where there is an *a priori* reason to believe that death or disabling injury will occur; except, perhaps, in those experiments where the experimental physicians also serve as subjects.
6. The degree of risk to be taken should never exceed that determined by the humanitarian importance of the problem to be solved by the experiment.
7. Proper preparations should be made and adequate facilities provided to protect the experimental subject against even remote possibilities of injury, disability, or death.
8. The experiment should be conducted only by scientifically qualified persons. The highest degree of skill and care should be required through all stages of the experiment of those who conduct or engage in the experiment.
9. During the course of the experiment the human subject should be at liberty to bring the experiment to an end if he has reached the physical or mental state where continuation of the experiment seems to him to be impossible.
10. During the course of the experiment the scientist in charge must be prepared to terminate the experiment at any stage, if he has probable cause to believe, in the exercise of the good faith, superior skill and careful judgment required of him that a continuation of the experiment is likely to result in injury, disability, or death to the experimental subject.

APPENDIX II

THE WILSON MEMORANDUM

26 Feb 1953

Memorandum for the SECRETARY OF THE ARMY
SECRETARY OF THE NAVY
SECRETARY OF THE AIR FORCE

SUBJECT: Use of Human Volunteers in Experimental Research

1. Based upon a recommendation of the Armed Forces Medical Policy Council, that human subjects be employed, under recognized safeguards, as the only feasible means for realistic evaluation and/or development of effective preventive measures of defense against atomic, biological or chemical agents, the policy set forth below will govern the use of human volunteers by the Department of Defense in experimental research in the fields of atomic, biological and/or chemical warfare.

2. By reason of the basic medical responsibility in connection with the development of defense of all types against atomic, biological and/or chemical warfare agents, Armed Services personnel and/or civilians on duty at installations engaged in such research shall be permitted to actively participate in all phases of the program, such participation shall be subject to the following conditions:

(a) The voluntary consent of the human subject is absolutely essential.

(1) This means that the person involved should have legal capacity to give consent; should be so situated as to be able to exercise free power of choice, without the intervention of any element of force, fraud, deceit, duress, overreaching, or other ulterior form of constraint or coercion; and should have sufficient knowledge and comprehension of the elements of the subject matter involved as to enable him to make an understanding and enlightened decision. This latter element requires that before the acceptance of an affirmative decision by the experimental subject there should be made known to him the nature, duration, and purpose of the experiment; the method and means by which it is to be conducted; all inconveniences and hazards reasonably to be expected: and the effects upon his health or person which may possibly come from his participation in the experiment.

(2) The concept [sic: consent] of the human subject shall be in writing, his signature shall be affixed to a written instrument setting forth substantially the aforementioned requirements and shall be signed in the presence of at least one witness who shall attest to such signature in writing.

(a) In experiments where personnel from more than one Service are involved the Secretary of the Service which is exercising primary responsibility for conducting the experiment is designated to prepare such an instrument and coordinate it for use by all the Services having human volunteers involved in the experiment.

(3) The duty and responsibility for ascertaining the quality of the consent rests upon each individual who initiates, directs or engages in the experiment. It is a personal duty and responsibility which may not be delegated to another with impunity.

(b) The experiment should be such as to yield fruitful results for the good of society, unprocurable by other methods or means of study, and not random and unnecessary in nature.

(c) The number of volunteers used shall be kept at a minimum consistent with item b. above.

(d) The experiment should be so designed and based on the results of animal experimentation and a knowledge of the natural history of the disease or other problem under study that the anticipated results will justify the performance of the experiment.

(e) The experiment should be so conducted as to avoid all unnecessary physical and mental suffering and injury.

(f) No experiment should be conducted where there is an a priori reason to believe that death or disabling injury will occur.

(g) The degree of risk to be taken should never exceed that determined by the humanitarian importance of the problem to be solved by the experiment.

(h) Proper preparation should be made and adequate facilities provided to protect the experimental subject against even remote possibilities of injury, disability, or death.

(i) The experiment should be conducted only by scientifically qualified persons. The highest degree of skill and care should be required through all stages of the experiment of those who conduct or engage in the experiment.

(j) During the course of the experiment the human subject should be at liberty to bring the experiment to an end if he has reached the physical or mental state where continuation of the experiment seems to him to be impossible.

(k) During the course of the experiment the scientist in charge must be prepared to terminate the experiment at any stage, if he has probable cause to believe, in the exercise of the good faith, superior skill and careful judgment required of him that a continuation of the experiment is likely to result in injury, disability, or death to the experimental subject.

(l) The established policy, which prohibits the use of prisoners of war in human experimentation, is continued and they will not be used under any circumstances.

3. The Secretaries of the Army, Navy and Air Force are authorized to conduct experiments in connection with the development of defenses of all types against atomic, biological and/or chemical warfare agents involving the use of human subjects within the limits prescribed above.

4. In each instance in which an experiment is proposed pursuant to this memorandum, the nature and purpose of the proposed experiment and the name of the person who will be in charge of such experiment shall be submitted for approval to the Secretary of the military department in which the proposed experiment is to be conducted. No such experiment shall be undertaken until such Secretary has approved in writing the experiment proposed, the person who will be in charge of conducting it, as well as informing the Secretary of Defense.

5. The addresses [sic] will be responsible for insuring compliance with the provisions of this memorandum within their respective Services.

/signed/
C. E. WILSON

Copies furnished:
Joint Chiefs of Staff
Research and Development Board

SEC. 1520A. RESTRICTIONS ON THE USE OF HUMAN SUBJECTS FOR TESTING OF CHEMICAL OR BIOLOGICAL AGENTS

(a) Prohibited activities
The Secretary of Defense may not conduct (directly or by contract)
(1) any test or experiment involving the use of a chemical agent or biological agent on a civilian population; or
(2) any other testing of a chemical agent or biological agent on human subjects.

(b) Exceptions
Subject to subsections (c), (d), and (e) of this section, the prohibition in subsection (a) of this section does not apply to a test or experiment carried out for any of the following purposes:
(1) Any peaceful purpose that is related to a medical, therapeutic, pharmaceutical, agricultural, industrial, or research activity.
(2) Any purpose that is directly related to protection against toxic chemicals or biological weapons and agents.
(3) Any law enforcement purpose, including any purpose related to riot control.

(c) Informed consent required
The Secretary of Defense may conduct a test or experiment described in subsection (b) of this section only if informed con-

sent to the testing was obtained from each human subject in advance of the testing on that subject.

(d) Prior notice to Congress
Not later than 30 days after the date of final approval within the Department of Defense of plans for any experiment or study to be conducted by the Department of Defense (whether directly or under contract) involving the use of human subjects for the testing of a chemical agent or a biological agent, the Secretary of Defense shall submit to the Committee on Armed Services of the Senate and the Committee on National Security of the House of Representatives a report setting forth a full accounting of those plans, and the experiment or study may then be conducted only after the end of the 30-day period beginning on the date such report is received by those committees.

(e) "Biological agent" defined
In this section, the term "biological agent" means any microorganism (including bacteria, viruses, fungi, rickettsiae, or protozoa), pathogen, or infectious substance, and any naturally occurring, bioengineered, or synthesized component of any such microorganism, pathogen, or infectious substance, whatever its origin or method of production, that is capable of causing
(1) death, disease, or other biological malfunction in a human, an animal, a plant, or another living organism;
(2) deterioration of food, water, equipment, supplies, or materials of any kind; or
(3) deleterious alteration of the environment.

APPENDIX IV

DECLARATION OF HELSINKI

1. Basic Principles

a. Biomedical research involving human subjects must conform to generally accepted scientific principles and should be based on adequately performed laboratory and animal experimentation and on a thorough knowledge of the scientific literature.

b. The design and performance of each experimental procedure involving human subjects should be clearly formulated in an experimental protocol which should be transmitted for consideration, comment, and guidance to a specially appointed committee independent of the investigator and the sponsor provided that this independent committee is in conformity with the laws and regulations of the country in which the research experiment is performed.

c. Biomedical research involving human subjects should be conducted only by scientifically qualified persons and under the supervision of a clinically competent medical person. The responsibility for the human subject must always rest with a medically qualified person and never rest on the subject of the research, even though the subject has given his or her consent.

d. Biomedical research involving human subjects cannot legitimately be carried out unless the importance of the objective is in proportion to the inherent risk to the subject.

e. Every biomedical research project involving human subjects should be preceded by careful assessment of the predictable risks in comparison with foreseeable benefits to the subjects or to others. Concern for the interests of the subject must always prevail over the interests of science and society.

f. The right of the research subject to safeguard his or her integrity must always be respected. Every precaution should be taken to respect the privacy of the subject and to minimize the impact of the study on the subject's physical and mental integrity and to the personality of the subject.

g. Physicians should abstain from engaging in research projects involving human subjects unless they are satisfied that the hazards involved are believed to be predictable. Physicians should cease any investigation if the hazards are found to outweigh the potential benefits.

h. In publication of the results of his or her research, the physician is obligated to preserve the accuracy of the results. Reports of experimentation not in accordance with the principles laid down in this declaration should not be accepted for publication.

i. In any research on human beings, each potential subject must be adequately informed of the aims, methods, anticipated benefits and potential hazards of the study and the discomfort it may entail. He or she should be informed that he or she is at liberty to abstain from participation in the study and that he or she is free to withdraw his or her consent to participation at any time. The physician should then obtain the subject's freely-given informed consent, preferably in writing.

j. When obtaining informed consent for the research project, the physician should be particularly cautious if the subject is in a dependent relationship to him or her or may consent under duress. In that case, the informed consent should be obtained by a physician who in not engaged in the investigation and who is completely independent of this official relationship.

k. In case of legal incompetence, informed consent should be obtained from the legal guardian in accordance with national legislation. Where physical or mental incapacity makes it impossible to obtain informed consent, or when the subject is a minor, permission from the responsible relative replaces that of the subject in accordance with national legislation.

l. The research protocol should always contain a statement of the ethical considerations involved and should indicate that the principles enunciated in the present Declaration are complied with.

2. Medical Research Combined with Professional Care (Clinical Research).

a. In the treatment of the sick person, the doctor must be free to use a new diagnostic and therapeutic measure, if in his or her judgement it offers hope of saving life, reestablishing health or alleviating suffering.

b. The potential benefits, hazards and discomforts of a new method should be weighed against the advantages of the best current diagnostic and therapeutic methods.

c. In any medical study, every patient—including those of a control group, if any—should be assured of the best proven diagnostic and therapeutic methods.

d. The refusal of the patient to participate in a study must never interfere with the physician-patient relationship.

e. If the physician considers it essential not to obtain informed consent, the specific reasons for this proposal should be stated in the experimental protocol for transmission to the independent committee.

f. The physician can combine medical research with professional care, the objective being the acquisition of new medical knowledge, only to the extent that medical research is justified by its potential diagnostic or therapeutic value for the patient.

3. Non-Therapeutic Biomedical Research Involving Human Subjects (Non-Clinical Biomedical Research).

a. In the purely scientific application of medical research carried out on a human being, it is the duty of the physician to remain the protector of the life and health of that person on whom biomedical research is being carried out.

b. The subjects should be volunteers—either healthy persons or patients for whom the experimental design is not related to the patient's illness.

c. The investigator or the investigating team should discontinue the research if in his or her or their judgement it may, if continued, be harmful to the individual.

d. In research on man, the interest of science and society should never take precedence over considerations related to the well being of the subject.

APPENDIX

V

EXCERPTS FROM:

BIOLOGICAL TESTING INVOLVING HUMAN SUBJECTS BY THE DEPARTMENT OF DEFENSE, 1977, NINETY-FIFTH CONGRESS, MARCH 8 AND MARCH 23, 1977;

DECLASSIFIED BY 056047, 15 SEPT 1975

Special Operations Division of Fort Detrick (page 245)

The Agency association with Fort Detrick involved the Special Operations Division (SOD) of that facility. This Division was apparently responsible for developing special applications for BW agents and toxins. Its principal customer appears to have been the US Army Special Forces. Its concern was with the development of both suitable agents and delivery mechanisms for special use in paramilitary situations. These applications clearly include one-on-one situations in which clandestine delivery was an objective. Both standard BW agents and biologically derived toxins were investigated by the Division. Discussions with former Fort Detrick employees indicate that SOD was first established as a distinct, highly secure activity within Fort Detrick in about 1948, though no records going back that far have been found. The Division was abolished in 1970 or 1971 as the Fort Detrick operation was terminated.

CIA Relationships with SOD (page 246)

The CIA relationship with SOD was formally established in May 1952 through a memorandum of agreement with the Army Chief Chemical Officer for the performance of certain research and development in the laboratory facilities of the Special Operations Division of the Army Biological Laboratory at Fort Detrick. The animus for establishing this relationship seems to have been a belief in OTS that the special capabilities of the Fort Detrick group and its access to biological materials of all sorts provided the Agency with expertise and capabilities which were appropriate to its function and not otherwise available. This experience included the development of two different types of agent suicide pills to be used in extremis and a successful operation using BW materials against a Nazi leader. In the latter case, Staph. enterotoxin (food poisoning) was administered to Hjalmar Schacht so as to prevent his appearance at a major economic conference during the war. This agent was included in the materials maintained for the Agency by SOD.

Activities of Peculiar CIA Interests (page 248)

Though discussions with people associated with the project reflect an overriding interest in incapacitants, particularly in later years, available records make it clear that CIA interests included maintaining a stockpile of lethal materials and delivery systems. The evidence indicates that the Agency relied upon the use of specific BW agents and toxins being investigated as a normal part of the Army's BW program.

A major early requirement to the Agency was to find a replacement for the standard cyanide L-Pill issued to agents in hazardous situations and U-2 pilots for suicide purposes in the event of capture. Work on this problem was done at Fort Detrick and ultimately centered on the coating of a number of 80 drill bit (the smallest made) with shellfish toxin. In conjunction with this project, a considerable amount of work was done in developing concealment schemes for the drill or pin to be used in the event suicide was necessary.

Primary Agency interest seemed to relate to the development of dissemination equipment to be used with a standard set of agents kept on the shelf. A number of such dissemination devices appear to be peculiarly suited for the type of clandestine use one might associate with Agency operations. Some of these were included among hardware stored for the Agency at Edgewood Arsenal subsequent to the closure of SOD: attaché case rigged to disseminate an agent into the air, a cigarette lighter rigged to disseminate an agent when lighted, a fountain pen dart launcher, an engine head bolt designed to release an agent when heated, a fluorescent light starter to activate the light and then release an agent, etc.

At a meeting in June 1952, at the very outset of the Agency's association with SOD when CIA representatives stated that they as yet had no specific requirements, a list of SOD priorities for work on dissemination devices was provided. This dissemination list included such things as cigarettes, chewing gum, cigarette lighters, wristwatches, fountain pens, rings, etc.

One development peculiarly associated with the CIA was the "microbioinoculator" which was an extremely small dart device which could be fired through clothing to penetrate the skin so as to inoculate the target with an agent without his perception of being hit. An added fillip to this development was the requirement that no indication of the use of such a device be discernible in the course of autopsy. A large amount of Agency attention was given to the problem of incapacitating guard dogs. Though most of the dart launchers used in these developments were developed for the Army, the Agency did request the development of a small hand-held launcher for its "peculiar" needs.

A lot of work was done on human incapacitation. OTS apparently received continuing requests for safe, effective and rapidly acting, incapacitating devices. Many of these related to requirements for incapacitating Viet Cong leaders before they could render themselves incapable of talking and terrorists before they could take retaliatory action. Substantial work was also done for the Agency in the development of spoilants for agricultural products, biological materials for the contamination of petroleum stores, and agents for use in the destruction of electronics, optical systems, structural materials, etc.

Shellfish Toxin (page 250)

By the late 1960s, a stockpile of some 15 to 20 different BW agents and toxins was maintained on a regular basis by SOD for possible Agency use. The supply included such agents as food poisons, infectious viruses, lethal botulism toxin, paralytic shellfish toxin, snake (krait) venom, *Microsporum gypseum* which produces severe skin disease, etc. Varying amounts of these materials ranging from 100 grams to 100 milligrams were maintained. The 11 grams of shellfish toxin—along with 8 milligrams of cobra venom—was found by the Chief of the OTS Chemistry Branch, in Vault B10. On 13 June the vault was put under 24-hour guard. The shellfish toxin was packaged in several different forms including two individual doses in tablet form.

APPENDIX VI

LETTER TO THE SECRETARY OF THE DEPARTMENT OF DEFENSE REGARDING GULF WAR SYNDROME

U.S. Senate
Committee on Banking, Housing, and Urban Affairs
Washington, DC

February 9, 1994

Hon. William Perry
Secretary, Department of Defense
Pentagon
Washington, DC.

Dear Secretary Perry:

After receiving complaints from a number of Michigan veterans who told me they were not receiving appropriate care from Department of Veterans Affairs' hospitals, I initiated an inquiry into the nature and scope of Gulf War Syndrome. This research uncovered a great deal of evidence that U.S. forces may have been exposed to chemical and possibly biological warfare agents as a result of the bombings of 18 chemical, 12 biological, and 4 nuclear facilities within Iraq during the Persian Gulf War. I have also listened to the compelling accounts by eyewitnesses, including chemical officers, of events which appear to be best explained as direct chemical agent attacks.

Disturbingly, I also began to receive reports of these illnesses being transmitted to the spouses and children of these veterans. Since I initiated this inquiry, several medical researchers have suggested that the origins of these illnesses might be biological. As Chairman of the Senate Committee on Banking, Housing, and Urban Affairs, with oversight responsibility for the Export Administration Act, I contacted the Centers for Disease Control and the U.S. Department of Commerce to determine what, if any, biological materials were exported to Iraq prior to the Gulf War. After receiving the export information from the U.S. Department of Commerce, my staff contacted the principal supplier of these materials, the American Type Culture Collection, to determine the genus, species, strain, and origins of these materials.

Records provided by the supplier show that, from at least 1985 through 1989, the period for which records were available, the United States government approved for sale to Iraq quantities of potentially lethal biological agents that could have been cultured or grown in large volume in an Iraqi biological warfare program. These exported materials were not attenuated or weakened and were capable of reproduction.

Materials shipped included: Bacillus anthracis, clostridium botulinum, clostridium perfringens, histoplasma capsulatum, brucella abortus, and brucella melitensis.

(A detailed listing of these materials is attached.)

I find it especially troubling that, according to the supplier's records, these materials were requested by and sent to Iraqi government agencies, including the Iraq Atomic Energy Commission, the Iraq Ministry of Higher Education, the State Company for Drug Industries, and the Ministry of Trade.

I have released this information to assist medical researchers seeking to diagnose and treat affected veterans and their families. During this session of Congress, the Committee on Banking,

Housing, and Urban Affairs will be reviewing the Export Administration Act, which is due for reauthorization. I have assured the veterans, their families, and the people of the United States that the policy under these licenses were granted will be examined and strengthened. The defense of the United States should not be undermined by export policies that allow this government to assist any pariah nation, such as Iraq, in the furtherance of nuclear, chemical, or biological weapons programs.

While it is extremely important to promote U.S. products and exports in international trade, it is also important to note that the average cost of each of these specimens was less than $60, and they were acquired from a 'not-from-profit' organization.

I ask that the Department of Veterans Affairs and the Department of Defense immediately establish disability rating systems for stricken Gulf War veterans that are dependent on the degree of individual disability rather than using some arbitrary point system. Further, the establishment of this disability rating must not be delayed because of an inability to arrive at a specific medical diagnosis.

I also call upon the newly created Persian Gulf Veterans Coordinating Board and the participating Secretaries of Veterans Affairs, Defense, and Health and Human Services, to expand their research to include the reported transmission of these illnesses to the spouses and children of these veterans, and to assess what, if any, public health hazard might exist.

In order to ensure that no information is being withheld, and consistent with the recommendation of the National Academy of Sciences in their investigation of the exposure of veterans to the Health Effects of Mustard Gas and Lewisite, the Secretary of Defense and the Secretary of Veterans Affairs should widely and publicly announce that personnel who believe they were exposed to chemical or biological warfare agents during the Persian Gulf War or who detected the presence of any chemical or biological warfare agents during the Gulf War are released from any oath of secrecy relative to these exposures or detections.

We must ensure that those men and women who served this country during the Gulf War, on active duty, in the reserves, and those who have since left the military services, receive proper medical attention. The National Archives has retained many letters, the unheard pleas and appeals of the veterans who returned home after World War I complaining of illnesses as a result of their exposure to mustard gas. Surely, we cannot tolerate turning a deaf ear on the thousands of veterans who served in the Gulf War. Without proper testing and treatment, their conditions will worsen. They cannot wait. Many are now destitute—their savings spent on medical care not being provided by the government. Others, unable to work, receive no pension or compensation because the Department of Veterans Affairs is unable to diagnose their illnesses.

I believe that this issue needs to be resolved, in order to ensure that our Armed Services are properly prepared for future conflicts that might involve the use of these weapons. I know that you share my concerns, both about the well-being of those who wear the uniforms of the United States Armed Forces, and about the preparedness of this nation to protect its forces in future conflict. I ask you to personally reply to these requests on or before March 31, 1994.

Sincerely,

Donald W. Riegle, Jr.
Chairman

APPENDIX VII

LETTER TO AUTHOR REGARDING BIOLOGICAL AGENTS AS POSSIBLE CAUSE OF GULF WAR SYNDROME

Professor Garth L. Nicolson
David Bruton Jr. Chair in Cancer Research
Department of Tumor Biology (108)
The University of Texas M.D. Anderson Cancer Center
Houston, Texas 77030 Phone (713) 792-7481 Fax (713) 794-0209

July 8, 1996

Dr. Andrew Goliszek
XXXXXXXX
XXXXXXXXXXXX Tel: XXXXXXXX

Dear Dr. Goliszek

For our interview I have attached some information, reprints and preprints on GWI and the possible use of CBW during Operation Desert Storm. For your information, we have found so far that about one-half of the GWI patients (and 2/2 British ODS veterans) have an invasive mycoplasma infection that can be successfully treated with antibiotics, such as doxycycline (Nicolson, G.L. and Nicolson, N.L. Doxycycline treatment and Desert

Storm, *JAMA* 273: 618–619, 1995; Nicolson, G.L. and Nicolson, N.L. Diagnosis and treatment of mycoplasma infections in Persian Gulf War Illness-CFIDS patients, *Int. J. Occup. Med Immunol. Tox.* 5: 69–78, 1996) (200 mg/d for 6 wk per course; several courses are usually required, similar to Lyme Disease), Cipro (1,000–1,5000 mg/d) or Zithromax (500 mg/d). We have developed new diagnostic procedures (Gene Tracking and forensic PCR) for analysis of the types of mycoplasmas found in GWI, and these may also be useful and informative for soldiers with GWI-CFIDS and some civilians with CFIDS (these diseases are essentially the same—(Nicolson, G.L. and Nicolson, N.L. Chronic fatigue illness and Operation Desert Storm, *J. Occup. Environ. Med.* 38: 14–16, 1996). I have included these reprints with this letter.

Currently, we are using a test called Gene Tracking to identify unusual DNA sequences unique to mycoplasmas in blood leukocytes (Nicolson, N.L. and Nicolson, G.L. The isolation, purification, and analysis of specific gene-containing nucleoprotein complexes, *Meth. Mol. Genet.* 5: 281–298, 1994). We have adapted forensic PCR procedures for the accurate determination of invasive mycoplasma infections, and this may be useful for clinical labs that are struggling with antibody approaches for detecting mycoplasmas. What is interesting about these mycoplasmas is that they contain retroviral DNA sequences (such as the HIV-1 *env* gene but not other HIV genes), suggesting that they may have been modified to make them more pathogenic and more difficult to detect. It's also interesting that we have been working with a support group of Texas Department of Corrections employees that were apparently exposed to the same unusual mycoplasma before ODS, possibly during a Defense Department-supported vaccine testing program in selected state prisons here in Texas. One of the biotech companies involved in this TDC study (in Houston, TX) had US Army contracts to study mycoplasmas (this is indicated in their publications on the subject) and has been named in lawsuits as selling or supplying CBW to Iraq.

We have been able to assist thousands of soldiers recover from a life-threatening disease that is caused by invasive mycoplasma infections. We have learned that over 6,000 US soldiers have died of infectious diseases and chemical exposure in Operation Desert Storm. I suspect that this is being hidden from the American public for political, economic and legal reasons. You have my permission to use any of the material that I have sent you in any way you see fit.

Sincerely,

Garth L. Nicolson, Ph.D.
David Bruton Jr. Chair in Cancer Research
Professor and Chairman
Department of Tumor Biology, UTMDACC
and
Professor of Pathology and Laboratory Medicine
Professor of Internal Medicine
The University of Texas Medical School at Houston

U.S. SUPREME COURT

BUCK V. BELL, 274 U.S. 200 (1927)

BUCK V. BELL, SUPERINTENDENT OF STATE COLONY

EPILEPTICS AND FEEBLE MINDED NO. 292

ARGUED APRIL 22, 1927

DECIDED MAY 2, 1927

Mr. Justice HOLMES delivered the opinion of the Court

This is a writ of error to review a judgment of the Supreme Court of Appeals of the State of Virginia affirming a judgment of the Circuit Court of Amherst County, by which the defendant in error, the superintendent of the State Colony for Epileptics and Feeble Minded, was ordered to perform the operation of salpingectomy upon Carrie Buck, the plaintiff in error, for the purpose of making her sterile. The case comes here upon the contention that the statute authorizing the judgment is void under the Fourteenth Amendment as denying to the plaintiff in error due process of law and the equal protection of the laws.

Carrie Buck is a feeble-minded white woman who was committed to the State Colony above mentioned in due form. She is the daughter of a feeble-minded mother in the same institution, and the mother of an illegitimate feeble-minded child. She was eighteen years old at the time of the trial of her case in the Circuit Court in the latter part of 1924. An Act of Virginia approved March 20, 1924 (Laws 1924, c. 394) recites that the health of the patient and the welfare of society may be promoted in certain cases by the sterilization of mental defectives, under careful safeguards, etc.; that the sterilization may be effected in males by vasectomy and in females by salpingectomy, without serious pain or substantial danger to life; that the Commonwealth is supporting in various institutions many defective persons who if now discharged would become a menace but if incapable of procreating might be discharged with safety and become self-supporting with benefit to themselves and to society; and that experience has shown that heredity plays an important part in the transmission of insanity, imbecility, etc. The statute then enacts that whenever the superintendent of certain institutions including the abovementioned State Colony shall be of opinion that it is for the best interest of the patients and of society that an inmate under his care should be sexually sterilized, he may have the operation performed upon any patient afflicted with hereditary forms of insanity, imbecility, etc., on complying with the very careful provisions by which the act protects the patients from possible abuse.

The superintendent first presents a petition to the board of directors of his hospital or colony, stating the facts and the grounds for his opinion, verified by affidavit. Notice of the petition and of the time and place of the hearing in the institution is to be served upon the inmate, and also upon his guardian, and if there is no guardian the superintendent is to apply to the Circuit Court of the County to appoint one. If the inmate is a minor notice is also to be given to his parents, if any, with a copy of the petition. The board is to see to it that the inmate will attend the hearings if desired by him or his guardian. The evidence is all to be reduced to writing, and after the board has made its order for or against the operation, the superintendent, or the inmate, or his guardian, may appeal to the Circuit Court of the County. The Circuit Court may consider the record of the board and the evidence before it and such other admissible evidence as may be offered, and may affirm, revise, or reverse the order of the board and enter such order as it deems just. Finally any party may apply to the Supreme Court of Appeals, which, if it grants the appeal, is to hear the case upon the record of the trial in the Circuit Court and may enter such order as it thinks the Circuit Court should have entered. There can be no doubt that so far as procedure is concerned the rights of the patient are most carefully considered, and as every step in this case was taken in scrupulous compliance with the statute and after months of observation, there is no doubt that in that respect the plaintiff in error has had due process at law.

The attack is not upon the procedure but upon the substantive law. It seems to be contended that in no circumstances could such an order be justified. It certainly is contended that the order cannot be justified upon the existing grounds. The judgment finds the facts that have been recited and that Carrie Buck 'is the probable potential parent of socially inadequate offspring, likewise afflicted, that she may be sexually sterilized without detriment to her general health and that her welfare and that of society will be promoted by her sterilization,' and therefore makes the order. In view of the general declarations of the Legislature and the specific findings of the Court obviously we cannot say as a matter of law that the grounds do not exist, and if they exist they justify the result. We have seen more than once that the public welfare may call upon the best citizens for their lives. **It would be strange if it could not call upon those who already sap the strength of the State for these lesser sacrifices, often not felt to be such by those concerned, in order to prevent our being swamped with incompetence. It is better for all the world, if instead of waiting to execute degenerate offspring for crime, or to let them starve for their imbecility, society can prevent those who are manifestly unfit from continuing their kind. The principle that sustains compulsory vaccination is broad enough to cover cutting the Fallopian tubes. Three generations of imbeciles are enough. But, it is said, however it might be if this reasoning were applied generally, it fails when it is confined to the small number who are in the institutions named and is not applied to the multitudes outside. It is the usual last resort of constitutional arguments to point out shortcomings of this sort.** But the answer is that the law does all that is needed when it does all that it can, indicates a policy, applies it to all within the lines, and seeks to bring within the lines all similarly situated so far and so fast as its means allow. Of course so far as the operations enable those who otherwise must be kept confined to be returned to the world, and thus open the asylum to others, the equality aimed at will be more nearly reached.

Judgment affirmed

Mr. Justice BUTLER dissents

APPENDIX IX

EXCERPTS FROM:

FINAL REPORT OF THE TUSKEGEE SYPHILIS STUDY

AD HOC ADVISORY PANEL (1973)

Background Data

The Tuskegee Study was one of several investigations that were taking place in the 1930's with the ultimate objective of venereal disease control in the United States. Beginning in 1926, the United States Public Health Service, with the cooperation of other organizations, actively engaged in venereal disease control work. In 1929, the United States Public Health Service entered into a cooperative demonstration study with the Julius Rosewald fund and state and local departments of health in the control of venereal disease in six southern states (2): Mississippi (Bolivar County); Tennessee (Tipton County); Georgia (Glynn County); Alabama (Macon County); North Carolina (Pitt County); Virginia (Albermarle County). These syphilis control demonstrations took place from 1930–1932 and disclosed a high prevalence of syphilis (35%) in the Macon County survey. Macon County was 82.4% Negro. The cultural status of this Negro population was low and the illiteracy rate was high.

During the years 1928–1942 the Cooperative Clinical Studies in the Treatment of Syphilis (3) were taking place in the syphilis clinics of Western Reserve University of Pennsylvania, and the University of Michigan. The Division of Venereal Disease, USPHS provided statistical support, and financial support was provided by the USPHS and a grant from the Milbank Memorial Fund. These studies included a focus on effects of treatment in latent syphilis which had not been clinically documented before 1932. A report issued in 1932 indicated a satisfactory clinical outcome in 35% of untreated latent syphilitics.

The findings of Bruusgaard of Oslo on the results of untreated syphilis became available in 1929. The Oslo study was classic retrospective study involving the analysis of 473 patients at three to forty years after infection. For the first time, as a result of the Oslo study, clinical data were available to suggest the probability of spontaneous cure, continued latency, or serious or fatal outcome. Of the 473 patients included in the Oslo study, 309 were living and examined and 164 were deceased. Among the 473 patients, 27.7 percent were clinically free from symptoms and Wassermann negative; 14.8 percent had no clinical symptoms with Wassermann positive; 14.1 percent had heart and vessel disease; 2.76 percent had general paresis and 1.27 percent had tabes dorsalis. Thus in 1932, as the Public Health Service put forth a major effort toward control and treatment, much was still unknown regarding the latent stages of the disease especially pertaining to its natural course and the epidemiology of late and latent syphilis.

REPORT ON CHARGE 1-B

Background Data

In 1932, treatment of syphilis in all stages was being provided through the use of a variety of chemotherapeutic agents including mercury, bismuth, arsphenamine, neoarsphenamine, iodides and various combinations thereof. Treatment procedures being used in the early 1930's extended over long periods of time (up to two years) and were not without hazard to the patient. As of 1932, also treatment was widely recommended and treatment schedules specifically for late latent syphilis were published and in use. (10). The rational for treatment at that time was based on the clinical judgment "that the latent syphilis patient must be regarded as potential carrier of the disease and should be treated for the sake of the Community's health." (3). The aims of treatment in the treatment of latent syphilis were stated to be: 1) to increase the probability of "cure" or arrest, 2) to decrease the probability of progression or relapse over the probable result if no treatment were given and 3) the control of potential infectiousness from contact of the patient with adults of either sex, or in the case of women with latent syphilis, for unborn children.

According to Pfeiffer (1935), (11) treatment of late syphilis is quite individualistic and requires the physician's best judgment based upon sound fundamental knowledge of internal medicine and experience, and should not be undertaken as a routine procedure. Thus, treatment was being recommended in the United States for all stages of syphilis as of 1932 despite the "spontaneous" cure concept that was being justified by interpretations of the Oslo study, the potential hazards of treatment due to drug toxicity and to possible Jarisch-Herxheimer reactions in acute late syphilis. (12)

Documented reports of the effects of penicillin in the 1940's and early 1950's vary from outright support and endorsement of the use of penicillin in late and latent syphilis, (13–15) to statements of possible little or no value, (16–17) to expressions of doubts and uncertainty (18–19) related to its value, the potency of penicillin, absence of control of the rate of absorption, and potential hazard related to severe Herxheimer effects.

Although the mechanism of action of penicillin is not clear from available scientific reports of late latent syphilis, the therapeutic benefits were clinically documented by the early 1950's and have been widely reported from the mid 1950's to the present. In fact, the Center for Disease Control of the USPHS has reported treatment of syphilitic mothers in all stages of infection with penicillin as of 1953 (20) and has demonstrated that penicillin is the most effective treatment yet known for neurosyphilis(1960).

TO: THE ASSISTANT SECRETARY FOR HEALTH AND SCIENTIFIC AFFAIRS
FROM: JAY KATZ, M.D.
TOPIC: RESERVATIONS ABOUT THE PANEL REPORT ON CHARGE 1

I should like to add the following findings and observation to the majority opinion:

(1) There is ample evidence in the records available to us that the consent to participation was not obtained from the Tuskegee Syphilis Study subjects, but that instead they were exploited, manipulated, and deceived. They were treated not as human subjects but as objects of research. The most fundamental reason for condemning the Tuskegee Study at its inception and throughout its continuation is not that all the subjects should have been treated, for some might not have wished to be treated, but rather that they were never fairly consulted about the research project, its consequences for them, and the alternatives available to them. Those who for reasons of intellectual incapacity could not have been so consulted should not have been invited to participate in the study in the first place.

(2) It was already known before the Tuskegee Syphilis Study was begun, and reconfirmed by the study itself, that persons with untreated syphilis have a higher death rate than those who have been treated. The life expectancy of at least forty subjects in the study was markedly decreased for lack of treatment.

(3) In addition, the untreated and the "inadvertently" (using the word frequently employed by the investigators) but inadequately treated subjects suffered many complications which could have been ameliorated with treatment. This fact was noted on occasion in the published reports of the Tuskegee Syphilis Study and as late as 1971. However, the subjects were not apprised of this possibility.

(4) One of the senior investigators wrote in 1936 that since "a considerable portion of the infected Negro population remained

untreated during the entire course of syphilis, an unusual opportunity (arose) to study the untreated syphilitic patient from the beginning of the disease to the death of the infected person." Throughout, the investigators seem to have confused the study with an "experiment in nature." But syphilis was not a condition for which no beneficial treatment was available, calling for experimentation to learn more about the condition in the hope of finding a remedy. The persistence of the syphilitic disease from which the victims of the Tuskegee Study suffered resulted from the unwillingness or incapacity of society to mobilize the necessary resources for treatment. The investigators, the USPHS, and the private foundations who gave support to this study should not have exploited this situation in the fashion they did. Unless they could have guaranteed knowledgeable participation by the subjects, they all should have disappeared from the research scene or else utilized their limited research resources for therapeutic ends. Instead, the investigators believed that the persons involved in the Tuskegee Study would never seek out treatment; a completely unwarranted assumption which ultimately led the investigators deliberately to obstruct the opportunity for treatment of a number of the participants.

(5) In theory if not in practice, it has long been "a principle of medical and surgical morality (never to perform) on man an experiment which might be harmful to him at any extent, even though the result might be highly advantageous to science" (Claude Bernard 1865), at least without the knowledgeable consent of the subject. This was one basis on which the German physicians who had conducted medical experiments in concentration camps were tried by the Nuremberg Military Tribunal for crimes against humanity. Testimony at their trial by official representatives of the American Medical Association clearly suggested that research like the Tuskegee Syphilis Study would have been intolerable in this country or anywhere in the civilized world. Yet the Tuskegee study was continued after the Nuremberg findings and the Nuremberg Code had been widely disseminated to the medical community. Moreover, the study was not reviewed in 1966 after the Surgeon General of the USPHS

promulgated his guidelines for the ethical conduct of research, even though this study was carried on within the purview of his department.

(6) The Tuskegee Syphilis Study finally was reviewed in 1969. A lengthier transcript of the proceedings, not quoted by the majority, reveals the tone of the five members of the reviewing committee repeatedly emphasized that a moral obligation existed to provide treatment for the "patients." His plea remained unheeded. Instead the Committee, which was in part concerned with the possibility of adverse criticism, seemed to be reassured by the observation that "if we established good liaison with the local medical society, there would be no need to answer criticism."

(7) The controversy over the effectiveness and the dangers of arsenic and heavy metal treatment in 1932 and of penicillin treatment when it was introduced as a method of therapy is beside the point. For the real issue is that the participants in this study were never informed of the availability of treatment because the investigators were never in favor of such treatment. Throughout the study the responsibility rested heavily on the shoulders of the investigators to make every effort to apprise the subjects of what could be done for them if they so wished. In 1937 the then Surgeon General of the USPHS wrote: "(f) or late syphilis no blanket prescription can be written. Each patient is a law unto himself. For every syphilis patient, late and early, a careful physical examination is necessary before starting treatment and should be repeated frequently during its course." Even prior to that, in 1932, ranking USPHS physicians stated in a series of articles that adequate treatment "will afford a practical, if not complete guaranty of freedom from the development of any late lesions." In conclusion, I note sadly that the medical profession, through its national association, its many individual societies, and its journals, has on the whole not reacted to this study except by ignoring it. One lengthy editorial appeared in the October 1972 issue of the *Southern Medical Journal* which exonerated the study and chastised the "irresponsible press" for bringing it to public attention. When will we take seriously our responsibilities, particular-

ly to the disadvantaged in our midst who so consistently throughout history have been the first to be selected for human research?

Respectfully submitted,

(sgd) Jay Katz, M.D.

APPENDIX X

AMERICAN EUGENICS PARTY PLATFORM

The following platform of the AMERICAN EUGENICS PARTY is presented to unite individuals and organizations in the United States and throughout the world in opposition to the dysgenic forces. With an increasing number of supporters for all, if not most, of the Platform, the eventual attainment of a Eugenic Society is assured:

1. RACE AND STOCK PURITY

National laws must be enacted to prohibit marriage between the races and to encourage stock (ethnic group) purity. The Negroes are too different genetically and will always be a source of conflict. Negroes must be resettled in Africa. The remaining genetic types are similar enough to associate amiably with one another, but the tendency for the different genetic types to live, work, and play in separate ghettos is to be encouraged.

2. NO PERSECUTION

No race or stock is to be harshly treated. All Caucasian stocks (Germans, Jews, Italians, Poles, etc.) are to remain separate and free from persecution or abuse and must unite to ward off the non-Caucasian genetic threat.

3. QUANTITY CONTROL

The United States is already overpopulated. We must stop all immigration and impose birth controls.

---

4. QUALITY CONTROL

Those genetic types within each race and stock having better traits will be encouraged to produce more offspring and those having the lesser qualities will be restricted in the number of their offspring.

5. EXPERTS RULE

History shows that equal vote for unequal people eventually destroys society. Experts are to be selected through various improved qualification tests; all are eligible to take the tests; more than one expert is to be in each important position; and rotation of leadership among the qualified. Experts make fewer errors in all phases of life. This "civil-service-test" selection of leaders will promote confidence and efficiency.

6. INCENTIVE ECONOMICS

A high incentive level is necessary in society because of acquired and hereditary differences. Those who are not capable are to be humanely treated and their basic needs guaranteed. To promote greater progress, significant creativity and other major contributions (societal necessities, acts of bravery, etc.) are to be specially rewarded. Depressions and recessions can be cured through control of negative incentives and regulation of the medium of exchange. Wars due to economic struggles between different genetic types will be solved by controlling the availability of markets.

GENETIC INEQUALITY + GENETIC STABILITY → GENETIC CONTROLLABILITY

7. ROLE OF WOMEN

The equalitarian and the overly-dependent outlooks toward women are incurred. Women are not completely similar or different but are a combination of mental factors which are described as "complementary, opposite, and equal." The similarities allow women to do all routine tasks in society (mental to professional) and some types of research, but the differences (which are appropriate for the home) cause women to become dysgenic and socialistic when they are allowed to vote and occupy leadership positions in government, or other society molding roles (most churches recognize male-female mental differences).

8. NEUTRAL ON GOD

The AMERICAN EUGENICS PARTY does not oppose any religion which upholds Eugenics Principles. As an organization, we do not take a position on the existence of God.

9. EDUCATION PROVISIONS

Each will be guaranteed the opportunity to attain a basic education. Those with similar abilities will be grouped together for efficient learning and teaching as well as to provide compatible social contact. The neighborhood school policy will be enforced. The curriculum will stress creativity to exercise the solving ability, and teaching methods which emphasize time-wasting, etc., answer withholding techniques will be discarded (no amount of environmental "brain stretching" can alter hereditary mental ability).

10. POSITIVE PUBLIC COMMUNICATION MEDIA

Dysgenic acts and ideas will not be conveyed over the public communication media. Such expression is negative to young and impressionable minds and prevents a pleasant atmosphere for the capable. Suggestions to improve the communication media will be encouraged and handled through responsible authorities.

11. WHITE DEFECTIVES

The Caucasians are the controlling race in the world but the reason we do not have a Eugenic Society today is because of interfering, mentally deficient Caucasians (White Defectives) in controlling positions (incorrect eugenics programs are a secondary cause). There are White Defectives who propose race mixing, while other White Defectives promote intercaucasian strife—both prevent Eugenics but in different ways. White Defectives must be removed from controlling positions.

12. DYSGENIC RECREATION

Violence-type sports such as boxing, bullfights, etc. are to be eliminated or suitably altered. These types of "recreations" help produce negative behavior in lesser mental types because they promote disregard for human and animal life.

13. ANTI-VICE

Laws are to be enacted and enforced to prohibit gambling, alcoholism, smoking in public places, prostitution, sexual perversion, narcotics, obscene display in literature, movies, etc. and other vices. Happiness, health, and progress are endangered by these dysgenic acts.

14. ANTI-POLLUTION

Pollution of water, air, and food is now such a serious problem that mental and physical well-being are affected. Strict laws must be enacted to stop pollution if hereditary potentials are to be fully realized (a deficient basic environment can retard or obstruct a hereditary potential).

15. EUGENIC MARRIAGE

Eugenic marriages will be encouraged through a nationwide classification of all marriageable individuals. Those with similar qualities will be encouraged to meet and eventually marry. Today, many do not marry because they are unable to find suitable partners. Those who marry unsuitable mates are not only unhappy in marriage (which often leads to divorce—an increasing tendency) but bring forth undesirable offspring. Unhappy marriages, divorces, and undesirable offspring can only be solved through a "mate-location-system."

APPENDIX XI MEMORANDUM ABOUT PROJECT ARTICHOKE

31 JAN 1975

MEMORANDUM FOR THE RECORD
SUBJECT: Project ARTICHOKE

ARTICHOKE is the Agency cryptonym for the study and/or use of "special" interrogation methods and techniques. These "special" interrogation methods have been shown to include the use of drugs and chemicals, hypnosis, and "total isolation," a form of psychological harassment.

Among the instances where details were located in which drugs were used in an operational environment under the auspices of Project ARTICHOKE, were the following:

(a) In 1954 three subjects were interrogated by a Project ARTICHOKE team utilizing drugs of an unspecified nature. The three subjects were identified as [deleted] in a memorandum dated 13 January 1955, with a cover sheet signed by Mr. [deleted]. The interrogations took place in [deleted] and the memorandum mentioned injections of "solution #1" and "solution #2" but these drugs were not further identified. It was noted in the memorandum that the cases were handled "under straight drug techniques—hypnosis or narco-hypnosis was not attempted."

(b) A memorandum dated 20 January 1959 to Mr. [deleted] from [deleted] indicated that a field request had been made for a "P-1 interrogation." The writer [deleted] identified a "P-1 interrogation" as one using LSD. Approval was granted on 27 January 1959 by the initials [deleted], presumably Mr. [deleted]. No further reference to the case could be found, thus no details were available.

(c) A series of cables between [deleted] and Headquarters in 1955 requested ARTICHOKE interrogations for nine persons. No disposition in this instance was found. However, a transmittal slip affixed to the materials dated in 1960 indicated that the ARTICHOKE interrogations probably did not actually take place in [deleted] at that time.

(d) A memo contained in the security file of [deleted] reflected that an ARTICHOKE team was dispatched to [deleted] in June 1952 to conduct ARTICHOKE interrogations on [deleted]. No further reference to this operation was noted, and no disposition could be found.

(e) In the case of [deleted] operation in [deleted] drugs were utilized in the interrogation which took place in [deleted]. Again, details of the operation were not available. However, an interview with the Office of Security representative who participated in the interrogation revealed that a form of LSD was used in this instance. In this case, approval was granted by Headquarters for the ARTICHOKE interrogation. A memorandum dated 6 July 1960, signed by Mr. [deleted], Deputy Director of Security, reflected that approval for use of drugs in this case was granted at a meeting of the Drug Committee on 1 July 1960 and cabled to [deleted].

As stated earlier, little detail was available in file information concerning the conduct of actual cases utilizing Project ARTICHOKE techniques. It appears obvious, however, that the few cases noted above were only a small part of the actual utilization of ARTICHOKE techniques in the field. For one thing, almost no information was available for the period prior to 1952, so that Project BLUEBIRD experiments and operations were not noted specifically. In addition, annual reports of accomplishments found in SRS log materials reflected a substantial amount of activity in the Project ARTICHOKE area. The review for 1953–1954 stated in part that SRS had "dispatched an ARTICHOKE team for permanent location in an overseas area." The review for 1954–1955 stated in part that SRS conducted numerous ARTICHOKE experiments and "prepared and dispatched an ARTICHOKE team to an overseas area to handle a number of sensitive cases."

In the review of file information contained in SRS materials, one incident which occurred in November 1953 appears worthy of note. Although it was not clear from file information whether or not the incident occurred under the auspices of Project ARTICHOKE, the incident did involve use of LSD in an experimental exercise. One Frank OLSON, a civilian employee of the Department of the ARMY, committed suicide a week or so after having been administered LSD by an Agency representative. Details concerning this incident apparently will be reported in a separate memorandum, but it appears that the drug was administered to several unwitting subjects by a Dr. GOTTLIEB, at that time a branch chief in TSS (now OTS). A short time after the LSD was administered, the subjects were told that they had been given LSD. On the day following the experiment, OLSON began to behave in a peculiar and erratic manner and was later placed under the care of a psychiatrist. A few days later, OLSON crashed through a window in a New York hotel in an apparent suicide.

A memorandum dated 1 December 1953 from the IG Staff caused the impoundment of all LSD materials. Information contained in the above mentioned files reflected that the drug had been administered without the prior knowledge or approval of the Office of Security or the Office of Medical Services.

APPENDIX XII

EXCERPT OF MEMORANDUM DATED DECEMBER 3, 1975 FROM LIGGETT & MYERS ABOUT RADIOACTIVE MATERIALS IN CIGARETTES

Subject: Radioactive Particles in Cigarette Smoke

Martell postulates that radioactive elements in the soil, or present in phosphate fertilizers undergo transformation to Radon (a gas) which is captured selectively by the small hair-like appendages on upper and lower surfaces of the tobacco plant . . . Absorption of Radon through inhalation of the atmosphere and ingestion of the other radioactive materials in food is constantly occurring. The uniqueness of exposure via cigarette smoke is that relatively higher concentrations of these radioactive materials could occur on smoke particles and these could then accumulate at bifurcations in the lung to provide a relatively long-term chronic exposure at localized sites.

It has been established that both radioactive ^{210}Po and ^{210}Pb are transferred from tobacco to smoke. We have carried out experiments to demonstrate that these materials reside in the particulate phase and, therefore, are not susceptible to selective filtration.

If we assume that it would be desirable to eliminate the ^{210}Pb and ^{210}Po from cigarette smoke, we have several options, some of which may not prove viable and some which may prove too expensive for practical application . . .

APPENDIX XIII

WAR CRIMES INDICTMENTS FOR HUMAN MEDICAL EXPERIMENTS

FROM NUREMBERG MILITARY TRIBUNALS UNDER CONTROL COUNCIL LAW NO. 10

Between September 1939 and April 1945 all of the defendants herein unlawfully, willfully, and knowingly committed war crimes, as defined by Article II of Control Council Law No. 10, in that they were principals in, accessories to, ordered, abetted, took a consenting part in, and were connected with plans and enterprises involving medical experiments without the subjects' consent, upon civilians and members of the armed forces of nations then at war with the German Reich and who were in the custody of the German Reich in exercise of belligerent control, in the course of which experiments the defendants committed murders, brutalities, cruelties, tortures, atrocities, and other inhuman acts. Such experiments included, but were not limited to, the following:

(A) High-Altitude Experiments. From about March 1942 to about August 1942 experiments were conducted at the Dachau concentration camp, for the benefit of the German Air Force, to investigate the limits of human endurance and existence at extremely high altitudes. The experiments were carried out in a low-pressure chamber in which atmospheric conditions and pressures prevailing at high altitude (up to 68,000 feet) could be duplicated. The experimental subjects were placed in the low-pressure chamber and thereafter the simulated altitude therein was raised. Many victims died as a result of these experiments and others suffered grave injury, torture, and ill treatment. The defendants Karl Brandt, Handloser, Schroeder, Gebhardt, Rudolf Brandt, Mrugowsky, Poppendick, Sievers, Ruff, Romberg, Becker-Freyseng, and Weltz are charged with special responsibility for and participation in these crimes.

(B) Freezing Experiments. From about August 1942 to about May 1943 experiments were conducted at the Dachau concentration camp, primarily for the benefit of the German Air Force, to investigate the most effective means of treating persons who had been severely chilled or frozen. In one series of experiments the subjects were forced to remain in a tank of ice water for periods up to 3 hours. Extreme rigor developed in a short time. Numerous victims died in the course of these experiments. After the survivors were severely chilled, rewarming was attempted by various means. In another series of experiments, the subjects were kept naked outdoors for many hours at temperatures below freezing. The victims screamed with pain as their bodies froze. The defendants Karl Brandt, Handloser, Schroeder, Gebhardt, Rudolf Brandt, Mrugowsky, [page 12] Poppendick, Sievers, Becker-Freyseng, and Weltz are charged with special responsibility for and participation in these crimes.

(C) Malaria Experiments. From about February 1942 to about April 1945 experiments were conducted at the Dachau concentration camp in order to investigate immunization for and treatment of malaria. Healthy concentration-camp inmates were infected by mosquitoes or by injections of extracts of the mucous glands of mosquitoes. After having contracted malaria the subjects were treated with various drugs to test their relative efficacy. Over 1,000 involuntary subjects were used in these experiments. Many of the victims died and others suffered severe pain and permanent disability. The defendants Karl Brandt, Handloser, Rostock, Gebhardt, Blome, Rudolf Brandt, Mrugowsky, Poppendick, and Sievers are charged with special responsibility for and participation in these crimes.

(D) Lost (Mustard) Gas Experiments. At various times between September 1939 and April 1945 experiments were conducted at Sachsenhausen, Natzweiler, and other concentration camps for the benefit of the German Armed Forces to investigate the most effective treatment of wounds caused by lost gas. Lost is a poison gas which is commonly known as mustard gas. Wounds deliberately inflicted on the subjects were infected with

Lost. Some of the subjects died as a result of these experiments and others suffered intense pain and injury. The defendants Karl Brandt, Handloser, Blome, Rostock, Gebhardt, Rudolf Brandt, and Sievers are charged with special responsibility for and participation in these crimes.

(E) **Sulfanilamide Experiments.** From about July 1942 to about September 1943 experiments to investigate the effectiveness of sulfanilamide were conducted at the Ravensbrueck concentration camp for the benefit of the German Armed Forces. Wounds deliberately inflicted on the experimental subjects were infected with bacteria such as streptococcus, gas gangrene, and tetanus. Circulation of blood was interrupted by tying off blood vessels at both ends of the wound to create a condition similar to that of a battlefield wound. Infection was aggravated by forcing wood shavings and ground glass into the wounds. The infection was treated with sulfanilamide and other drugs to determine their effectiveness. Some subjects died as a result of these experiments and others suffered serious injury and intense agony. The defendants Karl Brandt, Handloser, Rostock, Schroeder, Genzken, Gebhardt, Blome, Rudolf Brandt, Mrugowsky, Poppendick, Becker-Freyseng, Oberheuser, and Fischer are charged with special responsibility for and participation in these crimes.

(F) **Bone, Muscle, and Nerve Regeneration and Bone Transplantation Experiments.** From about September 1942 to about December 1943 experiments were conducted at the Ravensbrueck concentration camp, for the benefit of the German Armed Forces, to study bone, [page 13] muscle, and nerve regeneration, and bone transplantation from one person to another. Sections of bones, muscles, and nerves were removed from the subjects. As a result of these operations, many victims suffered intense agony, mutilation, and permanent disability. The defendants Karl Brandt, Handloser, Rostock, Gebhardt, Rudolf Brandt, Oberheuser, and Fischer are charged with special responsibility for and participation in these crimes.

(G) **Sea-water Experiments.** From about July 1944 to about September 1944 experiments were conducted at the Dachau concentration camp, for the benefit of the German Air Force and Navy, to study various methods of making sea water drinkable. The subjects were deprived of all food and given only chemically processed sea water. Such experiments caused great pain and suffering and resulted in serious bodily injury to the victims. The defendants Karl Brandt, Handloser, Rostock, Schroeder, Gebhardt, Rudolf Brandt, Mrugowsky, Poppendick, Sievers, Becker-Freyseng, Schaefer, and Beiglboeck are charged with special responsibility for and participation in these crimes.

(H) Epidemic Jaundice Experiments. From about June 1943 to about January 1945 experiments were conducted at the Sachsenhausen and Natzweiler concentration camps, for the benefit of the German Armed Forces, to investigate the causes of, and inoculations against, epidemic jaundice. Experimental subjects were deliberately infected with epidemic jaundice, some of whom died as a result, and others were caused great pain and suffering. The defendants Karl Brandt, Handloser, Rostock, Schroeder, Gebhardt, Rudolf Brandt, Mrugowsky, Poppendick, Sievers, Rose, and Becker-Freyseng are charged with special responsibility for and participation in these crimes.

(I) Sterilization Experiments. From about March 1941 to about January 1945 sterilization experiments were conducted at the Auschwitz and Ravensbrueck concentration camps, and other places. The purpose of these experiments was to develop a method of sterilization which would be suitable for sterilizing millions of people with a minimum of time and effort. These experiments were conducted by means of X-ray, surgery, and various drugs. Thousands of victims were sterilized and thereby suffered great mental and physical anguish. The defendants Karl Brandt, Gebhardt, Rudolf Brandt, Mrugowsky, Poppendick, Brack, Pokorny, and Oberheuser are charged with special responsibility for and participation in these crimes.

(J) Spotted Fever (Fleckfieber) Experiments. [It was definitely ascertained in the course of the proceedings, by both prosecution and defense, that the correct translation of "Fleckfieber" is typhus. A finding to this effect is contained in the judgment. A similar initial inadequate translation occurred in the case of "typhus" and "paratyphus" which should be rendered as typhoid and paratyphoid.] From about December 1941 to about February 1945 experiments were conducted at the Buchenwald and Natzweiler concentration camps, for the benefit [page 14] of the German Armed Forces, to investigate the effectiveness of spotted fever and other vaccines. At Buchenwald numerous healthy inmates were deliberately infected with spotted fever virus in order to keep the virus alive; over 90 percent of the victims died as a result. Other healthy inmates were used to determine the effectiveness of different spotted fever vaccines and of various chemical substances. In the course of these experiments 75 percent of the selected number of inmates were vaccinated with one of the vaccines or nourished with one of the chemical substances and after a period of 3 to 4 weeks, were infected with spotted fever germs. The remaining 25 percent were infected without any previous protection in order to compare the effectiveness of the vaccines and the chemical substances. As a result, hundreds of the persons experimented upon died. Experiments with yellow fever, smallpox, typhus, paratyphus [It was definitely ascertained in the course of the proceedings, by both prosecution ad defense, that the correct translation of "Fleckfieber" is typhus. A finding to this effect is contained in the judgment. A similar initial inadequate translation occurred in the case of "typhus" and "paratyphus" which should be rendered as typhoid and paratyphoid] A and B, cholera, and diphtheria were also conducted. Similar experiments with like results were conducted at Natzweiler concentration camp. The defendants Karl Brandt, Handloser, Rostock, Schroeder, Genzken, Gebhardt, Rudolf Brandt, Mrugowsky, Poppendick, Sievers, Rose, Becker-Freyseng, and Hoven are charged with special responsibility for and participation in these crimes.

(K) Experiments with Poison. In or about December 1943, and in or about October 1944, experiments were conducted at the Buchenwald concentration camp to investigate the effect of various poisons upon human beings. The poisons were secretly administered to experimental subjects in their food. The victims died as a result of the poison or were killed immediately in order to permit autopsies. In or about September 1944 experimental subjects were shot with poison bullets and suffered torture and death. The defendants Genzken, Gebhardt, Mrugowsky, and Poppendick are charged with special responsibility for and participation in these crimes.

(L) Incendiary Bomb Experiments. From about November 1943 to about January 1944 experiments were conducted at the Buchenwald concentration camp to test the effect of various pharmaceutical preparations on phosphorous burns. These burns were inflicted on experimental subjects with phosphorous matter taken from incendiary bombs, and caused severe pain, suffering, and serious bodily injury. The defendants Genzken, Gebhardt, Mrugowsky, and Poppendick are charged with special responsibility for and participation in these crimes. Between June 1943 and September 1944 the defendants Rudolf Brandt and Sievers unlawfully, willfully, and knowingly committed war crimes, as defined by article II of Control Council Law No. 10, in that they were principals in, accessories to, ordered, abetted, took a consenting part in, and were connected with plans and enterprises involving the murder of civilians and members of the armed forces of [page 15] nations then at war with the German Reich and who were in the custody of the German Reich in exercise of belligerent control. One hundred twelve Jews were selected for the purpose of completing a skeleton collection for the Reich University of Strasbourg. Their photographs and anthropological measurements were taken. Then they were killed. Thereafter, comparison tests, anatomical research, studies regarding race, pathological features of the body, form and size of the brain, and other tests, were made. The bodies were sent to Strasbourg and defleshed.

Between May 1942 and January 1944 (Indictment originally read "January 1943" but was amended by a motion filed with the Secretary General. See Arraignment, page 18) the defendants Blome and Rudolf Brandt unlawfully, willfully, and knowingly committed war crimes, as defined by Article II of Control Council Law No. 10, in that they were principals in, accessories to, ordered, abetted, took a consenting part in, and were connected with plans and enterprises involving the murder and mistreatment of tens of thousands of Polish nationals who were civilians and members of the armed forces of a nation then at war with the German Reich and who were in the custody of the German Reich in exercise of belligerent control. These people were alleged to be infected with incurable tuberculosis. On the ground of insuring the health and welfare of Germans in Poland, many tubercular Poles were ruthlessly exterminated while others were isolated in death camps with inadequate medical facilities.

Between September 1939 and April 1945 the defendants Karl Brandt, Blome, Brack, and Hoven unlawfully, willfully, and knowingly committed war crimes, as defined by Article II of Control Council Law No. 10, in that they were principals in, accessories to, ordered, abetted, took a consenting part in, and were connected with plans and enterprises involving the execution of the so-called "euthanasia" program of the German Reich in the course of which the defendants herein murdered hundreds of thousands of human beings, including nationals of German-occupied countries. This program involved the systematic and secret execution of the aged, insane, incurably ill, of deformed children, and other persons, by gas, lethal injections, and diverse other means in nursing homes, hospitals, and asylums. Such persons were regarded as "useless eaters" and a burden to the German war machine. The relatives of these victims were informed that they died from natural causes, such as heart failure. German doctors involved in the "euthanasia" program were also sent to Eastern occupied countries to assist in the mass extermination of Jews.

The said war crimes constitute violations of international conventions, particularly of Articles 4, 5, 6, 7, and 46 of the Hague Regulations, 1907, and Articles 2, 3, and 4 of the Prisoner-of-War Convention (Geneva, 1929), the laws and customs of war, the general principles of criminal law as derived from the criminal laws of all civilized nations, the internal penal laws of the countries in which such crimes were committed, and Article II of Control Council Law No. 10.

APPENDIX XIV

LETTER FROM SENATOR HOWARD METZENBAUM ABOUT THE DANGERS OF ASPARTAME

Howard M. Metzenbaum
United States Senate

140 Russell Senate Office Building
Washington, DC 20510

February 25, 1986

Dear Orrin:

I am at a loss to comprehend the thrust of your recent letters on my request for hearings on the safety concerns raised in the scientific community regarding NutraSweet.

When I sent you a 110-page report on February 6 on the failure of the U.S. Attorney to hold a grand jury investigation, you replied the same day that there were no health issues raised. You then asked that I share with you all information raising safety issues. Orrin, the report I sent you included a summary of current health concerns raised by nine different scientists. My report contained all the relevant references to medical journals and other citations. Now you have sent me another letter, dated February 18, in which you again request evidence.

As you know, I met last Thursday with Dr. Roger Coulombe

of Utah State University. You also had a conversation with Dr. Coulombe, as did your staff. Dr. Coulombe has informed both of us that his study of NutraSweet's effects on brain chemistry contains new and significant data.

All of the 12 mice tested showed brain chemistry changes after ingesting NutraSweet. Four other mice received no NutraSweet and showed no brain chemistry changes. Dr. Coulombe also informed us that the issues raised in his study were not tested prior to NutraSweet's approval. So, the FDA never reviewed this research prior to approving NutraSweet. It is critical to note that some of the lab animals which had reactions to NutraSweet were fed doses at levels currently being consumed by humans.

As you know, there have been many reports of seizures, headaches, mood alterations, etc., associated with NutraSweet. Dr. Coulombe's study, which has been accepted for publication in Toxicology and Applied Pharmacology, states: "It is therefore possible that Aspartame may produce neurobiochemical and behavioral effects in humans, particularly in children and susceptible individuals. Based on the foregoing, there is a need for additional research on the safety of this food additive."

Orrin, you have asked for new and significant scientific evidence about NutraSweet. Now you have it. Dr. Coulombe's research as well as the other research cited in my report raises new health concerns which have not been resolved. We need to hold hearings on NutraSweet—which is being used by over 100 million Americans. With an issue that is critical to the health of half the American population, how can you in good conscience say "no?"

We cannot rely upon the tests sponsored by the manufacturer of NutraSweet, G. D. Searle, and ignore the concerns being raised by independent studies. We don't need the company which is making hundreds of millions of dollars on this product telling us it's "safe," particularly when the credibility of that Company's testing on NutraSweet has been severely undermined. You know that the FDA recommended a criminal investigation of possible fraud in NutraSweet tests. The FDA has never before or since made such an investigation.

Although NutraSweet was later approved, credible scientific concerns continue to be raised. The Director of Clinical Research at M.I.T., Dr. Richard Wurtman, has recently published a letter in Lancet citing case reports suggesting a possible association between Aspartame and seizures. According to Dr. Wurtman, the reports are compatible with evidence that high Aspartame doses may produce neurochemical changes that, in laboratory animals, are associated with depressed seizure thresholds.

Dr. William Pardridge, U.C.L.A. School of Medicine, has also cited a possible link between one of Aspartame's principal components, phenylalanine, and the lowering of seizure thresholds in individual/individuals. He has also questioned the possible affects of NutraSweet on fetal development.

In July, 1985, Dr. Michael Mahalik of the Philadelphia College of Osteopathic Medicine, and Dr. Ronald Gautieri of Temple University School of Pharmacy, published a study on the potential of Aspartame to produce brain dysfunction in mouse neonates whose mothers were exposed to Aspartame in late gestation. They concluded that the possibility of brain dysfunction appears to be a viable sequela to excessive Aspartame exposure.

In June of last year, Dr. Adrian Gross, a former FDA Toxicologist, and member of the FDA Investigative Task Force which reviewed the Aspartame studies, sent me a letter stating that despite their serious shortcomings, at least those studies established beyond a reasonable doubt that Aspartame is capable of inducing brain tumors in experimental animals.

In February, 1985, letters were published in the Annals of Internal Medicine and the American Journal of Psychiatry, linking Aspartame to skin lesions and to severe headaches caused by chemical interactions with an anti-depressant drug, an M.A.O. inhibitor.

In December 1984, Dr. John Olney of Washington University published a study on excitotoxic amino acids including Aspartate, one of Aspartame's two constituent amino acids. He concludes that excitotoxins pose a significant hazard to the developing nervous systems of young children.

Dr. Louis Elsas, at Emory University, has raised concerns about Aspartame's other constituent amino acid, phenylalanine.

He has stated that if the mother's blood phenylalanine is raised to high concentrations in pregnancy, her child's brain development can be damaged. According to Dr. Elsas, it has not been determined how high the blood phenylalanine must be elevated to produce any of these effects. However, he believes that it has not been proven that all people can take as much Aspartame without fear of ill effects as they desire.

Appearing on the news program Nightline in May of last year, Dr. Elsas warned of a whitewashed scientific review, most of which has been supported by the industry itself, which has obvious conflict of interest.

All of these safety concerns have been raised after NutraSweet was approved for use in over 90 products. The FDA is currently considering petitions which would explain even further the dramatic increase in NutraSweet consumption. My staff has provided you with the references for all of the scientific concerns raised above. I strongly urge you to reconsider your decision and to convene oversight hearings in the Labor and Human Resources Committee as soon as possible to consider these issues.

By ignoring the safety concerns which have been raised, we are potentially jeopardizing the health and safety of over 100 million Americans who are ingesting NutraSweet in everything from soft drinks to pudding to children's cold remedies.

Very Sincerely Yours,

Howard M. Metzenbaum
United States Senator

APPENDIX XV

LETTER FROM THE EPA TO SENATOR METZENBAUM ABOUT THE DANGERS OF ASPARTAME

U.S. Environmental Protection Agency
Washington, DC 20460

3 November, 1987

Senator Howard M. Metzenbaum,
United States Senate,
140 Russell Senate Office Building,
Washington, DC, 20510

Dear Senator Metzenbaum,

The following represents a continuation of my letter to you of last week, October the 30th, 1987. In that letter I discussed the many serious problems with the quality or reliability of the experimental studies with aspartame carried out by or for G. D. Searle & Co. I noted there, that in 1976, the FDA Commissioner at that time, Dr. Alexander Schmidt, speaking for the FDA as an agency, publicly stated that he agreed with a set of conclusions, the first of which was that the FDA had no basis for reliance on the quality of studies generated by or for that firm.

[letter details pages of irregularities before coming a conclusion]

SUMMARY AND CONCLUSIONS.

From what has been discussed in my letter addressed to you last week as well as from what has been presented in the previous pages of this communication, I can conclude the following:

1. It is impossible for anyone to appreciate just how a determination by the FDA that the G. D. Searle & Co. experimental studies with aspartame were of an unacceptable quality in 1976 can be metamorphosed several years later into a view by that same Agency that essentially the same studies were sufficiently reliable for anyone to assess that this food additive is "reasonably certain" to be safe for consumption by humans.

2. Even if, contrary to the FDA's view in 1976, the quality of the conduct of those studies could be relied upon by the same agency to even begin making such a determination, at least one of those studies had reveled a highly significantly dose-related increase in the incidence of brain tumors as a result of exposure to aspartame. The full incidence of those brain tumors was not disclosed by G. D. Searle & Co. to the FDA prior to the initial approval for the marketing of aspartame in 1974; moreover, the review of that study in the FDA was so flawed that the Agency apparently did not even realize at that time that only a portion of the observations on brain tumors had in fact been submitted by G. D. Searle & Co. in their petition for that approval.

3. Quite aside from the remarkable significance of the increased incidence with dose of those brain tumors, the ADI of 50 mg/kg bodyweight recently set by the FDA for the human consumption of aspartame is alarmingly dangerous in that it involves an extremely high and, therefore, a totally unacceptable upper limit on the risk for those consuming aspartame: between 1/1,000 and 5/1,000 population to develop brain tumors as a result of such exposure.

4. Although in their report the GAO express the view that the FDA "followed its required process in approving aspartame (for marketing)" I would sharply disagree with such evaluation. Although the FDA may have gone through the motions

or it may have given the appearance of such a process being in place here, the people of this country expect and require a great deal more from that agency charged with protecting their public health: in addition to mere facade or window-dressing on the part of the FDA, they require a thorough and scientifically based evaluation by the Agency on the safety of the products it regulates.

Unfortunately this has clearly not been the case here. And without this kind of assurance, any such "process" or dance represents no more than a farce and a mockery of what is truly required.

Sincerely yours,

M. Adrian Gross,
Senior Science Advisor,
Benefits and Use Division
Office of Pesticide Programs
Sworn to be a true copy on 30 Oct., 1987.

PROJECT ARTICHOKE OPERATIONS DOCUMENT

TO	Director of Security
VIA	Deputy Director of Security
VIA	Chief, Security Research Staff
FROM	[Deleted]
SUBJECT:	Report of ARTICHOKE Operations, 20 to 23 January 1955.

1. Between Thursday, 20 January and Sunday, 23 January 1955, the ARTICHOKE team conducted a special operation [deleted]. In the opinion of team mentors and participating case officers of the [deleted], the ARTICHOKE operation was successful. Details follow:

2. It should be noted at this point that because these operations were the first ARTICHOKE operations undertaken in the United States the full names of those participating are omitted from this report and will not be revealed without consent of the Security Office. First names, titles or pseudonyms will be used throughout this report.

3. In view of the highly sensitive nature of the ARTICHOKE techniques and in view of the fact that this was the first ARTICHOKE operation carried out in the United States, the operation was conducted [deleted]. This safe house is far removed from surrounding neighbors in a large tract of land and is thoroughly isolated. A limited and Security-cleared household staff

maintained functions of the house and messing was by unwitting [deleted]. Actual ARTICHOKE operations were, as usual, carried out in a special area on the second floor of the house and neither the household staff nor the [deleted] were permitted in the area during any of the processing. SSD Division furnished one Security Officer during the entire period of the operation to act as a special guard and to handle any unusual situations which arose during the operation. This guard is hereinafter referred to as [deleted] in this report.

4. For matter of record, it should be noted that the subject was not a confinement problem and has been, at all times, fully cooperative. Guard detail was not present in connection with the Subject except in a general sense.

5. Technical matters in the case were handled entirely by the IS/FSD under the personal supervision of [deleted]. Full tape recordings were made of the entire case and tapes are to be turned over to the participating Division in the immediate future. It should be noted that during this particular operation, a special device was used in connection with the recorder. This device, which is easily concealable worked with remarkable efficiency and at no time during the entire recording was there any break due to technical failure. It should also be noted that a complex two-way transmitting-receiving unit was again used in this ARTICHOKE operation.

6. Cover for the actual operation followed standard procedure. The Subject was informed in general terms that before being sent for further work, it was necessary that certain tests be made on him physically and psychologically for our protection as well as his. Hence, a complete physical and psychiatric/psychological examination was required. Subject readily accepted this medical cover and the ARTICHOKE technique was introduced easily and with full consent of the SUBJECT.

THE CASE

7. Prior to the actual commencement of ARTICHOKE operations, a number of conferences had been held with the vari-

ous participating personnel involved. All hands had been briefed and procedures had been worked out; a general time schedule was prepared and operating instructions for ARTICHOKE were issued.

8. On the afternoon of 20 January, the Subject and Case Officer [deleted]. They were met by [deleted] of the interested Division. Using a covert car, Subject was taken to the [deleted], arriving there at approximately 9:30 PM. Prior to this, that is, during the day of Thursday, 20 January, the technical equipment had been checked out and installed and [deleted] had arrived at the covert area at approximately 8:00 AM for operational purposes. By previous arrangement, the [deleted] was picked up by [deleted] at approximately 9:30 PM. [Deleted] was brought to the safe house at 10:50 PM.

9. Shortly after the arrival of [deleted], a preliminary conference began at approximately 11:10 PM with the Subject, [deleted]. At 11:33 PM, the Subject, [deleted] went to the operations area and a few minutes later [deleted] started a general interrogation relative to Subject's background. This interrogation lasted until 12:25 when all except the Subject and [deleted] left the operations room. Tape recording was cut off at this time.

10. As a result of this interview, [deleted] stated that Subject's mental and physical condition was good and noted that the pulse at 12:25 PM was 120/80. Doctor also commented he had noted an increased amount of talk after a drink of whiskey and although there was some nervousness present, it was not excessive. [Deleted] stated he had given Subject two grams Phenobarbital to use in assisting the Subject to sleep and it was later confirmed the Subject had taken these prior to going to sleep.

11. Subject [deleted]. Because of this successful penetration and because of the extremely high quality of information which the Subject was obtaining, the case is regarded as most sensitive and important by the participating Division. Since the Subject's information had been checked and cross-checked many times by the operating Division's case officers, the Division was of the uniform opinion that the Subject was fully legitimate and fully coop-

erating with our efforts. They, however, desired ARTICHOKE to give added assurances to the Subject's story and to help them determine absolute suitability for further uses of the Subject in his work. For the record, it should be noted that no polygraph techniques had been applied in this case since a physical examination in [deleted] by apparently a cleared physician, had indicated too nervousness for successful polygraph testing.

12. Following established patterns and using medical cover as explained above, the [deleted] began the physical-psychological examination at 10:00 PM on the morning of Friday, 21 January. This examination continued until 1:00 PM when an hour was taken for lunch. At 2:00 PM [deleted] again continued a general examination of the Subject with [deleted] being used, as before lunch, as interpreter. This examination lasted until 3:00 PM when the [deleted] concluded the first medical session and a portable polygraph was taken in by [deleted] for the purpose of polygraph testing.

13. Between 3:00 PM and 4:25 PM [deleted]. At 5:00 PM, all work concluded for the day.

14. On Saturday, 22 January 1955, Subject had breakfast with [deleted] and [deleted]. At 9:35 AM, [deleted] arrived at the safe house and at 9:45 AM, [deleted] arrived. At 10:35 AM, the Subject, again with [deleted] acting as interpreter, was examined briefly by briefly by Dr. [deleted]. At 10:50 AM, [deleted] left the operations area and began polygraph testing. This examination lasted until 12:37 PM when it was concluded.

15. Subject was taken into the Special Operations Room with only the [deleted] present and at 2:36 PM the first intravenous infusion began. Slow injections were continued until 2:46 PM when the [deleted] signaled that the subject was fully affected by the chemicals and at this time special recording and transmitting equipment was brought into the Operations Room. Also at this time [deleted] left the room and [deleted] entered. From this point until approximately 4:15 PM when the interrogations ended, ARTICHOKE techniques were applied. These techniques, which followed a previously agreed-upon plan, were in three stages: A. A

fantasy which [deleted]. Results during this phase were good and subject had no control. Time approximately 15–20 minutes. B. A fantasy in which [deleted]. Results were again very good. Time approximately 40–45 minutes. C.

16. Following development of the fantasies as noted above, the subject was more or less directly interrogated by [deleted], and [deleted] introduced as [deleted]. Results only fair, although subject had little control. Time approximately 15 minutes.

17. Immediately following the conclusion of the ARTICHOKE treatments, a general conference was held with all hands present. It was agreed at this time that further ARTICHOKE treatments were unnecessary, that results were as conclusive, that in view of the subject's importance, additional work with chemicals or with the H technique might possibly antagonize the subject, hence would be unwarranted and unwise.

18. Following the conclusion of the general discussion all technical apparatus was removed from the premises, and all participating personnel with the exception of [deleted] left the area after the [deleted] had checked the subject.

19. On Sunday 23 January between approximately twelve noon and 1:30 PM, the [deleted] returned to the safe house and again re-examined the physical and mental condition of the subject. At this time the subject reported he had slept fairly well but he had a persistent headache. The [deleted] pointed out that the headache was a natural consequence of the "examination" and it would gradually disappear. In addition, the [deleted] wrote a prescription which was to be picked up in another name for future use by the subject as a general sedative.

20. At 1:50 PM approximately [deleted] left the safe house and the subject was turned over for handling to Case Officers of his participating Division.

CONCLUSIONS

21. In the opinion of the ARTICHOKE team, the operation was profitable and successful. In this case, the Subject was aware

that he had been given certain types of solutions but as to what he had been given or amounts given he had no knowledge. Checks made by [deleted] and later [deleted] apparently indicated that the Subject, although not having specific amnesia for the ARTICHOKE treatment, nevertheless was completely confused and memory was vague and faulty. This vagueness and failure of memory was intensified by the [deleted] explanation that the Subject had been dreaming—an opinion which it appears the Subject shared, at least in part.

SPECIAL COMMENTS

22. The work of the case officers of the participating Division in connection with this case was exceptionally good. Their understanding and appreciation of the ARTICHOKE techniques was extremely helpful.

23. The ARTICHOKE team wishes to commend [deleted] for the expert handling of the support function at the [deleted], which greatly assisted in the development of the ARTICHOKE work.

APPENDIX XVII PROJECT MKOFTEN DOCUMENTS

23 June 1970

MEMORANDUM FOR: [Deleted]
SUBJECT: Review of EA 3167 Study

1. In this study, nineteen subjects were divided into three groups which were treated with dosages of [deleted] units/kg of experimental agent 3167.

2. In the first group of six subjects, measurements of temperature, blood pressure, respiration, pulse, and pupil size, although showing some variation, did not reveal significant differences which could be related to drug symptomology. In every case, undesirable symptoms were noted, all six subjects experiencing "drowsiness" and "dry throat." Of the three cases of hallucination and mental incapacitation, only one was of a serious nature and this admittedly may have been due to an additional dose of the drug.

3. The second group ([deleted] units/kg) exhibited a variety of undesired side effects: drowsiness, dry or sore throat, nausea, loss of taste, blurred vision, heaviness in legs, lack of coordination. All seven subjects in this group experienced at least three of these symptoms. Four of the seven suffered severe mental incapacitation accompanied by heightened symptomology. In three of these seven cases a high pulse rate and dilated pupils could be

related to drug action though the papillary response was much stronger. In subjects not strongly affected by the drug, a lower pulse rate sometimes coincided with drowsiness and impairment of coordination.

4. The group which received the highest dosage proved as variable as the others. Although each subject exhibited the usual symptomology, only two of the six were strongly affected. Those two hallucinated and dropped to scores of zero on their numbers facilities tests with concomitant increases in pulse rate and pupil size. The four other subjects showed thought hindrance and lack of concentration but apparently as a consequence of extreme drowsiness.

5. In the majority of cases, the side effects appeared within 4 hours after injection. Their duration varied from about 4 hours to 19 days. The desirable primary effects generally did not appear till after the side effects were evident and in every case had a shorter duration, varying from 1 to 90 hours.

6. In the instance of mental incapacitation, the more pronounced effects appeared to be inability to relate to surroundings or time, inability to remember names, and poor performance on numbers facilities tested. Hallucinations were of both visual and auditory nature. Patients would see and hear persons not there and speak to them. Frequent complaints were blight lights or objects on the wall and roaches or flying insects in the room.

7. This study was somewhat unprofessional and a trifle slipshod. The results are inconclusive. Apparently, the drug is not reliable at the dosage levels tested. Only nine of the nineteen subjects experienced "desirable effects" (3 out of 6 at [deleted] units/kg; 4 out of 7 at [deleted] units/kg; 2 out of 6 at [deleted] units/kg, but all nineteen exhibited undesirable signs and/or symptoms.

---

13 October 1970

MEMORANDUM FOR THE RECORD
SUBJECT: Visit by Dr. [Deleted]

1. On 30 September, Dr. [deleted], Professor and Chairman of the Department of [deleted] visited Project OFTEN's screening facility prior to a subsequent meeting at [deleted]. Dr. [deleted] visit was arranged in accordance with the joint [deleted] steering committee's announced plans to have OFTEN's activities reviewed and assessed by at least two recognized outside experts in the field of pharmacology and behavioral science.

2. At the screening, Dr. [deleted] was given a complete briefing on OFTEN's objectives and the methods and procedures implemented to carry out these objectives. Those in attendance were Drs. [deleted]. In addition, Dr. [deleted] toured the laboratories where demonstrations were held and representative data discussed. At [deleted], Dr. [deleted] reviewed project OFTEN activities with Drs. [deleted] for FY 71, inspected the computer facility, and was briefed on [deleted], the file management program now used to search our toxicological data bank.

3. Dr. [deleted] expressed agreement with the design and operation of the OFTEN project as well as planned activities and made several suggestions for new agents that have potential as unique behavior modifiers. These will be screened immediately if available, together with other compounds suggested as criterion agents for standardization purposes.

4. Finally, Dr. [deleted] was invited to join the OFTEN project on a consulting basis and he accepted the role. It is planned to have him visit this installation quarterly with opportunity for more frequent consultation on a non-visit basis. He will also be present at some meetings of the joint [deleted] steering group. This aspect will be cleared with [deleted] in advance.

———

29 May 1973

MEMORANDUM FOR: [Deleted]
SUBJECT: Summary of Project OFTEN Clinical Tests at Edgewood

1. Funds in the amount of $37,000 were transferred to Edgewood Arsenal on 17 February 1971 for the purpose of determining the clinical effects of EA #3167, a glycolate class chemical previously developed by Edgewood. Analysis of Edgewood file data had flagged this item as possessing unusual potential as an incapacitant, strongly suggesting the possibility of [deleted].

2. The Soviets were known to be actively working in the glycolate area. Edgewood had partially investigated EA #3167 and found it to be effective [deleted] in animals. In addition, there had been several laboratory accidents in which the agent had produced prolonged psychotic effects in laboratory personnel.

3. Since the [deleted] were the routes of potential threat to U.S. VIPs and other key personnel, it was highly desirable that existing data on [deleted] in humans previously acquired by Edgewood be extended to include the [deleted]. Simultaneously, plans were developed to implement countermeasures as required.

4. Preliminary laboratory work was undertaken to determine the solubility and [deleted] of #3167. Additional work was undertaken to develop laboratory tests to identify the agent in blood. Further work was carried out on the masking effects of such common medicinals as aspirin, barbiturates, etc. The agent was found [deleted]. A good solvent was discovered. A detection test for #3167 was developed, but barbiturates were found to completely mask its presence.

5. Twenty human volunteer subjects, five prisoners (Holmesbury State Prison, Holmesbury, PA) and fifteen military volunteers in the Edgewood program were tested. Both the [deleted] were found to be effective with symptoms lasting up to six weeks.

6. Concerting countermeasures, certain [deleted].

7. In addition to the above project, in 1967, [deleted] established a contract through Edgewood with [deleted] for the collection of information on and samples of new psychopharmaceuticals developed in Europe and [deleted]. The focus was on unpublished data and unusual new developments. Agency support of this action consisted of [deleted] in 1967 and [deleted] in 1969. The Agency took advantage of a pre-existing contract between Edgewood and [deleted] for the collection of information on foreign chemical and pharmaceutical developments. Agency redirection, beginning in 1967, consisted of focusing on psychoactive drugs and on collection of samples.

8. Agency support of both the clinical testing of EA #3167 and of the collection of information on and samples of foreign developments was terminated in January 1973. The [deleted] transferred to Edgewood in 1972 for an enlarged foreign collection effort was withdrawn in January 1973. Expenditures for the human testing program were gradually reduced as subjects were cleared from the program during the necessary post-test follow-up observational and examination period. Agency involvement in the above activities was closely held at all times.

6 May 1974

MEMORANDUM FOR: Inspector General
SUBJECT: Project OFTEN

1. The purpose of this memorandum is to document to the best of my knowledge the activities associated with Project OFTEN. I am writing this at the request of Mr. [deleted]. I am writing it at this point in time because (a) in a recent telephone conversation with the Office of [deleted] it became apparent that there is very little written information available on the project; (b) all of the key people associated with the project are no longer with the Agency; (c) I am resigning from the Central Intelligence

Agency on 11 May 1974. I hope this memorandum and attachments will never be needed, but I believe it is in the interest of CIA to have the following information in case it should be required.

2. The project dealt with the behavioral effects of chemical compounds (drugs) on humans. Numerous sources of compounds and data bases were used including private industry, other U.S. Government agencies, and foreign sources. An entire research cycle was set up, from the discovery of new compounds or the development of hybrids, to animal screening, to clinical (human) testing. Numerous data bases were acquired to help refine our search for candidate compounds.

3. The following activities were conducted with the Edgewood Arsenal, Edgewood, Maryland. We obtained a large data base from them containing their animal toxicity screen data. They supplied U.S. Army volunteers for testing of our candidate compounds. We transferred funds to them for their efforts. As a result of this testing something called the "Boomer" was developed. After the project was cancelled one more data base was received containing their clinical data on humans. As the project had been cancelled this data base was not exploited but remained in storage. At a recent request of [deleted], I visited the [deleted] Building to help them determine the nature of all the stored computer data relating to the project. Upon examining a listing of the clinical data it became evident that the volunteers' names were incorporated in the data base. If the data base contains all of the information described on the forms used by the doctors at Edgewood, it seems this data base could be a severe invasion of privacy of these volunteers. One form, the biographical one, comes to mind immediately. This form contains questions about the volunteer's sex life, alcohol and drug use, parents' family life, and numerous personal questions. I believe the volunteers never intended for this information to leave the control of the U.S. Army. It should be noted that (to the best of my knowledge), [deleted] had no knowledge of the sensitivity of this data base until very recently since the data base was not being exploited. We also obtained their Wiswesser Line Notation (WLN) data base

which contained the WLN notation for the compounds they have researched.

4. The following activities were conducted with [deleted]. They did all the benchwork of animal screening. They took the candidate compounds and ran them through a series of screens on such animals as mice, rats, cats, and monkeys. It was a result of these screens that determined whether to go ahead with further testing in human Edgewood volunteers. It was also through this contractor that funds were used to pay university professors when needed.

5. Several contractual agreements were made with private industry to receive new compounds of possible interest to CIA. These companies include [deleted] and a few other organizations whose names I cannot remember. Numerous compounds were received from these organizations and the results of the screens, in the form of computer reports, were returned to the sources of the compounds. This was a delicate process because some of the compounds were under patent consideration by their companies. Several foreign sources were also used but I did not have access to which ones.

6. An effort was also put forth to develop our own WLN data base. Assistance was received from experts at Edgewood Arsenal and Fort Detrich.

7. It was my belief that the project had three primary operational purposes. First, it was hoped that new compounds could be derived that could be used offensively. An example would be to come up with a compound that could simulate a heart attack or a stroke in the targeted individual, or perhaps a new hallucinogen to cause the targeted individual to act bizarrely. Second, it was hoped that blockers or even immunizations could be developed for known drugs. Although this would be mainly for use by our people in hostile environments, any progress along those lines certainly would have been welcomed by conventional drug related agencies. Third, it was my understanding that we would use profiles of volunteers who had received known drugs for comparative analysis. For example, if one of our people suddenly started act-

ing peculiarly, a profile of his actions could be run through the data base to see if his particular combination of actions matched any known drug profiles. In addition to these three operational goals, other work was being done on a permitted search capability of the WLN data base. There were also plans to develop a file of all known Soviet research in the drug area. The basis for my understanding of these goals was direct conversations with the division chief who had control over the project, and the project officer who was running the project, and discussions at which I was present. To my knowledge only one compound was ever perfected. I have no knowledge of it or any other compound being used operationally.

8. I am also attaching a series of Activity Report forms I used over the years to document the progress of my efforts. While these forms represent only my participation in the project, I believe they do give precise information as to the who, when, and where of many of the activities that were involved.

9. If this memorandum or its attachments should raise further questions, I would be happy to assist in any way to get required answers. However, as I previously mentioned I will be leaving the Agency on 11 May 1974. If I do not receive any queries during this time, I will assume this memorandum and attachments were sufficient so that the Agency will not be caught by any surprises from this project in the future.

APPENDIX XVIII

EXECUTIVE ORDER 13139: IMPROVING HEALTH PROTECTION OF MILITARY PERSONNEL PARTICIPATING IN PARTICULAR MILITARY OPERATIONS

THE WHITE House
Office of the Press Secretary
For Immediate Release

September 30, 1999

By the authority vested in me as President by the Constitution and the laws of the United States of America, including section 1107 of title 10, United States Code, and in order to provide the best health protection to military personnel participating in particular military operations, it is hereby ordered as follows:

Section 1. Policy. Military personnel deployed in particular military operations could potentially be exposed to a range of chemical, biological, and radiological weapons as well as diseases endemic to an area of operations. It is the policy of the United States Government to provide our military personnel with safe and effective vaccines, antidotes, and treatments that will negate or minimize the effects of these health threats.

Sec. 2. Administration of Investigational New Drugs to Members of the Armed Forces.

(a) The Secretary of Defense (Secretary) shall collect intelligence on potential health threats that might be encountered in an area of operations. The Secretary shall work together with the Secretary of Health and Human Services to ensure appropriate countermeasures are developed. When the Secretary considers an investigational new drug or a drug unapproved for its intended use (investigational drug) to represent the most appropriate countermeasure, it shall be studied through scientifically based research and development protocols to determine whether it is safe and effective for its intended use.

(b) It is the expectation that the United States Government will administer products approved for their intended use by the Food and Drug Administration (FDA). However, in the event that the Secretary considers a product to represent the most appropriate countermeasure for diseases endemic to the area of operations or to protect against possible chemical, biological, or radiological weapons, but the product has not yet been approved by the FDA for its intended use, the product may, under certain circumstances and strict controls, be administered to provide potential protection for the health and well-being of deployed military personnel in order to ensure the success of the military operation. The provisions of 21 CFR Part 312 contain the FDA requirements for investigational new drugs.

Sec. 3. Informed Consent Requirements and Waiver Provisions.

(a) Before administering an investigational drug to members of the Armed Forces, the Department of Defense (DoD) must obtain informed consent from each individual unless the Secretary can justify to the President a need for a waiver of informed consent in accordance with 10 U.S.C. 1107(f). Waivers of informed consent will be granted only when absolutely necessary.

(b) In accordance with 10 U.S.C. 1107(f), the President may waive the informed consent requirement for the administration of an investigational drug to a member of the Armed Forces in connection with the member's participation in a particular military operation, upon a written determination by the President that obtaining consent.

(1) is not feasible;

(2) is contrary to the best interests of the member; or

(3) is not in the interests of national security.

(c) In making a determination to waive the informed consent requirement on a ground described in subsection (b)(1) or (b)(2) of this section, the President is required by law to apply the standards and criteria set forth in the relevant FDA regulations, 21 CFR 50.23(d). In determining a waiver based on subsection (b)(3) of this section, the President will also consider the standards and criteria of the relevant FDA regulations.

(d) The Secretary may request that the President waive the informed consent requirement with respect to the administration of an investigational drug. The Secretary may not delegate the authority to make this waiver request. At a minimum, the waiver request shall contain:

(1) A full description of the threat, including the potential for exposure. If the threat is a chemical, biological, or radiological weapon, the waiver request shall contain an analysis of the probability the weapon will be used, the method or methods of delivery, and the likely magnitude of its affect on an exposed individual.

(2) Documentation that the Secretary has complied with 21 CFR 50.23(d). This documentation shall include:

(A) A statement that certifies and a written justification that documents that each of the criteria and standards set forth in 21 CFR 50.23(d) has been met; or

(B) If the Secretary finds it highly impracticable to certify that the criteria and standards set forth in 21 CFR 50.23(d) have been fully met because doing so would significantly impair the Secretary's ability to carry out the particular military mission, a written justification that documents which criteria and standards have or have not been met, explains the reasons for failing to meet any of the criteria and standards, and provides additional justification why a waiver should be granted solely in the interests of national security.

(3) Any additional information pertinent to the Secretary's determination, including the minutes of the Institutional Review Board's (IRB) deliberations and the IRB members' voting record.

(e) The Secretary shall develop the waiver request in consultation with the FDA.

(f) The Secretary shall submit the waiver request to the President and provide a copy to the Commissioner of the FDA (Commissioner).

(g) The Commissioner shall expeditiously review the waiver request and certify to the Assistant to the President for National Security Affairs (APNSA) and the Assistant to the President for Science and Technology (APST) whether the standards and criteria of the relevant FDA regulations have been adequately addressed and whether the investigational new drug protocol may proceed subject to a decision by the President on the informed consent waiver request. FDA shall base its decision on, and the certification shall include an analysis describing, the extent and strength of the evidence on the safety and effectiveness of the investigational new drug in relation to the medical risk that could be encountered during the military operation.

(h) The APNSA and APST will prepare a joint advisory opinion as to whether the waiver of informed consent should be granted and will forward it, along with the waiver request and the FDA certification to the President.

(i) The President will approve or deny the waiver request and will provide written notification of the decision to the Secretary and the Commissioner.

Sec. 4. Required Action After Waiver is Issued.

(a) Following a Presidential waiver under 10 U.S.C. 1107(f), the DoD offices responsible for implementing the waiver, DoD's Office of the Inspector General, and the FDA, consistent with its regulatory role, will conduct an ongoing review and monitoring to assess adherence to the standards and criteria under 21 CFR 50.23(d) and this order. The responsible DoD offices shall also adhere to any periodic reporting requirements specified by the President at the time of the waiver approval.

The Secretary shall submit the findings to the President and provide a copy to the Commissioner.

(b) The Secretary shall, as soon as practicable, make the congressional notifications required by 10 U.S.C. 1107(f)(2)(B).

(c) The Secretary shall, as soon as practicable and consistent with classification requirements, issue a public notice in the Federal Register describing each waiver of informed consent determination and a summary of the most updated scientific information on the products used, as well as other information the President determines is appropriate.

(d) The waiver will expire at the end of 1 year (or an alternative time period not to exceed 1 year, specified by the President at the time of approval), or when the Secretary informs the President that the particular military operation creating the need for the use of the investigational drug has ended, whichever is earlier. The President may revoke the waiver based on changed circumstances or for any other reason. If the Secretary seeks to renew a waiver prior to its expiration, the Secretary must submit to the President an updated request, specifically identifying any new information available relevant to the standards and criteria under 21 CFR 50.23(d). To request to renew a waiver, the Secretary must satisfy the criteria for a waiver as described in section 3 of this order.

(e) The Secretary shall notify the President and the Commissioner if the threat countered by the investigational drug changes significantly or if significant new information on the investigational drug is received.

Sec. 5. Training for Military Personnel.

(a) The DoD shall provide ongoing training and health risk communication on the requirements of using an investigational drug in support of a military operation to all military personnel, including those in leadership positions, during chemical and biological warfare defense training and other training, as appropriate. This ongoing training and health risk communication shall include general information about 10 U.S.C. 1107 and 21 CFR 50.23(d).

(b) If the President grants a waiver under 10 U.S.C. 1107(f), the DoD shall provide training to all military personnel conducting the waiver protocol and health risk communication to all military personnel receiving the specific investigational drug to be administered prior to its use.

(c) The Secretary shall submit the training and health risk communication plans as part of the investigational new drug protocol submission to the FDA and the reviewing IRB. Training and health risk communication shall include at a minimum:

(1) The basis for any determination by the President that informed consent is not or may not be feasible;

(2) The means for tracking use and adverse effects of the investigational drug;

(3) The benefits and risks of using the investigational drug; and

(4) A statement that the investigational drug is not approved (or not approved for the intended use).

(d) The DoD shall keep operational commanders informed of the overall requirements of successful protocol execution and their role, with the support of medical personnel, in ensuring successful execution of the protocol.

Sec. 6. Scope.

(a) This order applies to the consideration and Presidential approval of a waiver of informed consent under 10 U.S.C. 1107 and does not apply to other FDA regulations.

(b) This order is intended only to improve the internal management of the Federal Government. Nothing contained in this order shall create any right or benefit, substantive or procedural, enforceable by any party against the United States, its agencies or instrumentalities, its officers or employees, or any other person.

WILLIAM J. CLINTON
THE WHITE HOUSE,
September 30, 1999.

APPENDIX XIX

EXCERPTS FROM: NSDA DRAFT REPORT CONCERNING ASPARTAME, CONGRESSIONAL RECORD S5507-S5511, MARCH 7, 1985

The National Soft Drink Association ultimately supported the approval of aspartame for use in beverages. However, in 1983, the NSDA had its attorneys prepare a draft document objecting to the introduction of the product, citing certain unresolved public health concerns. The document was never filed with the FDA. However, the document's existence was made known in the course of the hearings concerning aspartame.

The following brief excerpts are drawn from the draft document, entitled "Objections of the National Soft Drink Association to a Final Rule Permitting the Use of Aspartame in Carbonated Beverages and Carbonated Beverage Syrup Bases and a Request for a Hearing on the Objections." Readers are urged to examine the complete document, which was published in the 1985 Congressional Record, in order to assess the significance of these excerpts in context.

"G. D. Searle and Company has not demonstrated to a reasonable certainty that the use of aspartame in soft drinks, without quantitative limitations, will not adversely affect human health."

"Searle has not met its burden of demonstrating to a reasonable certainty that the unlimited use of aspartame, especially in combination with carbohydrates, will not adversely affect human health."

"Searle has not met its burdens under section 409 of the Federal Food, Drug and Cosmetic Act, 21 U.S.C. 348 ("FDC Act") to demonstrate that aspartame is safe and functional for use in soft drinks."

"Collectively, the extensive deficiencies in the stability studies conducted by Searle to demonstrate that APM and its degradation products are safe in soft drinks intended to be sold in the United States, render those studies inadequate and unreliable. It is not possible on the basis of these studies to conclude that the petitioner has demonstrated that, notwithstanding its inherent instability, APM is safe for use in soft drinks."

"FDA has underestimated the amount of aspartame that can be consumed through its use in soft drinks because the agency has focused on adult users assumed to average 60 kilograms in weight."

"There is scientific evidence suggesting that increases in brain PHE and TYR levels of the order seen in the rat studies can effect synthesis of neurotransmitters, which themselves can affect important physiological functions and potentially behavior."

APPENDIX XX

LETTER FROM THE EPA REGARDING DANGERS OF ASPARTAME

30 October, 1987

Senator Howard M. Metzenbaum,
United States Senate,
140 Russell Senate Office Building
Washington, DC, 20510

Dear Senator Metzenbaum,

The following is in response to a request for comments addressed to me by Mr. James C. Wagoner of your Office in reference to the safety of the artificial sweetener aspartame, known commercially as Nutrasweet.

As you may know, during my service with the FDA from 1964 to 1979 I participated along with others in the extensive investigation of the quality of experimental studies carried out by or for the G. D. Searle & Co. of Skokie, Ill. Inasmuch as I had participated both in the "on-site" investigations at G. D. Searle & Co., as well as in the evaluation of the findings that emerged, my signature along with those of others appears on the final report of that FDA investigation (known also as the Searle Task Force Report) which was dated March the 24th, 1976.

In early 1979 I was transferred for duty from the FDA to the EPA to assume a position involving a promotion for me. My comments here ought not to be taken to imply in any way that they represent the views of the EPA since this agency has no regulatory concerns whatsoever in the area of food additives; rather, such comments of mine represent strictly my own views.

During that 1975 FDA investigation at G. D. Searle & Co. and at a number of their contractors, a total of 25 distinct experimental studies were intensively audited. Almost half of those 25 studies (11, to be exact) were carried out for aspartame with the remaining 14 studies having been distributed amongst 6 drug products manufactured by G. D. Searle & Co. It is worthy of note that the conduct of *all* experimental studies by that firm, regardless whether they entailed food additives or drug products, was the responsibility of a single group in the G. D. Searle & Co.'s organization: the Pathology-Toxicology or Path-Tox Department. Practices that were noted in connection with any given such study were quite likely to have been noted also for other studies that were audited, and this was a situation which was in no way unexpected: after all, the set of all such studies executed by that firm from about 1968 to the mid 1970's were conducted in essentially the same facilities, by virtually the same technicians, professional workers and supervisors, and the nature of such studies does not differ much whether a food additive or a drug product is being tested for safety in laboratory animals. It is in this sense, therefore, that the overall conclusions summarized at the beginning of the Searle Task Force Report have relevance to *all* the studies audited in 1975 (whether they had reference to aspartame or to any of the six drug products of Searle's) and, by extension, to the totality of experimental studies carried out by that firm around that time—1968 to 1975.

The FDA's Task Force Report starts at the top of its page 1 with:

"At the heart of the FDA's regulatory process is its ability to rely upon the integrity of the basic safety data submitted by sponsors of regulated products. Our investigation clearly demonstrates that, in the (case of the) G. D. Searle Company, we have no basis for such reliance now.

"Reliance on a sponsor is justified when FDA has reasonable assurance that the sponsor will: (1) inform the agency of *all* material results, observations, and conclusions of an experiment, (2) report fully and completely *all* of the conditions and circumstances under which an experiment was conducted, and (3) submit its reports to the FDA in a timely fashion so that measures to protect the public health and safety can be taken promptly when warranted. Through our efforts, we have uncovered serious deficiencies in Searle's operations and practices which undermine the basis for reliance on Searle's integrity in conducting high quality animal research to accurately determine or characterize the toxic potential of its products."

"Searle has not met the above criteria on a number of occasions and in a number of ways. We have noted that Searle has not submitted *all* of the facts of experiments to FDA, retaining unto itself the unpermitted option of filtering, interpreting, and not submitting information which we would consider material to the safety evaluation of the product. Some of our findings suggest an attitude of disregard for FDA's mission of protection of the public health by selectively reporting the results of studies in a manner which allays the concerns of questions of an FDA reviewer. Finally, we have found instances of irrelevant or unproductive animal research where experiments have been poorly conceived, carelessly executed, or inaccurately analyzed or reported."

"While a single discrepancy, error, or inconsistency in any given study may not be significant in and of itself, the cumulative findings of problems within and across the studies we investigated reveal a pattern of conduct which compromises the scientific integrity of the studies. We have attempted to analyze and characterize the problems and to determine why they are so pervasive in the studies we investigated."

"Unreliability in Searle's animal research does not imply, however, that its animal studies have provided no useful information on the safety of its products. Poorly controlled experiments containing random error blur the differences between treated and control animals and increase the difficulty of discriminating between the two populations to detect a product-induced effect. A positive finding of toxicity in the test animals in a poorly controlled study provides a reasonable lower bound on the true toxicity of the substance. The agency must be free to conclude that the results from such a study, while admittedly imprecise as to incidence or severity of the untoward effect, cannot be overlooked in arriving at a decision concerning the toxic potential of the product."

In addition to those general comments and references to no basis for reliance on reports generated by the G. D. Searle Company, serious deficiencies in Searle's operations and practices, Searle's integrity, Searle's selectively reporting the results, poorly conceived, carelessly executed and inaccurately analyzed or reported experiments at Searle's, a pattern of conduct which compromises the scientific integrity of the studies, pervasive problems in the Searle's studies, their unreliability, etc., which apply across the board to all studies investigated, there are a number of additional problems that attach specifically to the aspartame studies.

It should be pointed out that the Task Force Report detailing those general conclusions as well as those that relate specifically to the *aspartame* studies are not merely the views of the members of the Task Force itself. That Task Force operated under the direction of a Steering Committee composed of a number of FDA Bureau Directors as well as others and the Chairman of that Committee was none other then the FDA Commissioner himself. In fact the Task Force Report was addressed to the Commissioner in his capacity as Chairman of the Steering Committee, and, it seems clear that both the Committee and the Commissioner accepted that report and transmitted it to the United States Senate as an *institutional FDA report* without changing in it as much as a semicolon. The following are quotes from pages 3 and 4 of the record of hearings of April 8–9 and July 10, 1976, held by Sen. Edward Kennedy, Chairman, Subcommittee on Administrative Practice and Procedure, Committee on the Judiciary, and Chairman, Subcommittee on Health, Committee on Labor and Public Health.

Given those conclusions reached on the quality of Searle experimental studies in general and of the aspartame studies in particular, as we have seen above, by both the FDA *as an institution* and its Commissioner in 1976, how is it possible for another Commissioner in July, 1981, to reapprove the use of aspartame being marketed in dry foods? How is it possible for yet another Commissioner two years later, in July, 1983, to have extended such approval for marketing aspartame also in carbonated beverages? Such approvals were based on largely the very same studies that were examined by the Task Force in 1975–76.

It seems to me that no amount of additional examinations of pathology material such as undertaken by the UAREP and others, no additional statistical analyses carried out on the data, and no judgmental evaluations or interpretations of any data arising from those studies can in any way rectify the basic problem expressed by the Task Force, i.e., the FDA itself: in the absence of reasonable expectation that the experimental animals were administered the correct dosages of the test agent, any observational data carried out on those animals must be regarded as questionable or flawed. This is to say nothing of all the myriad of other problems involving the competence of those conducting such studies, and the care they exercised in their execution. Once a study is carried out and the test animals are disposed of, all that remains are the number of tiny bits of fissure preserved from their organs for microscopic examination and the written records of observations made by those who actually carried out that study. While the tissues themselves *can* be examined by others long after the remains of those animals no longer exist, the reliability of the written records has already been found to be unacceptable in a great variety of ways. Clearly, there is no way that even the most competent scientists can make any new observations on those animals at a time subsequent to the conduct of the study. Once a study is compromised in its executions, it is beyond salvation by anyone.

Even with respect to those small portions of tissue preserved for microscopic examination for an indefinite period of time after any study is completed there are serious problems as presented in the 1976 FDA report with respect to Searle studies in general and for the aspartame studies in particular: there is little if any assurance that such samples of tissues as were preserved actually originate from the specific animals said by Searle or Hazleton to have been their source (see the discussion on page 57 paragraph 2 et seq.) Furthermore, due to the unacceptably high rate of postmortem autolysis, a great many such tissues were not collected at all from the experimental animals. In any such study of even a few hundred test animals, it takes no more than a dozen or so of them to exhibit a particular lesion (such as brain tumors, for instance) where missing no more than one or two animals manifesting such tumors in any given exposure group may well make the difference whether that particular lesion is or is not significantly associated with the test agent, i.e., aspartame or any of its related chemicals.

Following the Senate hearing in the Spring and Summer of 1976, during the winter beginning in that year the FDA began negotiating with G. D. Searle & Co. on retaining the UAREP (Universities Associated with Research and Education in Pathology), a private organization, on the feasibility of investigating a number of other Searle studies with aspartame. When I heard of those negotiating being in effect, I wrote a memorandum to Mr. Carl Sharp, the chairman of the FDA's Searle Task Force, on November the 4th, 1976. A copy of it is given here as *Attachment 1*, and my apprehensions over such plans is clearly evident there. Basically, they amounted to the fact that the UAREP was totally unsuited for such a task since it had never before engaged in anything like it and I also objected to the idea that Searle was to fund that particular activity by the UAREP. As mentioned there, the FDA had just received a supplemental appropriation from the US Congress for the express purpose of expanding its own activities in that very area of investigating the conduct of such experimental studies by the regulated industry. Under that appropriation (which came to some $16,000,000) a great number of additional investigators were hired and trained for this particular task by the FDA.

A few months prior to the UAREP beginning its investigations in August of 1977, in April of that same year, yet another FDA investigation of three aspartame studies conducted at G. D. Searle & Co. was undertaken. The 76-page report of that investigation (also known as the Bressler report, after the name of the leader of the investigative team, Mr. Jerome Bressler, a compliance officer in the FDA's Chicago District) reveals the reference to a single one of those studies (the 115-week experiment in rats exposed to DKP or diketopiperazine, a breakdown product of aspartame) the following:

- substitutions of some of the animals in that study;

- the presence of intercurrent disease amongst the test animals and the administration of drugs to combat this, neither of which were completely reported to the FDA;
- incomplete examinations of tissues from the experimental animals;
- excision of tissue masses likely to be tumors from live animals during the study;
- absence of batch records for the mixing of the test substance into the diet of the test animals;
- incomplete stability studies for the agent on test;
- absence of homogeneity studies for the agent on test;
- deficiencies in the methods of chemical assay for the actual DKP that was mixed into the diet of the experimental rats;
- problems with the dosage of the DKP that was given to those rats;
- problems with the fixation-in-toto and autolysis;
- failure to report to the FDA all tissue masses (likely to be tumors) which were found in the experimental rats;
- failure to report to the FDA all internal tumors present in the experimental rats, e.g., polyps in the uterus (animal K9MF), ovary neoplasms (Animals H19CF, H19CF, and H7HF) as well as other lesions (Animal D29CF);
- inconsistencies between different parts of the report on this study submitted by G. D. Searle & Co. to the FDA on the precise nature of the lesions manifested by the test rats;
- numerous transcription errors in that report.

Interestingly and most important, the Bressler investigating group found not only that no homogeneity tests were conducted by G. D. Searle & Co. on the mixture of the test agent within the animals' diet, but they actually obtained *direct* evidence that in fact the distribution of the test agent in that diet was clearly *not* homogeneous due to failure to have the test agent ground in a sufficiently fine manner. Descriptive remarks on this issue were found by the FDA investigators in a notebook kept by Searle personnel on observations made during the study, as was a Polaroid photograph taken by the same Searle technicians and which clearly shows the test agent in the form of coarse particles with the animals' diet. If follows that the experimental rats could have consumed their feed without actually touching the DKP and, consequently, no-one can state with any assurance whatsoever just how much DKP (if any) those rats were actually exposed to in the course of that study. Evidence such as this obtained by the FDA investigators seems to me to have been crucial to the interpretation of any findings or observations by Searle.

On page 32 of the GAO report one can read the view of the FDA's Center for Food Safety and Applied Nutrition (CFSAN) on the findings of the investigators. To me these read like a script written for Abbott and Costello in the sense of their having their perceptions inside-out or upside-down—"the diets *may* have been homogeneous because of a dose-related increase in the incidence of uterine polyps and decrease in blood cholesterol levels" (a clear non-sequitur, such as one almost never encounters in real life); on the problem with autolysis of the tissues the CFSAN felt "they could not determine whether the results would have been altered if these tissues had been obtained before autolysis (an obvious instance of placing the burden of proof that a study is unsound on the Government rather than requiring the petitioner for approval of a food additive to demonstrate, as the Law requires, that any study is of sound quality); the observation by the investigators that 329 fetuses were examined in two days by a single person (a clear impossibility) was laid aside by the CFSAN with another non-sequitur: that "the Searle scientist who performed these examinations estimated that he examined about 30 fetuses a day . . ."; on the fact that an insufficient number of sections were made out of the heart, the CFSAN observed: ". . . while there was no evidence that the study was compromised by this issue, the practice of not making enough sections through the organs, as specified in the protocol, did not preclude a possible failure to observe abnormalities which may have occurred."

Despite all these problems, at least some of which undermined or compromised the study in an unredeeming manner, apparently the CFSAN and the FDA Commissioner found the quality of those three studies reported by the Bressler investigating group as being in fact of an acceptable nature and G. D. Searle & Co. was notified to this effect in September, 1977.

The investigation undertaken by the UAREP began in August 1977. After reading the report of that group, it became painfully clear to me that the misgivings which I foresaw in November 1976 (see *Attachment 1* here) were indeed justified and my worst fears were eventually realized. If one compares the kind of detailed and painstaking findings made by the professional investigators from the FDA both in 1975 and in 1977 with the rather amateurish activities by the UAREP outlined in their report, the contrast between these could hardly have been greater. Of course, inasmuch as G. D. Searle had paid for the UAREP investigation, the cost of it for the FDA was nil; what the FDA got in return for its money, was not worth much more than this.

Perhaps the most disappointing aspect of this entire fiasco with the quality or reliability of the experimental studies with aspartame was the failure of the Public Board of Inquiry (PBOI) to consider these aspects in their deliberations. The PBOI expressly declined to do so even after the principal objectors to the approval of aspartame for marketing, Mr. James Turner and Dr. John Olney, asked for such consideration. To me it seems almost beyond belief that a collection of scientists can sit in judgment over the interpretations to be made on a set of results arising from certain studies, not only failing to consider the adequacy of the conduct of those studies but actually *refusing* to do so.

Given this sort of circumstance, it should not come as a surprise to anyone that eventually the Commissioner of the FDA finally reapproved aspartame for marketing even though his own panel of experts was divided over the issue whether this particular food additive had been shown in a reasonable manner to be safe.

As mentioned in the GAO report (page 12 there) "The Federal Food, Drug, and Cosmetic Act does not specifically define 'safety'. However, the legislative history of the Food Additives Amendment indicates that safety means 'proof of a reasonable certainty that no harm will result from the proposed use of an additive'." It is intuitively clear to anyone that no "reasonable certainty" can attach to any results emanating from studies as profoundly flawed as the Commissioner of the FDA had determined in 1976 and as amply reconfirmed since then.

This concludes my remarks on the quality or reliability of the experimental studies with aspartame carried out by the G. D. Searle & Co. or by the contractors working under the direction of that firm.

Since Mr. Wagoner of your Office has requested my comments in a very short period of time, I am expediting this letter to you now; however, I plan to send you in the very near future an additional communication where two other issues are discussed in some detail:-the problem with the brain tumors induced by aspartame and that the FDA's having set a very high (and, to my view, clearly dangerous) level of Acceptable Daily Intake, or ADI, for this particular food additive in the diet of humans.

Finally, I wish to state here that, quite aside from my professional background as a scientist and speaking merely as an individual citizen, I am grateful for the concern you have had over the safety of aspartame for many years now; as such, I wish to thank you for having given me this opportunity of being of some service to you. With best wishes for the future, I remain, Senator Metzenbaum,

Sincerely yours,

M. Adrian Gross,
Senior Science Advisor,
Benefits and Use Division,
Office of Pesticide Programs
Sworn to be a true copy on 30 Oct., 1987.

BIBLIOGRAPHY

1 The Chemical Revolution: Bringing Bad Things to Life

America's Children and the Environment: A First View of Available Measures, Environmental Protection Agency, EPA 240-R-00-006, December 2000.

Army Regulation 70-25. *Research and Development, Use of Volunteers As Subjects of Research*, Department of the Army, Washington, DC, March 26, 1968.

Batt, S. and Gross, L. The estrogen connection: how synthetic chemicals may increase breast cancer, *Sierra Magazine,* September/October 1999.

Century of biological and chemical weapons. *BBC News*, September 25, 2001.

Colborn, T. et al. *Our Stolen Future*, Penguin Books, New York, 1996.

Daniels, J. L. et al. Pesticides and childhood cancers, *Environmental Health Perspectives*, 105(10): 1068, 1997.

Davidson, L. Lethal breeze, *Deseret News*, June 5, 1994.

Davidson, L. and Bauman, J. Toxic Utah: a land littered with poisons, *Deseret News*, February 12, 2001.

Davis, D. L. et al. Is brain cancer mortality increasing in industrial countries? *American Journal of Industrial Medicine*, 19:421, 1991.

Davis, D. L. et al. Agricultural exposures and cancer trends in developed countries, *Environmental Health Perspectives*, 100:42, 1992.

Department of Defense Fact Sheet. Office of the Assistant Secretary of Defense (Health Affairs) Deployment Health Support Directorate, *Deseret Test Center Project SHAD: Big Tom.*, October 9, 2002.

Department of Defense Fact Sheet. Office of the Assistant Secretary of Defense (Health Affairs) Deployment Health Support Directorate, *Deseret Test Center: Devil Hole, Phase II*, October 9, 2002.

Department of Defense Fact Sheet. Office of the Assistant Secretary of Defense (Health Affairs) Deployment Health Support Directorate, *Deseret Test Center: Elk Hunt, Phase II,* October 9, 2002.

Department of Defense Fact Sheet. Office of the Assistant Secretary of Defense (Health Affairs) Deployment Health Support Directorate, *Deseret Test Center: Tall Timber*, October 9, 2002.

Department of Defense Fact Sheet. Special Assistant to the Under Secretary of Defense (Personnel and Readiness) for Gulf War Illnesses, Medical Readiness and Military Deployments, *Project Shipboard Hazard and Defense (SHAD): Fearless Johnny*, October 9, 2002.

Department of Defense Fact Sheet. Special Assistant to the Under Secretary of Defense (Personnel and Readiness) for Gulf War Illnesses, Medical Readiness and Military Deployments. *Project Shipboard Hazard and Defense (SHAD): Flower Drum, Phase I*, October 9, 2002.

Devesa, S. S. et al. Recent cancer trends in the United States, *Journal of the National Cancer Institute*, 87(3):175, 1995.

Dewer, H. Chemical weapons all around the world, *Sun News*, November 15, 1997.

Dillon, L. and Snyder, B. Toxic Utah: trash, troubles are piling up, *Deseret News*, February 16, 2001.

Environmental dangers of open-air testing of lethal chemicals, Hearing before the Subcommittee on Conservation and Natural Resources, Committee on Government Operations, U.S. House of Representatives, May 20–21, 1969.

Environmental Working Group. *The English Patients: Human Experiments and Pesticide Policy*. Washington, DC, July, 1998.

Fagin, D. and Lavelle, M. *Toxic Deception: How the Chemical Industry Manipulates Science, Bends the Law and Endangers Your Health*, Common Courage Press, Monroe, MA, 1999.

Garthoff, R. L. Polyakov's run, *Bulletin of the Atomic Scientists*, 56(5):37.

Goodwin, B. *Keen as Mustard: Britain's Horrific Chemical Warfare Experiments in Australia,* University of Queensland Press, Queensland, Australia, 1998.

Gray, L. E. Xenoendocrine disruptors: laboratory studies on male reproductive effects, *Toxicology Letters*, 102:331, 1998.

Haynes, M. *Pesticides Companies Using Humans in Lab Studies*, Environmental Working Group, Washington, DC, July 27, 1998.

Hoffman, D. Soviets reportedly built weapon despite pact, *Washington Post*, August 16, 1998.

Hogendoorn, E. J. A chemical weapons atlas, *Bulletin of the Atomic Scientists*, 53(5):1, 1997.

Hunter, D. J. and Kelsey, K. T. Pesticide residues and breast cancer: the harvest of a silent spring? *Journal of the National Cancer Institute*, 85:598.

Jensen, T. K. et al. Do environmental estrogens contribute to the decline in male reproductive health? *Clinical Chemistry*, 41:1896, 1995.

Kaplan, D. E. Terrorism's next wave: nerve gas and germs are the new weapons of choice, *U.S. News and World Report,* pp. 27–31, November 17, 1997.

Keeler, J. R., Hurst, C. G., and Dunn, M. A. Pyridostigmine used as a nerve agent pretreatment under wartime conditions, *Journal of the American Medical Association,* 266(5):693, 1991.

Kelley, M. Chemical weapons tests by US in the 60s, *Associated Press,* October 8, 2002.

Leiss, J. K. and Savitz, D. A. Home pesticide use and childhood cancer: a case controlled study, *American Journal of Public Health,* 85(2):249, 1995.

Klassen, C. D. *Casarrett and Doull's Toxicology. The Basic Science of Poisons,* 5th edition. McGraw-Hill, New York, 1996.

Knox, E. G. and Gilman, E. A. Hazard proximities of childhood cancers in Great Britain from 1953–80, *Journal of Epidemiology and Community Health,* 51:151, 1997.

Koch, E. and Fleming, D. Bizarre experiments at SADF research farms, *Weekly Mail and Guardian,* December 15, 1994.

Koop, C. E. and Lundberg, G. D. Violence in America: a public health emergency, *Journal of the American Medical Association,* 267(22):3075, 1992.

Marshall, E. Search for a killer: focus shifts from fats to hormones, *Science,* 259: 618, 1993.

Masters, R., Hone, B., and Doshi, A. *Environmental Toxicology,* Gordon and Breach, New York, 1997.

Moline, J. M. et al. Exposure to hazardous substances and male reproductive health: a research framework, *Environmental Health Perspectives,* 108:9, September 2000.

Moscow Placed Loopholes in Chemical Weapons Convention; Treaty Won't Ban Two Classes of Binary Nerve Agents, Foreign Policy Alert No. 23, American Foreign Policy Council, Washington, DC, September 9, 1996.

Motluck, A. Pollution may lead to a new life of crime, *New Scientist,* 154(2084): 4, 1997.

Nitrate in Drinking Water Associated with Increased Risk for Non-Hodgkin. NCI press release, September 6, 1996.

Pechura, C. M. and Rall, D. P. *Veterans at Risk: The Health Effects of Mustard Gas and Lewisite,* Institute of Medicine, National Academy Press, Washington, DC, 1993.

Pogoda, J. M. and Preston-Martin, S. Household pesticides and risk of pediatric brain tumors, *Environmental Health Perspectives,* 105(11):1214, 1997.

Porter, W. P. et al. Endocrine, immune and behavioral effects of aldicarb (carbamate), atrazine (triazine) and nitrate (fertilizer) mixtures at groundwater concentrations, *Toxicology and Industrial Health,* 15:133, 1999.

Proctor, R. N. *Cancer Wars: How Politics Shapes What We Know and Don't Know About Cancer*, Basic Books, New York, 1995.

Repetto, R. and Sanjay, B. *Pesticides and the Immune System: The Public Health Risks*, World Resources Institute, Washington, DC, 1996.

Riegle Report: U.S. Chemical and Biological Warfare-Related Dual Use Exports to Iraq and Their Possible Impact on the Health Consequences of the Gulf War, United States Senate, 103rd Congress, 2nd Session, May 25, 1994.

Rockefeller Report: Is Military Research Hazardous to Veterans' Health? Lessons Spanning Half a Century, United States Senate, 103rd Congress, 2nd Session, December 8, 1994.

Schettler, T. et al. *In Harm's Way: Toxic Threats for Child Development*, Greater Boston Physicians for Social Responsibility, May 2000.

Schmidt, C. W. Childhood cancer: a growing problem, *Environmental Health Perspectives*, 106(1):A18, 1998.

Schmidt, C. W. The promise and pitfalls of human testing, *Chemical Innovation*, 30(5):14, 2000.

Sharabi, Y. et al. Survey of symptoms following intake of pyridostigmine during the Persian Gulf War, *Israeli Journal of Medical Science*, 27:656, 1991.

Sidell, F. R. *Clinical Considerations in Nerve Agent Intoxication*, Academic Press, San Diego, 1992.

Simcox, N. et al. Pesticides in household dust and soil: exposure pathways for children of agricultural families, *Environmental Health Perspectives*, 103(12):1127, 1995.

Sokolski, H. *Rethinking bio-chemical dangers*, Foreign Policy Research Institute, February 2000.

Somia, G. et al. Accumulation of chlorpyrifos on residential surfaces and toys accessible to children, *Environmental Health Perspectives*, 106(1):9, 1998.

Toppari, J. et al. Male reproductive health and environmental xenoestrogens. *Environmental Health Perspectives*, 104:4, 1996.

UK Porton Down—a sinister air? *BBC News*, August 20, 1999.

U.K. probes chemical tests on soldiers, *United Press International*, May 3, 2001.

U.S. Congress. *Proliferation of Weapons of Mass Destruction: Assessing the Risks*, Office of Technology Assessment, Washington, DC, 1993.

U.S. Congress. Hearing before the Subcommittee on Conservation and Natural Resources, Committee on Government Operations, U.S. House of Representatives. *Environmental Dangers of Open-Air Testing of Lethal Chemicals*, May 20–21, 1969.

U.S. General Accounting Office. *Human Experimentation: An Overview on Cold War Era Programs*, September 28, 1994.

Voccia, I. et al. Immunotoxicology of pesticides: a review, *Toxicology and Industrial Health*, 15:119, 1999.

Waller, J. M. The chemical weapons coverup, *Wall Street Journal*, February 13, 1997.

Wolff, M. S. and Weston, A. Breast cancer risk and environmental exposures, *Environmental Health Perspectives*, 105:891, 1997.

Zumwalt, E. R. *Report to Secretary of the Department of Veterans Affairs on the Association Between Adverse Health Effects and Exposure to Agent Orange*, Department of Veterans Affairs, May 5, 1990.

2 Nature's Weapons: Man and Biological Warfare

A Study of the Vulnerability of Subway Passengers in New York City to Covert Action with Biological Agents. Miscellaneous Publication 25, Department of the Army, Fort Detrick, MD, January 1968.

Alibek, K. A. *Behind the Mask: Biological Warfare*, Institute for the Study of Conflict, Ideology, and Policy, Boston, MA, September–October, 1998.

Alibek, K. A. Russia's deadly expertise, *New York Times*, March 27, 1998.

Alibek, K. A. Terrorist and intelligence operations: potential impact on the U.S. economy, statement before the Joint Economic Committee, statement to the United States Congress, May 20, 1998.

Alibek, K. A. and Handelman, *S. Biohazard: The Chilling True Story of the Largest Covert Biological Weapons Program in the World—Told from Inside by the Man Who Ran It*, Random House, New York, 1999.

Alibek, K. A. Combating terrorism: assessing the threat of a biological weapons attack, statement before the House Subcommittee on National Security, Veterans Affairs, and International Relations, October 12, 2001.

Anderson, J. Soviets push biological weapons work, *Washington Post*, p. B-15, December 4, 1984.

Armstrong, F. B. et al. *Recombinant DNA and the Biological Warfare Threat*, U.S. Army Test and Evaluation Command, Dugway Proving Ground, May 1981.

Arostegul, M. Castro weaponizes West Nile virus, *Insight Magazine*, September 16, 2002.

Behavior of Aerosol Clouds Within Cities. Joint Quarterly Report No. 3, Chemical Corps, U.S. Army, January–March, 1953.

Behavior of Aerosol Clouds Within Cities. Joint Quarterly Report No. 4, Chemical Corps, U.S. Army, April–June 1953.

Berstein, B. J. America's biological warfare program in the Second World War, *Journal of Strategic Studies 2*, March 1992.

Biological Weapons Join Pesticides in Misguided War on Drugs. Pesticide Action Network North America Updates Service, August 1, 2000.

Block, S. M. The growing threat of biological weapons, *American Scientist*, 89:28.

Brown, C. E. Human retention from single inhalations of *Bacillus subtilis* spore aerosols, in *Inhaled Particles and Vapors*, Proceedings of an International Symposium of the British Occupational Hygiene Society, Oxford, Pergamon Press, New York, 1961.

Brown, P. O. Integration of retroviral DNA, *Current Topics in Microbiology and Immunology*, 157:19.

Carruth, G. A. *U.S. Army Activity in the U.S. Biological Warfare Programs*, Volume 1, Washington, D.C., February 24, 1977.

Carter, G. B. *Porton Down: 75 Years of Chemical and Biological Research*, HMSO, London, 1992.

Cole, L. A. *Clouds of Secrecy: The Army's Germ Warfare Tests over Populated Areas*, Rowman and Littlefield, Totowa, NJ, 1988.

Cole, L. A. The specter of biological weapons, *Scientific American*, p. 60, December 1996.

Cole, L. A. *The Eleventh Plague: The Politics of Biological and Chemical Warfare*, W. H. Freeman and Company, New York, 1998.

Cousins, N. How the U.S. used its citizens as guinea pigs, *Saturday Review*, p. 10, November 10, 1979.

Covert, N. *Cutting Edge: A History of Fort Detrick, Maryland, 1943–1993*, U.S. Army Garrison, Public Affairs Office, Fort Detrick, MD, 1993.

Dando, M. *Biological Warfare in the 21st Century: Biotechnology and the Proliferation of Biological Weapons*, London, 1994.

Davis, B. R. et al. *Microbiology*, 3rd ed., Harper and Row, Hagerstown, MD, 1980.

Davis, C. J. Nuclear blindness: an overview of the biological weapons programs of the former Soviet Union and Iraq, *Emerging Infectious Diseases*, 5(4): 509, 1999.

Derbes, V. J. De Mussis and the Great Plague of 1348, *Journal of the American Medical Association*, 196(1):60, 1966.

Eitzen, E. M. and Takafuji, E. T. Historical overview of biological warfare, in *Textbook of Military Medicine, Medical Aspects of Chemical and Biological Warfare*, Office of the Surgeon General, Department of the Army, 1997.

Endicott, S. and Hagerman, E. *The United States and Biological Warfare*, Indiana University Press, Bloomington, IN, 1998.

Franz, D. *International Biological Warfare Threat*, statement before the United States Senate Joint Committee on Judiciary and Intelligence, Second Session, 105th Congress, March 4, 1998.

Garrett, L. *The Coming Plague: Newly Emerging Diseases in a World Out of Balance*, Penguin Books, New York, 1995.

Gold, H. *Unit 731 Testimony: Japan's Wartime Human Experimentation Program*, Yenbooks, Tokyo, 1996.

Gower, R., Powell, R. and Rolling, B. Japan's biological weapons, 1930–1945, *Bulletin of the Atomic Scientists*, October 1988.

Harris, R. and Paxman, J. *A Higher Form of Killing*, Hill and Wang, New York, 1982.

Harris, S. H. Japanese biological warfare research on humans: a case study of microbiology and ethics. *Annals of the New York Academy of Sciences*, 666:21, 1992.

Harris, S. H. *Factories of Death*, Routledge, New York, 1994.

Hersh, S. M. *Chemical and Biological Warfare: America's Hidden Arsenal*, Bobbs-Merrill, Indianapolis, IN, 1968.

Hogshire, J. U.S. biological roulette on drugs, *Covert Action Quarterly*, June 2000.

Horowitz, L. G. *Emerging Viruses: AIDS and Ebola*, Tetrahedron, Rockport, MA, 1996.

Inglesby, T. V. et al. Anthrax as a biological weapon, *Journal of the American Medical Association*, 281(18):1735, 1999.

Istock, C. A. Bad medicine, *Bulletin of the Atomic Scientists*, November/December 1998.

Jelsma, J. Military implications of biotechnology in M. Fransman et al., eds. *The Biotechnology Revolution*, Blackwell, Oxford, 1995.

Leahy, P. Interview with Mark Johnson about West Nile Virus on WKDR, Burlington, VT.

Lederberg, J. *Biological Weapons: Limiting the Threat*, MIT Press, Cambridge, MA, 1999.

Lema, M. J. Death in a droplet, *American Society of Anesthesiologists Newsletter*, 65(3):1, 2001.

Letter to author from Dr. Garth Nicolson, University of Texas M. D. Anderson Cancer Center, about Gulf War illness.

Livingstone, N. C. and Douglass, J. *CBW: The Poor Man's Atomic Bomb*, Institute for Foreign Policy Analysis, Cambridge, MA, 1984.

Mangold, T. and Goldberg, J. *Plague Wars: A True Story of Biological Warfare*, St. Martins Press, New York, 2000.

McNeill, W. H. *Plagues and Peoples*, Anchor Books, New York, 1977.

Miller, J. et al. *Germs: Biological Weapons and America's Secret War*, Simon and Schuster, New York, 2001.

Mobley, J. A. Biological warfare in the twentieth century: lessons from the past, challenges for the future, *Military Medicine*, 160:547, 1995.

Moehringer, J. R. Gulf war ailment called contagious, *Los Angeles Times*, A-1, March 9, 1997.

Nicolson, G. L. and Nicolson, N. L. Testimony about Gulf War illness. Committee on Government Reform and Oversight, Subcommittee on Human Resource and Intergovernmental Relations, U.S. House of Representatives, June 26, 1997.

Open air testing with simulated biological and chemical warfare agents. Testimony by Leonard Cole before the Committee on Veterans Affairs, May 6, 1994.

Perretta, B. Historic structure stands as reminder of biological warfare, *Capital News Service*, April 20, 1999.

Poupard, J. A. and Miller, L. A. History of biological warfare: catapults to capsomeres, *Annals of the New York Academy of Sciences*, 666:9, 1992.

Powell, J. W. Japan's germ warfare: the U.S. coverup of a war crime, *Bulletin of Concerned Asian Scholars*, 12(4):3, 1980.

Powell, J. W. A hidden chapter in history, *Bulletin of the Atomic Scientists*, 37(8):45, 1981.

Preston, R. Chemical and biological weapons threats to America: are we prepared? A statement for the record before the Senate Judiciary Subcommittee on Technology, Terrorism and Government Information and the Senate Select Committee on Intelligence, April 22, 1998.

Preston, R. The bioweaponeers, *The New Yorker*, March 9, pp. 52–65, 1998.

Preston, R. The demon in the freezer, *The New Yorker*, July 12, pp. 44–61, 1999.

Preston, R. West Nile mystery, *The New Yorker*, October 18, 1999.

Principi, A. J. Secretary of veterans affairs press conference on GWI and ALS, Washington, DC, December 10, 2001.

Regis, E. *The Biology of Doom: The History of America's Secret Germ Warfare Project*, Henry Holt and Company, New York, 1999.

Report on special BW operations. Memorandum for the Research and Development Board of the National Military Establishment by I. L. Baldwin, Washington, DC, October 5, 1948.

Riegle Report: U.S. Chemical and Biological Warfare–Related Dual Use Exports to Iraq and their Possible Impact on the Health Consequences of the Gulf War, United States Senate, 103rd Congress, 2nd Session, May 25, 1994.

Ringle, K. Army sprayed germs on unsuspecting travelers, *Washington Post*, p. B-1, December 5, 1984.

Roberts, G. The deadly legacy of Anthrax Island, *The Sunday Times Magazine*, London, February 15, 1981.

Robodonirina, M. et al. Fusarium infections in immunocompromised patients: case reports and literature review, *European Journal of Clinical Microbiology and Infectious Diseases*, 13:152, 1994.

Rose, A. H. *Chemical Microbiology*, 3rd ed., Plenum Press, New York, 1976.

Rothschild, J. H., Brigadier General, U.S.A. (Ret.), *Tomorrow's Weapons, Chemical and Biological*, McGraw-Hill, New York, 1964.

Satcher, D. Letter to Senator Donald Riegle with CDC shipments to Iraq, including West Nile virus, June 21, 1995.

"Secret Germ Warfare Experiment?" CBS News Special Report, May 15, 2000.

Special Report No. 142. Biological Warfare Trials at San Francisco, California, 20–27 September 1950, U.S. Chemical Corps Biological Laboratory, January 22, 1951.

Stearn, W. E. and Stearn, A. *The Effect of Smallpox on the Destiny of the Amerindian*, Bruce Humphries Publishers, Boston, 1945.

U.S. Congress. *Proliferation of Weapons of Mass Destruction: Assessing the Risks*, Office of Technology Assessment, Washington, DC, 1993.

U.S. Congress. Hearings before the Subcommittee on Health and Scientific Research of the Committee on Human Resources, United States Senate, *Biological Testing Involving Human Subjects by the Department of Defense, 1977*, March 8 and May 23, 1977, Government Printing Office, Washington, DC, 1977.

Use of Human Subjects in Research: Operational Guidelines. Department of the Army, U.S. Army Medical Research Institute of Infectious Diseases, Fort Detrick, MD, USAMRIID Regulation No. 70–25, August 7, 2000.

"VA alert on biowar tests." CBS News Special Report, December 7, 2000.

Vegar, J. Terrorism's new breed. Are today's terrorists more likely to use chemical and biological weapons? *Bulletin of the Atomic Scientists*, March/April 1998.

Vulliamy, E. U.S. prepares to spray genetically-modified herbicides on Colombians, *London Observer*, July 2, 2000.

Williams, P. and Wallace, D. *Unit 731: The Japanese Army's Secret of Secrets*, Hodder and Stoughton, London, 1989.

Wright, S. and Sinsheimer, R. Recombinant DNA and biological warfare, *Bulletin of the Atomic Scientists*, 39(9):23, 1983.

Zilinskas, R. A. New biotechnology: potential problems, likely promises, *Politics and the Life Sciences*, 2(1):43, 1983.

3 The Eugenics Movement: Past, Present, and Future

Aarons, M. and Loftus, J. *Unholy Trinity: The Vatican, the Nazis, and Soviet Intelligence*, St. Martin's Press, New York, 1991.

Adams, M. B. *The Wellborn Science: Eugenics in Germany, France, Brazil, and Russia*, Oxford University Press, New York, 1990.

Alexander, L. Medical science under dictatorship, *New England Journal of Medicine*, 241:39, 1949.

Annas, G. J. and Grodin, M. A. *The Nazi Doctors and the Nuremberg Code: Human Rights in Human Experimentation,* Oxford University Press, New York, 1992.

Barber, B. The ethics of experiments with human subjects, *Scientific American,* 234(2):25, 1976.

Barondess, J. A. Medicine against society. Lessons from the Third Reich, *Journal of the American Medical Association,* 276:1657, 1966.

Beyerchen, A. D. *Scientists Under Hitler,* Yale University Press, New Haven, CT, 1977.

Buchanan, P. *The Death of the West: How Mass Immigration, Depopulation and a Dying Faith Are Killing Our Culture and Country,* Thomas Dunne Books, New York, 2001.

Cole, L. A. *Clouds of Secrecy: The Army's Germ Warfare Tests over Populated Areas,* Rowman and Littlefield, Totowa, N.J. 1988.

Copper, W. *Behold a Pale Horse,* Light Technology Publications, Flagstaff, AZ, 1991.

Deposition by Dr. Miklos Nyiszli before the Budapest Commission for the Welfare of Deported Hungarian Jews, July 28, 1945.

Dowbiggin, I. R. *Keeping America Sane: Psychiatry and Eugenics in the United States and Canada 1880–1940,* Cornell University Press, Ithaca, NY, 1997.

DuBois, J. *The Devil's Chemists,* Beacon Press, Boston, MA, 1952.

Ferencz, B. *Less Than Slaves,* Harvard University Press, Cambridge, MA, 1979.

Friedman, M. J. Eugenics and the new genetics, *Perspectives in Biology and Medicine,* 35(1):145, 1991.

Gallagher, N. L. *Breeding Better Vermonters: The Eugenics Project in the Green Mountain State,* University Press of New England, Hanover, NH, 1999.

Galton, F. *Essays in Eugenics,* Garland Press, New York, 1985.

Garver, K. L. and Garver, B. Eugenics, euthanasia, and genocide, *Linacre Quarterly,* 59(3):24, 1992.

Garver, K. L. and Garver, B. The Human Genome Project and eugenic concerns, *American Journal of Human Genetics,* 54(1):148, 1994.

Glannon, W. Genes, embryos, and future people, *Bioethics,* 12(3):187, 1998.

Global 2000 Report to the President: Entering the Twenty-First Century, U.S. Department of State, 1977.

Gould, S. J. The smoking gun of eugenics, *Natural History,* 100(12):8, 1991.

Graham, L. Science and values: the eugenics movement in Germany and Russia in the 1920s, *American Historical Review,* 82:1148, 1977.

Grant, G. *Grand Illusions: The Legacy of Planned Parenthood,* Wolgemuth and Hyatt, Brentwood, TN, 1988.

Hatchett, R. Brave new worlds: perspectives on the American experiences of eugenics, *Pharos,* 54(4):13, 1991.

Heller, J. R. and Bruyere, P. T. Untreated syphilis in the male Negro. II. Mortality during 12 years of observation, *Journal of Venereal Disease Information,* 27:34, 1946.

Herken, G. and David, J. Doctors of death, *New York Times,* January 13, 1994.

Hilts, P. Medical experts testify on tests done without consent, *New York Times,* June 3, 1991.

Hitler, A. *Mein Kampf,* Houghton Mifflin, Boston, 1971.

Holtzman, N. A. and Rothstein, M. A. Eugenics and genetics discrimination, *American Journal of Human Genetics,* 50(3):457, 1992.

Hunt, L. *The United States Government, Nazi Scientists and Project Paperclip, 1945–1990,* St. Martin's Press, New York, 1991.

Jaroff, L. Vaccine jitters, *Time Magazine,* September 13, 1999.

Jones, J. H. *Bad Blood: The Tuskegee Syphilis Experiments,* Free Press, New York, 1993.

Kevles, D. J. *In the Name of Eugenics: Genetics and the Uses of Human Heredity,* Knopf, New York, 1985.

Kohn, M. *The Race Gallery: The Return of Racial Science,* Jonathan Cape, London, 1995.

Kuhl, S. *The Nazi Connection: Eugenics, American Racism, and German National Socialism,* Oxford University Press, New York, 1994.

Larson, E. J. *Sex, Race, and Science: Eugenics in the Deep South,* John Hopkins University Press, Baltimore, MD, 1995.

Lasby, C. *Project Paperclip,* Atheneum Books, New York, 1971.

Ledley, F. D. Distinguishing genetics and eugenics on the basis of fairness, *Journal of Medical Ethics,* 20(3):157, 1994.

Lerner, B. H. and Rothman, D. J. Medicine and the holocaust: learning more of the lessons, *Annals of Internal Medicine,* 122:793, 1995.

Lifton, R. J. *The Nazi Doctors: Medical Killing and the Psychology of Genocide,* Basic Books, New York. 1986.

Lubinsky, M. S. Scientific aspects of early eugenics, *Journal of Genetic Counseling,* 2(2):77, 1993.

Ludmerer, K. M. *Genetics and American Society: A Historical Appraisal,* Johns Hopkins University Press, Baltimore, MD, 1972.

Mayer, A. J. *Why Did the Heavens Not Darken: The Final Solution in History,* Pantheon Books, New York, 1990.

Naumann, B. *Auschwitz,* Pall Mall Press, London, 1966.

Nazi Conspiracy and Aggression, U.S. Government Printing Office, Washington, DC, 1946.

Nyiszli, M. *Auschwitz: A Doctor's Eyewitness Account,* Frederick Fell, New York, 1960.

Paul, D. B. *Controlling Human Heredity: 1865 to the Present,* Humanities Press International, Atlantic Highlands, NJ, 1995.

Pellegrino, E. D. The Nazi doctors and Nuremberg: some moral lessons revisited, *Annals of Internal Medicine,* 127:307, 1997.

Pesare, P. J. et al. Untreated syphilis in the male Negro: observation of abnormalities over 16 years, *American Journal of Syphilis, Gonorrhea, and Venereal Diseases,* 34:201, 1950.

Postgate, J. Eugenics returns, *Biologist,* 42(2):96, 1995.

Prenger, U. The development of anti-fertility vaccines, *Biotechnology and Development Monitor,* 25:2, 1995.

Proctor, R. *Racial Hygiene: Medicine Under the Nazis,* Harvard University Press, Cambridge, MD, 1988.

Rafter, N. H. *White Trash: The Eugenic Family Studies, 1877–1919,* Northeastern University Press, Boston, MA, 1988.

Ravnitzky, M. J. Statement before the Nazi war criminal records interagency working group, U.S. District Courthouse, New York, September 27, 1999.

Reilly, P. R. *The Surgical Solution: A History of Involuntary Sterilization in the United States,* Johns Hopkins University Press, Baltimore, MD, 1991.

Rivers, E. et al. Twenty years of follow-up experience in a long-range medical study, *Public Health Report,* 68(4):391, 1953.

Rockwell, D. H. et al. The Tuskegee study of untreated syphilis: the 30th year of observation, *Archives of Internal Medicine,* 144:792, 1964.

Rudin, E. An urgent need, *Birth Control Review,* 17(4):102, 1933.

Sanger, M. *Women and the New Race,* Brentano's Press, New York, 1920.

Sanger, M. *The Pivot of Civilization,* Brentano's Press, New York, 1922.

Sanger, M. Plan for peace, *Birth Control Review,* 16(4):107, 1932.

Smith, D. J. *The Eugenic Assault on America: Scenes in Red, White and Black,* George Mason University Press, Fairfax, VA, 1993.

Smith, D. J. and Nelson, K. R. *The Sterilization of Carrie Buck,* New Horizon Press, Far Hills, NJ, 1989.

Trials of War Criminals Before the Nuremberg Military Tribunals, 1949–1953, U.S. Government Printing Office, Washington, DC.

United States Supreme Court, *Buck v. Bell, Superintendent of State Colony Epileptics and Feeble Minded,* No. 292, 274 U.S. 200, May 2, 1927.

Vondelehr, R. A. et al. Untreated syphilis in the male Negro: comparative study of treated and untreated cases, *Journal of Venereal Disease Information,* 17:260, 1936.

Wilkerson, I. Medical experiment still haunts blacks, *New York Times,* June 3, 1991.

Winterbotham, F. W. *The Nazi Connection,* Weidenfeld and Nicolson, London, 1979.

4 Human Radiation Experiments

Advisory Committee on Human Radiation Experiments, Final Report, Government Printing Office, Washington, DC, 1995.

Allen, K. D. *Radiology in World War II*, Medical Department, U.S. Army, Government Printing Office, Washington, DC, 1966.

American Nuclear Guinea Pigs: Three Decades of Radiation Experiments on U.S. Citizens. Hearings on Human Total Body Irradiation (TBI), Committee on Energy and Commerce and Committee on Science and Technology, U.S. House of Representatives, September 23, 1981.

America's Children and the Environment: A First View of Available Measures, Environmental Protection Agency, EPA 240-R-00-006, December 2000.

Bassett, S. H. et al. *The Excretion of Hexavalent Uranium Following Intravenous Administration. II. Studies on Human Subjects*, University of Rochester, URB37, June 1948.

Bordes, P. A. et al. *Desert Rock I: A Psychological Study of Troop Reactions to an Atomic Explosion*, HumRRO-TR-1, February 1953.

Byrnes, V. A. et al., *Ocular effects of thermal radiation from atomic detonations—flashblindness and chorioretinal burns, project 4.5*, School of Aviation Medicine, November 30, 1955.

Ducoff, H. S. et al. Biological studies with arsenic-76. II. Excretion and tissue localization, *Proceedings of the Society for Experimental Biology and Medicine*, 69: 548, 1948.

Effect of Light from Very-Low-Yield Nuclear Detonation on Vision (Dazzle) of Combat Personnel, Operation Hardtack, Project 4.3, Headquarters Field Command, Defense Atomic Support Agency, April–October 1958.

Exposure of the American People to Iodine-131 from the Nevada Nuclear Bomb Tests: Review of the National Cancer Institute Report and Public Health Implications, National Academy Press, Washington, DC, 1999.

Fink, R. M. *Biological Studies with Polonium, Radium, and Plutonium*, McGraw-Hill, New York, 1950.

Gallagher, C. and Schneider, K. *American Ground Zero: The Secret Nuclear War*, MIT Press, Cambridge, MA, 1993.

Hawkins, D. *Project Y: The Los Alamos Story*, Tomash Publishers, Los Angeles, CA, 1983.

Hawley, C. A. et al. *Controlled Environmental Radioiodine Tests at the National Reactor Testing Station, Idaho Operations Office*, U.S. Atomic Energy Commission, IDOB12035, June 1964.

Heller, C. G. *Effects of Ionizing Radiation on the Testicular Function of Man: 9 Year Progress Report*, Pacific Northwest Research Foundation, May 1972.

Hewlett, R. G. and Duncan, F. *Atomic Shield: A History of the Atomic Energy*

Commission, Vol. 2, 1974–1952. U.S. Atomic Energy Commission, Washington, DC, 1972.

History of Air Force Atomic Cloud Sampling, AFSC Historical Publications, Series 61-142-1.

Human Experimentation: An Overview on Cold War Era Programs, U.S. General Accounting Office, GAO/T-NSIAD-94-266, September 28, 1994.

Human Radiation Experiments: The Department of Energy Roadmap to the Story and the Records, U.S. Department of Energy, Office of Scientific and Technical Information, Oak Ridge, TN, 1995.

Joint Panel on the Medical Aspects of Atomic Warfare, Department of Defense, Research and Development Board, Committee on Medical Sciences, 8th meeting, Washington, DC, February 24, 1951.

Jones, V. C. *Manhattan: The Army and the Atomic Bomb, Special Studies, U.S. Army in World War II.* Government Printing Office, Washington, DC, 1985.

Katz, J. The Nuremberg Code and the Nuremberg Trial, *Journal of the American Medical Association*, 276:1662, 1996.

Langham, W. H. et al. Distribution and excretion of plutonium administered intravenously to man, Los Alamos Scientific Laboratory, *Health Physics*, 38: 1031, 1980.

Matell, E. A. Radioactivity of tobacco trichomes and insoluble cigarette smoke particles, *Nature*, 249(454):215, 1974.

Martmer, E. E. et al. A study of the uptake of iodine (I-131) by the thyroid of premature infants, *American Journal of Diseases of Children*, 17:503, 1955.

Middlesworth, L. V. Radioactive iodine uptake of normal newborn infants, *American Journal of Diseases of Children*, 88:439, 1954.

Military Application of Radioactive Materials, Radiological Warfare Study Group, Second Report, June 17, 1948.

Morrison, R. T. et al. Radioiodine uptake studies in newborn infants, *Journal of Nuclear Medicine*, 4:162, 1963.

Niedenthal, J. A. history of the people of Bikini following nuclear weapons testing in the Marshall Islands: with recollections and views of elders of Bikini Atoll, *Health Physics*, 73(1):28, 1997.

Nishimi, R. Y. Research involving human subjects, Hearing before the Subcommittee on Energy, Committee on Science, Space, and Technology, U.S. House of Representatives, February 10, 1994.

Nuclear Health and Safety: Examples of Post World War II Radiation Releases at U.S. Nuclear Sites, U.S. General Accounting Office, GAO/RCED-94-51FS, November 1993.

Oberhansly, C. and Oberhansly, D. *Downwinders: An Atomic Tale*, Black Ledge Press, Salt Lake City, UT, 2001.

Ogborn, R. E. et al. Radioactive-iodine concentration in thyroid glands of newborn infants, *Pediatrics*, 26:771, 1960.

Operation Crossroads, 1946. U.S. Defense Nuclear Agency, Washington, DC 1984.

Operation Crossroads Fact Sheet. Department of the Navy, Naval Historical Center, Washington, DC, 1999.

Panel to investigate safety of civilian populations from tests of radiological warfare agents. Memorandum to military liaison committee, May 2, 1949.

Paulsen, C. A. *The Study of Irradiation Effects on the Human Testis: Including Histologic, Chromosomal and Hormonal Aspects, Terminal Report*, AEC Contract #AT(45B1)B225, University of Washington School of Medicine, January 31, 1973.

Radioactive fallout from nuclear testing at Nevada test site, 1950–60. Hearing before the Subcommittee on Labor, Health and Human Services, and Education, and Related Agencies, Committee on Appropriations, October 1, 1997.

Radioactive warfare. Memorandum from Joseph G. Hamilton, M.D., December 31, 1946.

Rahman, S. M. et al. Tobacco's radiation: its sources and potential hazards, *Ohio Medicine*, 83(2):113, 1987.

Recommendation that the armed services conduct experiments on human subjects to determine effects of radiation exposure. Memorandum from Major General Harry G. Armstrong, USAF, April 13, 1950.

Rowley, M. J. et al. Effects of graded doses of ionizing radiation on the human testis, *Radiation Research*, 59:665, 1974.

Shurcliff, W. A. *Bombs at Bikini: The Official Report of Operation Crossroads*, William H. Wise, New York, 1947.

Stannard, J. N. *Radioactivity and Health: A History*, Office of Scientific and Technical Information, Washington, DC, 1988.

Titus, C. A. *Bombs in the Backyard: Atomic Testing and American Politics*, University of Nevada Press, Las Vegas, NV, 1987.

Unauthorized medical experimentation on service personnel. Memorandum from Secretary of the Navy Frank Knox to all ships and sailors, April 7, 1943.

United States Nuclear Tests, July 1945 Through September 1992, DOE/NV-209 (Rev.14), December 1994.

Welsome, E. *The Plutonium Files: America's Secret Experiments in the Cold War*, Dial Press, 1999.

---

5 The CIA and Human Experiments

Abramson, H. *The Use of LSD in Psychotherapy*, Josiah Macy Foundation, New York, 1960. Advisory Committee on Human Radiation Experiments, minutes of meeting of March 15–17, 1995, Executive Chambers, Madison Hotel, Washington, DC.

Bamford, J. *The Puzzle Palace: Inside the National Security Agency, America's Most Secret Intelligence Organization*, Penguin Books, New York, 1982.

Barker, E. T. and Buck, M. F. LSD in a coercive milieu therapy program, *Canadian Psychiatric Journal*, 22:311, 1977.

Bimmerle, G. Truth drugs in interrogation, *Studies in Intelligence*, 2(5): A1, 1961.

Blum, W. *The CIA: A Forgotten History*, Zed Books, London, 1986.

Bowart, W. H. *Operation Mind Control: Our Secret Government's War Against Its Own People*, Dell Books, New York, 1978.

Bower, T. *The Paperclip Conspiracy: The Hunt for the Nazi Scientists*, Little, Brown and Company, Boston, 1987.

Buckman, J. Brainwashing, LSD, and the CIA: Historical and ethical perspectives, *International Journal of Social Psychology*, 23:8, 1977.

Cameron, D. E. Red light therapy in schizophrenia, *British Journal of Physical Medicine*, 10:11, 1936.

Cameron, D. E. Psychic driving, *American Journal of Psychiatry*, 112:502, 1956.

Cameron, D. E. Production of different amnesia as a factor in the treatment of schizophrenia, *Comprehensive Psychiatry*, 1:26, 1960.

Cameron, D. E. et al. The depatterning treatment of schizophrenia, *Comprehensive Psychiatry*, 3:65, 1962.

Central Intelligence Agency. *Factbook on Intelligence*, December 1992.

Collins, A. *In the Sleep Room: The Story of the CIA Brainwashing Experiments in Canada*, Lester and Orpen Dennys, Toronto, 1988.

Condon, R. *The Manchurian Candidate*, McGraw-Hill, New York, 1959.

Foreign and Military Intelligence, Book I, United States Select Committee on Government Operations with Respect to Intelligence Activities, 94th Congress, 2nd Session, April 26, 1976.

Frazer, H. *Uncloaking the CIA*, Free Press, New York, 1975.

Freund, P. A. *Experiments with Human Subjects*, George Brazillier, New York, 1969.

Gillmor, D. *I Swear by Apollo: Dr. Ewen Cameron, the CIA, and the Canadian Mind-Control Experiments*, Eden Press, Fountain Valley, CA, 1986.

Horowitz, L. G. *Emerging Viruses: Aids and Ebola*, Tetrahedron, Rockport, MA, 1996.

Hudler, G. *Magical and Mischievous Molds*, Princeton University Press, Princeton, NJ, 1998.

Hunt, L. *Secret Agenda: The United States Government, Nazi Scientists, and Project Paperclip, 1945 to 1990*, St. Martin's Press, New York, 1991.

Jeffreys-Jones, R. *The CIA and American Democracy*, Yale University Press, New Haven, CT, 1989.

Kessler, R. *Inside the CIA: Revealing the Secrets of the World's Most Powerful Spy Agency*, Pocket Books, New York, 1992.

Lawrence, K. The CIA and the mad scientist: drugs, psychiatry, and mind control in Canada, *Covert Action Information Bulletin*, 28:29, 1987.

Lee, M. and Shlain, B. *Acid Dreams: The CIA, LSD, and the Sixties Rebellion*, Grove Press, New York, 1986.

Marks, J. *The Search for the Manchurian Candidate: The CIA and Mind Control*, Time Books, New York, 1979.

Matossian, M. K. *Poisons of the Past: Molds, Epidemics and History*, Yale University Press, New Haven, CT, 1989.

McRae, R. M. *Mind Wars: The True Story of Government Research into the Military Potential of Psychic Weapons*, St. Martin's Press, New York, 1984.

Mendelbaum, W. A. *The Psychic Battlefield*, St. Martin's Press, New York, 2000.

Richelson, J. T. *The Wizards of Langley: Inside the CIA's Directorate of Science and Technology*, Westview Press, Boulder, CO, 2001.

Rockefeller, N. A. et al. *Report to the President by the Commission on CIA Activities Within the United States*, Rockefeller Commission, New York, 1975.

Ross, C. A. *Bluebird: Deliberate Creation of Multiple Personality by Psychiatrists*, Manitou Communications, Richardson, TX, 2000.

Schnabel, J. *Remote Viewers: The Secret History of America's Psychic Spies*, Dell Books, New York, 1997.

Smith, R. H. *The Secret History of America's First Central Intelligence Agency*, University of California Press, Berkeley, 1972.

Smith, R. J. *The Unknown CIA: My Three Decades with the Agency*, Berkeley Books, New York, 1989.

Thomas, G. *Journey into Madness: The True Story of CIA Mind Control and Medical Abuse*, Bantam Books, New York, 1989.

Trento, J. J. *The Secret History of the CIA*, Prima Publishing, Rocklin, CA, 2001.

U.S. Congress. Hearings before the Select Committee to Study Governmental Operations with Respect to Intelligence Activities, *Final Report*, Report No. 94-755, 6 vols., Government Printing Office, Washington, DC, 1976.

U.S. Congress. Joint hearings before the Subcommittee on Health, Subcommittee on Administrative Practice and Procedure of the Committee on the Judiciary, and the Committee on Labor and Public Welfare, United States Senate, *Biomedical and Behavioral Research. Human-Use Experimentation Programs of the Department of Defense and Central Intelligence Agency, Sep-*

tember 10, 12 and November 7, 1975, Government Printing Office, Washington, DC, 1976.

U.S. Congress. Hearings before the Subcommittee on Health and Scientific Research of the Committee on Human Resources, United States Senate, *Biological Testing Involving Human Subjects by the Department of Defense, 1977*, March 8 and May 23, 1977, Government Printing Office, Washington, DC, 1977.

U.S. Congress. Joint Hearing before the Select Committee on Intelligence and the Subcommittee on Health and Scientific Research of the Committee on Human Resources, United States Senate, *Project MKULTRA, the CIA's Program of Research in Behavior Modification*, August 3, 1977, Government Printing Office, Washington, DC, 1977.

U.S. Congress. Hearings before the Subcommittee on Health and Scientific Research of the Committee on Human Resources, United States Senate, *Human Drug Testing by the CIA, 1977*, September 20 and 21, 1977, Government Printing Office, Washington, DC, 1977.

Vankin, J. *Conspiracies, Cover-ups, and Crimes: Political Manipulation and Mind Control in America*, Paragon House, New York, 1991.

Weberman, A. J. Mind control: the story of mankind research unlimited, inc., *Covert Action Information Bulletin*, 9:15, 1990.

Weinstein H. M. *Psychiatry and the CIA: Victims of Mind Control*, American Psychiatric Press, Washington, DC, 1990.

Westerfield, B. H. *Inside CIA's Private World: Declassified Articles from the Agency's Internal Journal, 1955–1992*, Yale University Press, New Haven, CT, 1997.

6 Silent Conspirators: The Government–Industry Connection from Aspartame to AZT

Aspartame—not for the dieting pilot? *Aviation Safety Digest*. Spring 1989.

Bessler Report. *FDA Report on D. G. Searle Pharmaceuticals*, 1977.

Blaylock, R. L. *Excitotoxins: The Taste That Kills*, Health Press, Albuquerque, NM, 1997.

Cathcart, R. F. Vitamin C in the treatment of acquired immune deficiency syndrome (AIDS), *Medical Hypothesis*, 14(4):423, 1984.

Cauchon, D. FDA advisers tied to industry, *USA Today*, September 25, 2000.

DHHS. *Report on All Adverse Reactions in the Adverse Reaction Monitoring System*, April 20, 1995.

DHHS Memorandum. Summary of adverse reactions attributed to aspartame, 1980–1996, DHHS Technical Information Specialist, HFS-728, June 26, 1997.

DHHS. *FDA Consideration of Codex Alimentarius Standards*, Docket No. 97N-0218, October 6, 1997.

Duesberg, P. H. and Ellison, B. J. Is the AIDS virus a science fiction? Immunosuppresive behavior, not HIV, may be the cause of AIDS, *Policy Review*, Summer 1990.

Duesberg, P. H. *Inventing the AIDS Virus*, Regnery Publishing, Washington, DC, 1996.

Fujimaki, H. et al. Mast cell response to formaldehyde, *International Archives of Allergy and Immunology*, 2:401, 1992.

Guiso, G. et al. Effect of aspartame on seizures in various models of experimental epilepsy, *Toxicology and Applied Pharmacology*, 96(3):485, 1988.

Horowitz, L. G. *Emerging Viruses: AIDS & Ebola, Nature, Accident or Intentional?* Tetrahedron, Rockport, MA, 1996.

John, D. R. Migraine provoked by aspartame, *New England Journal of Medicine*, 314:456, 1986.

Lauristen, J. AZT on trial: Did the FDA rush to judgment and thereby further endanger the lives of thousands of people? *New York Native*, 235, October 1987.

Lauristen, J. and Duesberg, P. *Poison by Prescription: The AZT Story*, Pagan Press, 1990.

Lazarou, J. et al. Incidence of adverse drug reactions in hospitalized patients, *Journal of the American Medical Association*, 279:1200, 1998.

Letter to Senator Orrin Hatch from Senator Metzenbaum calling for aspartame hearings, February 26, 1986.

Letter to Senator Howard Metzenbaum from EPA regarding dangers of aspartame, October 30, 1987.

Liesivuori, J. and Savolainen, H. Methanol and formic acid toxicity: biochemical mechanisms, *Pharmacology and Toxicology*, 69:157, 1991.

Mahler, T. J. and Wurtman, R. Possible neurologic effects of aspartame, a widely used food additive, *Environmental Health Perspectives*, 75:53, 1987.

Moser, R. H. Aspartame and memory loss, *Journal of the American Medical Association*, 272(19):153, 1975.

NSDA Trade Secret Report on Aspartame: Food and Drug Sweetener Strategy, part of Congressional Record, S5507-S5511, March 7, 1985.

Olney, J. W. et al. Brain damage in mice from voluntary ingestion of glutamate and aspartate, *Neurobehavioral Toxicology and Teratology*, 2:125, 1980.

Rath, M. Why animals don't get heart attacks but people do, *Health News*, 2000.

Rath, M. and Pauling, L. Hypothesis: lipoprotein(a) is a surrogate for ascorbate, *Proceedings of the National Academy of Sciences*, 87:6204, 1990.

Rath, M. and Pauling, L. Immunological evidence for the accumulation of lipoprotein(a) in the atherosclerotic lesion of the hypoascorbemic guine pig, *Proceedings of the National Academy of Sciences*, 87:9388, 1990.

Rath, M. and Pauling, L. Solution to the puzzle of human cardiovascular disease: its primary cause is ascorbate deficiency leading to the deposition of lipoprotein(a) and fibrinogen/fibrin in the vascular wall, *Journal of Orthomolecular Medicine*, 6:125, 1991.

Roberts, H. J. Reactions attributed to aspartame-containing products: 551 cases, *Journal of Applied Nutrition*, 40:85, 1988.

Ross, W. E. et al. Comparison of DNA damage by methylmelamines and formaldehyde, *Journal of the National Cancer Institute*, 67:217, 1981.

Rowen, J. A. et al. Aspartame and seizure susceptibility: results of a clinical study in reportedly sensitive individuals, *Epilepsia*, 36(3):270, 1995.

Sonnabend, J. A. et al. The acquired immunodeficiency syndrome: a discussion of etiological hypotheses, *AIDS Research*, 1(1):10, 1984.

Trocho, C. et al. Formaldehyde derived from dietary aspartame binds to tissue components in vivo, *Life Sciences*, 63(5):337, 1998.

U.S. Air Force. Aspartame alert, *Flying Safety*, 48(5):20, 1992.

United States Senate Report. *NutraSweet Health and Safety Concerns*, U.S. Senate Committee on Labor and Human Resources, November 3, 1987.

Walton, R. G. Seizure and mania after high intake of aspartame, *Psychosomatics*, 27:218, 1986.

Walton, R. G. et al. Adverse reactions to aspartame: double-blind challenge in patients from a vulnerable population, *Biological Psychiatry*, 34:13, 1993.

World Trade Agreement Application of Sanitary and Phytosanitary Measures (SPS Agreement).

7 Organized Medicine: A Century of Human Experimentation

Barber, B. et al. *Research on Human Subjects: Problems of Social Control in Medical Experimentation*. Russell Sage Foundation, New York, 1973.

Beecher, H. K. Experimentation on man, *New England Journal of Medicine*, 274:1383, 1966.

Bochetta, M. et al. Human mesothelial cells are unusually susceptible to SV40-mediated transformation and asbestos cocarcinogenicity, *Proceedings of the National Academy of Sciences*, 97(18):10214, 2000.

Bogomolny, R. L. *Human Experimentation*, SMU Press, Dallas, TX, 1976.

Caplan, A. L. *When Medicine Went Mad*, Humana Press, Totowa, NJ, 1992.

Charen, M. Body parts for sale: fetal harvesting, *The Washington Times*, November 15, 1999.

Childress, J. et al., *Experimentation with Human Subjects*, University Publications of America, Frederick, MD, 1984.

DHHS Report. *Institutional Review Boards: A Time for Reform*, June 1998.

Donnelly, M et al. Fenfluramine and dextroamphetamine treatment of childhood hyperactivity, *Archives of General Psychiatry*, 46:205, 1989.

Ervin, M. Guinea pigs don't get to say no, *Disability News Service*, 1998.

Freund, P. A. *Experimentation with Human Subjects*, George Brazillier, New York, 1969.

Golby, S. Experiments at the Willowbrook State School, *The Lancet*, April 10, 749, 1971.

Gray, B. H. *Human Subjects in Medical Experimentation*, John Wiley and Sons, New York, 1975.

Greenberg, D. S. What ever happened to the war on cancer? *Discover*, pp. 47–66, March 1991.

Grodin, M. A. et al. *Children as Research Subjects*, Oxford University Press, New York, 1994.

Halperin, J. M. et al. Serotonin, aggression, and parental psychopathology in children with attention-deficit hyperactivity disorder, *Journal of the American Academy of Child and Adolescent Psychiatry*, 36:1391, 1997.

Hilts, P. J. Experiments on children are reviewed, *New York Times*, April 15, 1998.

Hornblum, A. M. *Aces of Skin: Human Experiments at Holmesburg Prison*, Routledge, New York, 1998.

James, W. *Immunization: The Reality Behind the Myth*, Bergin and Garvey, New York, 1988.

Katz, J. *Experimentation with Human Beings: The Authority of the Investigator, Subject, Professions, and State in the Human Experimentation Process*, Russell Sage Foundation, New York, 1972.

Kerr, K. Parents never knew of test risks: study didn't warn of deadly side effects, *Newsday*, A04 1998.

Knox, R. A. Study finds conflicts in medical reports, *Boston Globe*, p. A12, January 8, 1998.

Kong, D. Still no solution in the struggle on safeguards, *Boston Globe*, A01 November 1998.

Kops, S. Oral polio vaccine and human cancer: a reassessment of SV40 as a contaminant based upon legal documents, *Anticancer Research*, 20:4745, 2000.

Lederer, S. E. Orphans as guinea pigs: American children and medical experimenters, 1890–1930, in Roger Cooter (ed.), *The Name of the Child: Health and Welfare, 1880–1940*, Routledge, New York, 1992.

Lederer, S. E. *Subjected to Science: Human Experimentation in America Before the Second World War*, Johns Hopkins University Press, Baltimore, MD, 1995.

McCann, U. D. et al. Brain serotonin neurotoxicity and primary pulmonary hypertension from fenfluramine and dexenfluramine: a systematic review of evidence, *Journal of the American Medical Association*, 278:66, 1997.

Meyer, P. B. *Drug Experiments on Prisoners*, Lexington Books, Lexington, MA, 1976.

Miller, N. Z. *Immunization: Theory vs. Reality—An Expose on Vaccinations*, New Atlantean Press, Santa Fe, NM, 1996.

Moreno, J. D. *Undue Risk: Secret State Experiments on Humans*, W. H. Freeman, New York, 2000.

Pappworth, M. H. *Human Guinea Pigs: Experimentation on Man*, Beacon Press, Boston, MA, 1967.

Rothman, D. J. *The Willowbrook Wars*, Harper and Row, New York, 1984.

Scheibner, V. *Vaccination: Over 100 Years of Orthodox Research Shows That Vaccines Represent a Medical Assault on the Immune System*, Australian Print Group, Victoria, Australia, 1993.

Stelfox, H. T. et al. Conflict of interest in the debate over calcium-channel antagonists, *New England Journal of Medicine*, 338(2):101, 1998.

Stix, G. The ties that bind: a study of journal publications details investigators' financial interests in their research, *Scientific American*, March 17, 1997.

Strickler, H. Simian virus 40 and human cancers, *Einstein Quarterly Journal of Biology and Medicine*, 18:14, 2001.

Torch, W. C. Diphtheria-pertussis-tetanus (DPT) immunization: a potential cause of the sudden infant death syndrome, *Neurology*, 32(4): 334.

U.S. Congress. Hearings before the Subcommittee on Human Resources of the Committee on Government Reform and Oversight, United States House of Representatives, *Institutional Review Boards: A System in Jeopardy*, June 11, 1998, Government Printing Office, Washington, DC, 1998.

Whitaker, R. Testing takes a human toll, *Boston Globe*, p. AO1, November 15, 1998,

Wilson, D. and Heath, D. He saw the tests as a violation of trusting desperate human beings, *Seattle Times*, March 12, 2001.

Wilson, D. and Heath, D. Patients never knew the full danger of trials they staked their lives on, *The Seattle Times*, March 11, 2001.

8 Ethnic Weapons: The New Genetic Warfare

Adams, J. *AIDS:The HIV Myth*, St. Martin's Press, New York, 1989.

Balter, M. Montagnier pursues the Mycoplasma–AIDS link, Science, 251: 271, 1991.

British Medical Association. *Biotechnology, Weapons and Humanity*, Harwood Academic, New York, 1999.

Cantwell, A. R. *AIDS: The Mystery and the Solution*, Aries Rising Press, Los Angeles, CA, 1984.

Cantwell, A. R. *Queer Blood: The Secret AIDS Genocide Plot*, Aries Rising Press, Los Angeles, CA, 1993.

Chase, A. *Magic Shots*, William Morrow, New York, 1982.

Curtis, T. The origin of AIDS, *Rolling Stone*, 626:54, 1992.

Duesberg, P. H. HIV and AIDS: correlation but not causation, *Proceedings of the National Academy of Science*, 86:755, 1989.

Duesberg, P. H. AIDS epidemiology: inconsistencies with human immunodeficiency virus and with infectious disease, *Proceedings of the National Academy of Science*, 88:1575, 1991.

Duesberg, P. H. *Inventing the AIDS Virus*, Regnery Publishing, Washington, DC, 1996.

Elswood, B. F. and Stricker, R. B. Polio vaccines and the origin of AIDS, *Medical Hypothesis*, 42:347, 1994.

Ethnic weapons: apartheid's final solution? *Resister: Bulletin of the Committee on South African War Resistance*, No. 23: *Chemical Warfare Threat,* pp. 13–17, 1983.

Gallo, R. C. et al. Isolation of human T-cell leukemia virus in acquired immune deficiency syndrome (AIDS), *Science*, 220:865, 1983.

Gonda, M. A. et al. Characterization and molecular cloning of a bovine lentivirus related to human immunodeficiency virus, *Nature*, 330:388, 1987.

Hatch, R. Cancer warfare, *Covert Action*, 39:14, 1991.

Hooper, E. *The River: A Journey to the Source of HIV and AIDS*, Little, Brown and Company, Boston, MA, 1999.

Horowitz, L. G. *Emerging Viruses: AIDS and Ebola, Nature, Accident or Intentional?* Tetrahedron, Rockport, MA, 1996.

Lander, E. S. The new genomics: global views of biology, *Science*, 274:25, 536, 1996.

Larson, C. A. Ethnic weapons, *Military Review*, November 3–11, 1970.

McDonald, M. I. et al. Hepatitis B surface antigen could harbour the infective agent of AIDS, *Lancet*, 2:882, 1983.

Mohan, C. R. Ethnic weapons? *Strategic Analysis*, 7:555, 1984.

Moss, A. R. et al. Incidence of the acquired immunodeficiency syndrome in San Francisco, 1980–1983, *Journal of Infectious Diseases*, 152:152, 1985.

Nassar, G. An Israeli ethnic bomb? *Al-Ahram Weekly*, 404:19, 1998.

O'Brien, S. J. and Dean, M. In search of AIDS-resistant genes. *Scientific American*, p. 28, September 1997.

O'Brien, S. J. et al. Evidence for concurrent epidemics of human herpes virus 8 and human immunodeficiency virus type I in U.S. homosexual men: rates, risk factors, and relationships to Kaposi's sarcoma, *Journal of Infectious Disease*, 180:1010, 1999.

Piller, C. and Yamamoto, K. R. *Gene Wars: Military Control over the New Genetic Technologies,* Willam Morrow, New York, 1988.

Progress Report No. 8. *Special Virus Cancer Program*, Etiology Area–National Cancer Institute, U.S. Department of Health, Education and Welfare, Public Health Service, and National Institutes of Health, August 1971.

Race weapon is possible. *Defense News*, March 23, P. 2, 1992.

Seale, J. Origins of the AIDS viruses, HIV-1 and HIV-2: Fact or Fiction? *British Journal of the Royal Society of Medicine*, 81:617, 1988.

Shilts, R. *And the Band Played On*, St. Martin's Press, New York, 1987.

Sonnabend, J. A. The Etiology of AIDS, *AIDS Research*, 1:1, 1983.

Starr, B. Cohen warns of new terrors beyond CW, *Jane's Defense Weekly*, June 27, 1997.

Starr, B. Interview: US Secretary of Defense. William Cohen, *Jane's Defense Weekly*, August 13, 1997.

Szmuness, W. et al. Hepatitis-B vaccine: demonstration of efficacy in a controlled clinical trial in a high-risk population in the United States, *New England Journal of Medicine*, 303:833, 1980.

Thatcher, G. Genetic weapons: Is it on the horizon? *Christian Science Monitor*, p. B11, December 15, 1988.

Thomas, P. *Big Shot: Passion, Politics, and the Struggle for an AIDS Vaccine*, Public Affairs, New York, 2001.

Wright, K. Mycoplasmas in the AIDS spotlight, *Science*, 248:682, 1990.

9 What the Future Holds: Human Experimentation in the Twenty-first Century

Anderson, W. F. Human gene therapy, *Nature*, 392:25, 1998.

Begich, N. and Manning, J. HAARP: High frequency vandalism in the sky, *Nexus Magazine*, January 1996.

Begich, N. and Manning, J. *Angels Don't Play This Haarp: Advances in Tesla Technology,* Earthpulse Press, Anchorage, AL 1997.

Breslau, K. Tiny weapons with giant potential, *Newsweek,* June 24, 2002.

Cantor, C. *Genomics: The Science and Technology Behind the Human Genome Project*, John Wiley and Sons, New York, 1999.

Charen, M. Body parts for sale—fetal harvesting, *Washington Times*, November 15, 1999.

Cohen, P. Dozens of human embryos cloned in China, *New Scientist*, March 6, 2002.

Drexler, K. E. Molecular engineering: an approach to the development of general capabilities for molecular manipulation, *Proceedings of the National Academy of Sciences*, 78:5275, 1981.

Elias, P. Saudi Arabia starts stem cell program, *Associated Press*, June 13, 2002.

Freitas, R. *Nanomedicine: Basic Capabilities*, Landes Bioscience, Austin, TX, 1999.

Halperin, J. M. et al. Serotonin, aggression, and parental psychopathology in children with attention-deficit hyperactivity disorder, *Journal of the American Academy of Child and Adolescent Psychiatry*, 36:1391, 1997.

Hook, C. C. Testimony before the U.S. House of Representatives, Committee on Government Reform of the Subcommittee on Criminal Justice, Drug Policy, and Human Resources; Hearing on embryonic stem cell research, July 17, 2001.

Human cloning experiments underway, *BBC News*, June 17, 1999.

Huxley, A. *Brave New World*, Harper, New York, 1932.

Kresina, T. F. *An Introduction to Molecular Medicine and Gene Therapy*, Wiley-Liss, New York, 2000.

Lauritzen, P. *Cloning and the Future of Human Embryo Research*, Oxford University Press, New York, 2001.

Lemoine, N. R. and Vile, R. G. *Understanding Gene Therapy*, Springer-Verlag, New York, 2000.

Lyon, J. and Gorner, P. *Altered Fates: Gene Therapy and the Retooling of Human Life*, W. W. Norton, New York, 1996.

Madrazo, I. et al. Fetal homotransplants (ventral mesencephalon and adrenal tissue) to striatum of parkinsonian subjects. *Archives of Neurology*, 47:1281, 1990.

McCann, U. D. et al. Brain serotonin neurotoxicity and primary pulmonary hypertension from fenfluramine: a systematic review of evidence, *Journal of the American Medical Association*, 278:666.

Merkle, R. C. Nanotechnology and medicine, in *Advances in Anti-Aging Medicine*, Volume 1, edited by Ronald M. Klatz, Mary Ann Liebert, Inc., Larchmont, 1996.

Molecular Medicine for the 21st Century, Dateline Los Alamos, Los Alamos National Laboratory, November 1995.

Muldoon, M. F. et al. D,L-Fenfluramine challenge test: experience in nonpatient sample, *Biological Psychiatry*, 39:761, 1996.

NIH Guide: *Availability of Human Fetal Tissue*, Volume 23, Number 10, March 10, 1994.

Pine, D. S. et al. Neuroendocrine response to fenfluramine challenge in boys, *Archives of General Psychiatry*, 54:839, 1997.

Potter, R. A. and Leder, P. *Exploring the Biomedical Revolution*, Johns Hopkins University Press, Baltimore, MD, 2000.

Quesenberry, A. *Stem Cell Biology and Gene Therapy*, John Wiley and Sons, New York, 1998.

Rovner, J. U.S. Senate sidesteps vote on cloning patent ban, *Reuters Health*, June 18, 2002.

Seth, P. *Adenoviruses: Basic Biology to Gene Therapy*, Landes Bioscience, Austin, TX, 1999.

Thompson, L. Human gene therapy: harsh lessons, high hopes, *FDA Consumer Magazine*, September–October 2000.

U.S. Congress. Hearings before the Subcommittee on Health and Environment of the Committee on Commerce, United States House of Representatives, *Fetal Tissue: Is It Being Sold in Violation of Federal Law?* March 9, 2000, Government Printing Office, Washington, DC, 2000.

Varmus, H. Statement for the record on human cloning. House Committee on Commerce of the Subcommittee on Health and Environment, United States House of Representatives, February 12, 1998.

Watson, J. *Human Genome*, St. Martin's Press, New York, 2001.

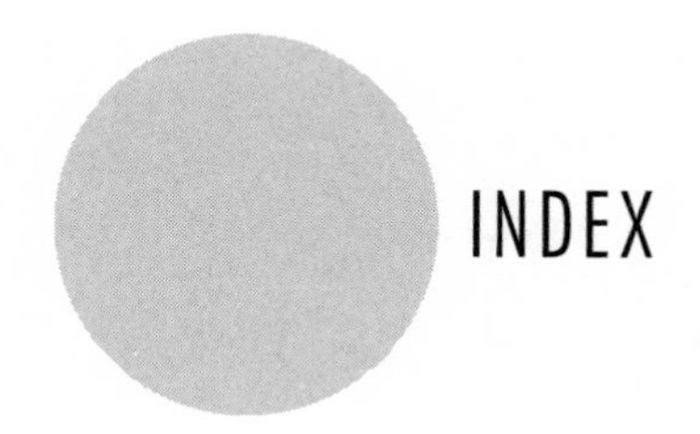

INDEX